Lecture Notes in Computer Science 4952

Commenced Publication in 1973
Founding and Former Series Editors:
Gerhard Goos, Juris Hartmanis, and Jan van Leeuwen

Editorial Board

Christian Floerkemeier   Marc Langheinrich
Elgar Fleisch   Friedemann Mattern
Sanjay E. Sarma (Eds.)

The Internet
of Things

First International Conference, IOT 2008
Zurich, Switzerland, March 26-28, 2008
Proceedings

 Springer

Volume Editors

Christian Floerkemeier
Sanjay E. Sarma
Massachusetts Institute of Technology
Auto-ID Labs
77 Massachusetts Avenue, Cambridge, MA 02139, USA
E-mail: {floerkem,sesarma}@mit.edu

Marc Langheinrich
Friedemann Mattern
ETH Zurich
Inst. for Pervasive Computing
Clausiusstr. 59, 8092 Zurich, Switzerland
E-mail: {langhein,mattern}@inf.ethz.ch

Elgar Fleisch
ETH Zurich and
University of St. Gallen
Dufourstrasse 40a, 9000 St. Gallen, Switzerland
E-mail: elgar.fleisch@ethz.ch

Library of Congress Control Number: 2008922619

CR Subject Classification (1998): C.2, C.3, B.7, E.3, D.4.6, K.6.5, J.1

LNCS Sublibrary: SL 3 – Information Systems and Application, incl. Internet/Web
and HCI

ISSN   0302-9743
ISBN-10   3-540-78730-5 Springer Berlin Heidelberg New York
ISBN-13   978-3-540-78730-3 Springer Berlin Heidelberg New York

Springer is a part of Springer Science+Business Media

springer.com

© Springer-Verlag Berlin Heidelberg 2008

Typesetting: Camera-ready by author, data conversion by Scientific Publishing Services, Chennai, India
Printed on acid-free paper   SPIN: 12244263   06/3180   5 4 3 2 1 0

Preface

This volume contains the proceedings of the Internet of Things (IOT) Conference 2008, the first international conference of its kind. The conference took place in Zurich, Switzerland, March 26–28, 2008. The term 'Internet of Things' has come to describe a number of technologies and research disciplines that enable the Internet to reach out into the real world of physical objects. Technologies such as RFID, short-range wireless communications, real-time localization, and sensor networks are becoming increasingly common, bringing the 'Internet of Things' into industrial, commercial, and domestic use. IOT 2008 brought together leading researchers and practitioners, from both academia and industry, to facilitate the sharing of ideas, applications, and research results.

IOT 2008 attracted 92 high-quality submissions, from which the technical program committee accepted 23 papers, resulting in a competitive 25% acceptance rate. In total, there were over 250 individual authors from 23 countries, representing both academic and industrial organizations. Papers were selected solely on the quality of their blind peer reviews. We were fortunate to draw on the combined experience of our 59 program committee members, coming from the most prestigious universities and research labs in Europe, North America, Asia, and Australia. Program committee members were aided by no less than 63 external reviewers in this rigorous process, in which each committee member wrote about 6 reviews. The total of 336 entered reviews resulted in an average of 3.7 reviews per paper, or slightly more than 1000 words of feedback for each paper submitted. To ensure that we had quality reviews as well as substantive deliberation on each paper, a subsequent discussion phase generated 270 discussion items. As a result, some 40 submissions were selected for discussion at the meeting of the program chairs.

The term 'Internet of Things' describes an area with tremendous potential; where new sensing and communication technologies, along with their associated usage scenarios and applications, are driving many new research projects and business models. Three major themes pervade the technical discussions collected in this volume: novel sensing technologies to capture real-world phenomena; the evaluation of novel applications using both new and existing technologies; and the appropriate infrastructure to facilitate communication with (and the localization of) billions of networked real-world objects. While all these technological developments are exciting, they also bear profound challenges from a social, legal, and economic perspective. The research areas covered at IOT 2008 were thus not only technical in nature, but reflected the diverse angles from which to approach this emerging research field.

The scientific papers presented at IOT 2008 were not its only highlight. In addition to the technical sessions, the conference featured keynote speeches by leading figures from industry and academia, such as Bob Iannucci (Nokia), Gerd

Wolfram (Metro Group), Peter Zencke (SAP), and Haruhisa Ichikawa (UEC Tokyo). IOT 2008 also included an industrial track where industry experts presented challenges and lessons learned from current technology deployments. The conference offered a demo reception, as well as a full day of workshops and tutorials.

Several organizations provided financial and logistical assistance in putting IOT 2008 together, and we would like to acknowledge their support. We thank ETH Zurich for the conference organization and for managing the local arrangements. We very much appreciate the support of our Platinum Sponsor SAP, along with the generous donations from Siemens, Metro Group, Google, ERCIM, and IBM. We would also like to thank the keynote speakers and industrial experts who provided a fascinating commercial perspective on current developments towards an 'Internet of Things'. Lastly, we would like to thank both the authors who submitted their work to IOT 2008 and the program committee members and our external reviewers, who spent many hours reviewing submissions, shepherding papers, and providing the feedback that resulted in the selection of the papers featured in these proceedings.

March 2008   Christian Floerkemeier
                                                    Marc Langheinrich
                                                       Elgar Fleisch
                                                  Friedemann Mattern
                                                       Sanjay Sarma

Organization

Conference Committee

Conference Chairs	Elgar Fleisch (ETH Zurich & University of St. Gallen, Switzerland) Friedemann Mattern (ETH Zurich, Switzerland) Sanjay Sarma (MIT, USA)
Program Chairs	Christian Floerkemeier (MIT, USA) Marc Langheinrich (ETH Zurich, Switzerland)
Industrial Program	Ulrich Eisert (SAP Research, Germany)
Workshops & Demos	Florian Michahelles (ETH Zurich, Switzerland)

Conference Management

Financial	Marc Langheinrich (ETH Zurich, Switzerland)
Local Arrangements	Steve Hinske (ETH Zurich, Switzerland)
Press	Steve Hinske (ETH Zurich, Switzerland)
Proceedings	Philipp Bolliger (ETH Zurich, Switzerland)
Publicity	Benedikt Ostermaier (ETH Zurich, Switzerland)
Registration	Marc Langheinrich (ETH Zurich, Switzerland) Steve Hinske (ETH Zurich, Switzerland)
Sponsoring	Steve Hinske (ETH Zurich, Switzerland)
Student Volunteers	Steve Hinske (ETH Zurich, Switzerland)
Web	Marc Langheinrich (ETH Zurich, Switzerland)

Program Committee

Karl Aberer (EPFL, Switzerland)
Manfred Aigner (TU Graz, Austria)
Michael Beigl (TU Braunschweig, Germany)
Alastair Beresford (University of Cambridge, UK)
Peter Cole (The University of Adelaide, Australia)
Nigel Davies (Lancaster University, UK)
Jean-Pierre Émond (University of Florida, USA)
Alois Ferscha (University of Linz, Austria)
Elgar Fleisch (ETH Zurich & University of St. Gallen, Switzerland)
Anatole Gershman (Carnegie Mellon University, USA)
Bill Hardgrave (University of Arkansas, USA)

Mark Harrison (Cambridge University, UK)
Ralf Guido Herrtwich (Daimler, Germany)
Lutz Heuser (SAP Research, Germany)
Lorenz Hilty (EMPA, Switzerland)
Thomas Hofmann (Google, Switzerland)
Ryo Imura (Hitachi & University of Tokyo, Japan)
Sozo Inoue (Kyushu University, Japan)
Yuri Ivanov (Mistubishi Electric Research Laboratories, USA)
Behnam Jamali (The University of Adelaide, Australia)
Guenter Karjoth (IBM Zurich Research Lab, Switzerland)
Wolfgang Kellerer (DoCoMo Euro-Labs, Germany)
Daeyoung Kim (Information and Communications University, Korea)
Kwangjo Kim (Information and Communication University, Korea)
Tim Kindberg (Hewlett-Packard Laboratories, UK)
Gerd Kortuem (Lancaster University, UK)
Anthony LaMarca (Intel Research Seattle, USA)
Friedemann Mattern (ETH Zurich, Switzerland)
Hao Min (State Key Lab of ASIC and System, Fudan University, China)
Jin Mitsugi (Keio University, Japan)
Paul Moskowitz (IBM T. J. Watson Research Center, USA)
Jun Murai (Keio University, Japan)
Osamu Nakamura (Keio University, Japan)
Paddy Nixon (University College Dublin, Ireland)
Thomas Odenwald (SAP Research, USA)
Ravi Pappu (Thingmagic, USA)
Joseph Paradiso (MIT, USA)
Aaron Quigley (University College Dublin, Ireland)
Hartmut Raffler (Siemens, Germany)
Matt Reynolds (Georgia Institute of Technology, USA)
Antonio Rizzi (University of Parma, Italy)
Sanjay Sarma (MIT Auto-ID Center, USA)
Albrecht Schmidt (University of Duisburg-Essen, Germany)
James Scott (Microsoft Research, UK)
Ted Selker (MIT Media Lab, USA)
Andrea Soppera (BT Research, UK)
Sarah Spiekermann (Humboldt-University Berlin, Germany)
Michael ten Hompel (Fraunhofer Logistik, Germany)
Frédéric Thiesse (University of St. Gallen, Switzerland)
Khai Truong (University of Toronto, Canada)
Kristof Van Laerhoven (TU Darmstadt, Germany)
Harald Vogt (SAP Research, Germany)
Wolfgang Wahlster (DFKI, Germany)
Kamin Whitehouse (University of Virginia, USA)
John Williams (MIT, USA)

Reviewers

Gildas Avoine (Université catholique de Louvain, Belgium)
John Barton (IBM Research, USA)
Paul Beardsley (MERL, USA)
Richard Beckwith (Intel Research, USA)
Mike Bennett (University College Dublin, Ireland)
Martin Berchtold (University of Karlsruhe, Germany)
Jan Beutel (ETH Zurich, Switzerland)
Urs Bischoff (Lancaster University, UK)
Philipp Bolliger (ETH Zurich, Switzerland)
Gregor Broll (Ludwig-Maximilians-University Munich, Germany)
Michael Buettner (University of Washington, USA)
Tanzeem Choudhury (Intel Research Seattle, USA)
Lorcan Coyle (University College Dublin, Ireland)
Jonathan Davies (University of Cambridge, UK)
Benessa Defend (University of Massachusetts, USA)
Christos Efstratiou (Lancaster University, UK)
Michael Fahrmair (DoCoMo Euro-Labs, Germany)
Martin Feldhofer (Graz University of Technology, Austria)
Guido Follert (Dortmund University, Germany)
Adrian Friday (Lancaster University, UK)
Benoit Gaudin (University College Dublin, Ireland)
Stephan Haller (SAP Research, Switzerland)
Timothy Hnat (University of Virginia, USA)
Paul Holleis (University of Duisburg-Essen, Germany)
Elaine Huang (RWTH Aachen, Germany)
Xavier Huysmans (Isabel, Belgium)
Tatsuya Inaba (Keio University, Japan)
Lenka Ivantysynova (Humboldt-University Berlin, Germany)
Michael Jones (MERL, USA)
Oliver Kasten (SAP Research, Switzerland)
Hans Kastenholz (EMPA, Switzerland)
Dagmar Kern (University of Duisburg-Essen, Germany)
Matthias Lampe (ETH Zurich, Switzerland)
Carsten Magerkurth (SAP Research, Switzerland)
Bill Manning (USC/ISI, USA)
David Merrill (MIT Media Laboratory, USA)
Maximilian Michel (DoCoMo Euro-Labs, Germany)
Masateru Minami (Shibaura Institute of Technology, Japan)
Maxime Monod (EPFL, Switzerland)
Luciana Moreira Sà de Souza (SAP Research, Germany)
Steve Neely (University College Dublin, Ireland)
Britta Oertel (IZT, Germany)
Matthias Ringwald (ETH Zurich, Switzerland)
Christof Roduner (ETH Zurich, Switzerland)

Kay Roemer (ETH Zurich, Switzerland)
Enrico Rukzio (Lancaster University, UK)
Sajid Sadi (MIT Media Laboratory, USA)
Ross Shannon (University College Dublin, Ireland)
Paolo Simonazzi (University of Parma, Italy)
Timothy Sohn (University of California, San Diego, USA)
Tamim Sookoor (University of Virginia, USA)
Joao Pedro Sousa (George Mason University, USA)
Graeme Stevenson (University College Dublin, Ireland)
Martin Strohbach (NEC Europe Ltd., Germany)
Shigeya Suzuki (Keio University, Japan)
Thorsten Teichert (University Hamburg, Germany)
Toshimitsu Tsubaki (NTT Network Innovation Laboratories, Japan)
Keisuke Uehara (Keio University, Japan)
Rossano Vitulli (University of Parma, Italy)
Leonardo Weiss F. Chaves (SAP Research, Germany)
Raphael Wimmer (University of Munich, Germany)
Koji Yatani (University of Toronto, Canada)

Platinum Sponsor

SAP

Gold Sponsors

Metro Group
Siemens

Silver Sponsor

Google
Ambient Systems B.V.

Bronze Sponsors

European Research Consortium for Informatics and Mathematics (ERCIM)
IBM

Table of Contents

EPC Network

Middleware

Business Aspects

RFID Technology and Regulatory Issues

Applications

Sensing Systems

Multipolarity for the Object Naming Service

Sergei Evdokimov, Benjamin Fabian, and Oliver Günther

Institute of Information Systems
Humboldt-Universität zu Berlin
Spandauer Str. 1, 10178 Berlin, Germany
{evdokim,bfabian,guenther}@wiwi.hu-berlin.de

Abstract. The Object Naming Service (ONS) is a central lookup service of the EPCglobal Network. Its main function is the address retrieval of manufacturer information services for a given Electronic Product Code (EPC) identifier. This allows dynamic and globally distributed information sharing for items equipped with RFID tags compatible to EPCglobal standards. However, unlike in the DNS system, the ONS Root is unipolar, i.e., it could be controlled or blocked by a single country. This could constitute a major acceptance problem for the use of the EPCglobal Network as a future global business infrastructure. In this article we propose a modification to the ONS architecture called MONS, which offers multipolarity for ONS and corresponding authentication mechanisms.

> The people who can destroy a thing, they control it

Dune
FRANK HERBERT

1   Introduction

One of the central applications of Radio Frequency Identification (RFID) is efficient identification of physical objects. As compared to its predecessor, the barcode, RFID provides extended reading range, does not require a line of sight between a reader and an RFID tag, and allows for fine-grained identification due to larger amounts of data that can be stored on a tag. However, since most RFID tags still have very modest technical characteristics, it will often be more efficient to let the tag itself only store an identification number. All the data corresponding to this number is stored in a remotely accessible datastore. By taking advantage of the Internet this approach renders such data globally available and allows several parties all over the world to benefit from it.

The future global use of RFID and RFID-related data makes it pivotal to provide common standards for data formats and communication protocols. Currently the primary provider of such standards is EPCglobal – a consortium of companies and organizations set up to achieve worldwide standardization and

C. Floerkemeier et al. (Eds.): IOT 2008, LNCS 4952, pp. 1–18, 2008.

adoption of RFID. According to already developed standards [1], the global availability of RFID related data is achieved by having the RFID tags store an Electronic Product Code (EPC) identifier, while related data is stored in remote datastores accessible via EPC Information Services (EPCIS). For locating a manufacturer EPCIS that can provide data about a given EPC identifier, EPCglobal proposes the Object Naming Service (ONS) [2] that resolves this identifier to the address of the corresponding EPCIS. Based on the same principles as the Domain Name System (DNS), the ONS relies on a hierarchy of namespaces. EPCglobal is delegating control of the root of this hierarchy to VeriSign [3] – a U.S.-based company, also known as a major certification authority for SSL/TLS, one of the DNS root operators, and maintainer of the very large .com domain.

Since RFID tags are foreseen by many to become ubiquitous and play a vital role in supply chains worldwide, such concentration of power in hands of a single entity can lead to mistrust in the ONS, and may involve the introduction of proprietary services, increase in fixed costs, and loss of the benefits that an open, freely accessible, global system could bring. A similar trend can be observed for Global Navigation Satellite Systems: In spite of the fact that the U.S.-operated Global Positioning System (GPS) is globally available, free of charge, and even though deployment and maintenance costs are extremely high, various nations start or plan to introduce their own navigation systems. To prevent a similar fragmentation scenario for the ONS, it seems reasonable to modify the initial design to take the distribution of control between the participating parties into account, and make the ONS *multipolar* – in contrast to the existing unipolar design. In this article we document the unipolar nature of ONS and propose several modifications to allow for multipolarity without radically changing the existing design (unlike e.g. [4]). In addition, we discuss approaches that could make the proposed architecture more secure by ensuring integrity and authenticity of the data delivered.

Our article is structured as follows. First we discuss the current ONS specification from the viewpoint of multipolarity in section 2. Next, in section 3 we discuss DNS principles and procedures, which are also relevant for ONS operations, followed by a comparison of ONS Root vs. DNS Root multipolarity. In section 4 we present MONS, our proposal for multipolar ONS, followed in section 5 by a corresponding outlook on multipolarity of ONSSEC, the use of DNSSEC for ONS data authentication. In section 6 we give a conclusion and discuss future research.

2   ONS – State of the Art

The Object Naming Service (ONS) is the central name service of the EPCglobal Network [2], [1]. It is based on DNS to alleviate efforts required for ONS introduction and operation because DNS is the widely established protocol for name resolution on the Internet [5]. In this section we describe briefly the specifics of ONS, followed by a discussion of this protocol from the viewpoint of multipolarity.

Header 8 Bits	Filter 3 Bits	Partition 3 Bits	Company Prefix 20-40 Bits	Object Class 4-24 Bits	Serial Number 38 Bits
00110000 "SGTIN-96"	001 "Retail"	101 "24:20 Bits"	200452	5742	5508265

Fig. 1. Electronic Product Code (SGTIN-96 Example)

2.1 ONS Principles

The task of ONS is the retrieval of dynamic lists of Web addresses of (usually) manufacturer EPC Information Services (EPCIS) [6] for specific EPC identifiers. Those identifiers, e.g. the 96 bit SGTIN-96 variant in Fig. 1, uniquely identify items and are stored on attached RFID tags compatible to the EPC standard [7] (which we will call EPC tags in the following). The most important parts of such an EPC are Company Prefix, which corresponds to an EAN.UCC Company Prefix and identifies the owner of the prefix – the EPC Manger (usually the item manufacturer), Object Class, which can be assigned by the manufacturer and describes the item category, and Serial Number, which differentiates between similar objects of the same category.

Besides SGTIN-96, the EPC standard also defines several other encoding schemes: GID-96, SGTIN-198, SGLN-96 etc. The choice of a scheme may depend on the application scenario and a company's preferences. In the rest of paper we will be referring to the SGTIN-96 scheme, however, due to the structural similarity of the EPC encoding schemes, proposed solutions are applicable to all the schemes described in the EPC specification.

The ONS and the related, but not yet fully specified EPCIS Discovery Services [1], allow for high flexibility in the linking of physical objects equipped with simple EPC tags and the information about those objects. This information can be stored in various internal or external databases, and can be shared over the Internet using EPCIS, especially those offered by the object manufacturer or by various stakeholders in the supply chain. The list of information sources can easily be updated to include new EPCIS or to change addresses of existing ones, without any change to the anticipated masses of EPC tags deployed in the field.

The inner workings of the ONS are described in [2], for an example query procedure see Fig. 2. Since EPCglobal standards make use of general roles to describe system functionality, we give a short specific example – the arrival of a new RFID-tagged good in a shop. An RFID reader located in the delivery area of the shop reads out the tag and receives an EPC identifier in binary form. Then it forwards the EPC identifier to a local inventory system. This inventory system needs to retrieve item information from the manufacturer's database on the Internet, e.g. to verify the item is fresh and genuine, and to enhance smart advertisement throughout the shop. The system hands the EPC identifier over to a specific software library, the local ONS resolver, which translates the

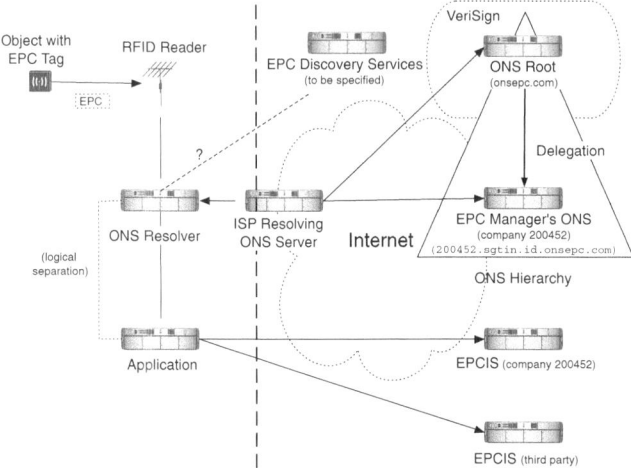

Fig. 2. ONS Resolution

identifier into a domain name compatible with the Domain Name System (DNS, for details of its working see Section 3), e.g. 5742.200452.sgtin.id.onsepc.com. This name, which does not make use of the EPC Serial Number as of now, is an element of the DNS domain onsepc.com that has been reserved for ONS and is used for delegation purposes. The resolver queries the resolving ONS server of its organization or Internet Service Provider (ISP). If the EPCIS address list is not known yet (as in our example of a new item) or has been retrieved and cached before, but is now considered as potentially out-of-date, the ONS Root is contacted. This ONS Root, a service run exclusively by the company VeriSign [3], recognizes the Company Prefix part of the DNS-encoded EPC identifier, and delegates the query to the EPC Manager's ONS server, which has the authoritative address of the manufacturer EPCIS stored in a DNS record called Naming Authority Pointer (NAPTR). Once this address has been determined, the shop inventory system can contact the manufacturer EPCIS directly, e.g. by the use of Web services. To locate different EPCIS for additional information, the use of so-called EPCIS Discovery Services is planned, which are not specified at the time of this writing. However, as is indicated by [1], these search services will (at least in part) be run by EPCglobal.

2.2   ONS and Multipolarity

The ONS Root will formally be under control of the international consortium EPCglobal, but practically run by the U.S.-based company VeriSign. We abstract from these particular circumstances to a more general scenario. Let the ONS Root, as it is designed today, be controlled by a single company C belonging to a nation or group of closely allied nations N. At any given time and state of global politics, there exists the possibility for the government(s) of N to influence those actions of C that concern international relationships – this influence can be exerted either directly via laws, or indirectly via political or economic pressure.

Attack Model: Unilateral ONS Blocking. The current design of the ONS would allow N the following unilateral blocking attack against another nation F: The ONS Root can be easily configured to formally deny any information to clients originating in F (compliant to the ONS protocol), or simply ignore any query from IP addresses belonging to F. An even more efficient way would be to drop inbound ONS packets from F at border routers of N. The result of this attack would be stalled information at all companies in F. Cached addresses of EPCIS could still be used, but cannot be easily updated anymore. To recover, F may consider building its own version of an ONS Root answering all local queries. However, to feed this new root information from alternative external sources would be tedious and probably very time-consuming. There would be serious business drawbacks for companies in F during that time. Companies outside of F, for example in N, would only (and in the worst case for N) be affected if they heavily rely on business with F (due to retaliate blocking of EPCIS access from N by F or stale data at the ONS Root) – this corresponds to a virtual embargo situation. All other companies would not directly be affected, leading to a comparatively low risk for N. In a highly connected global economy based on the EPCglobal network this kind of attack, or even its threat, could be highly effective and more efficient than a simple general disruption of the global system. This should be prevented already by a design that spreads out the control of the ONS Root more evenly.

Attack Model: Traffic Eavesdropping and Analysis. ONS queries and responses are transmitted in plaintext and can easily be read by an adversary who is able to intercept them [8]. The control over the ONS Root allows N to eavesdrop on all ONS queries reaching the root nameservers and to gather global business intelligence about location and movements of items tagged with EPC tags virtually for free and without risk. Such attacks are relatively easy to launch, both technically and legally[1], and could force parties concerned with their privacy to refuse ONS adoption and to look for alternative solutions.

Before we discuss our design proposals to mitigate these attacks in section 4, we first have to take a deeper look at the origin and inner workings of DNS in the next section.

3   ONS vs. DNS Root Control

3.1   DNS Principles

The basic application of the DNS is the resolution of human-memorizable, alphanumerical hostnames into the corresponding purely numerical Internet Protocol

---

[1] According to a recently accepted amendment to Foreign Intelligence Surveillance Act (FISA), U.S. intelligence is allowed to intercept electronic communication between U.S. and non-U.S. bodies if the communication passes across U.S.-based networks (Protect America Act of 2007).

(IP) addresses used for datagram routing. At an early stage of the Internet, the ARPANET, name resolution was performed by referring to a flat text file that stored mappings between the hostnames and the IP addresses. Obviously, maintaining and synchronizing copies of the hosts files on all computers connected to ARPANET was extremely inefficient. To address this issue, the name resolution protocol was updated to introduce a central distribution of the master hosts file via an online service maintained by the Network Information Center. This architecture worked successfully for about a decade. However, the rapid growth of the Internet rendered this centralized approach impractical. The increasing number of changes introduced to the hosts file and its growing size required hosts to regularly download large volumes of data and often led to propagation of network-wide errors.

As a reaction, shortly after deployment of TCP/IP, the new Domain Name System (DNS) was introduced (classical RFCs include 1034, 1035, see [9]). A hostname now has a compound structure and consists of a number of labels separated by dots, e.g. `www.example.com.` (the final dot is often omitted). The labels specify corresponding domains: the empty string next to the rightmost dot corresponds to the *root domain*, the next label to the left to the *top-level domain (TLD)*, followed by the *second-level domain (SLD)* and so forth. The resolution of the hostname into the corresponding IP address is carried out by a tree-like hierarchy of DNS nameservers. Each node of the hierarchy consists of DNS nameservers that store a list of *resource records* (RRs) mapping domain names into IP addresses of Internet sites belonging to a *zone* for which the DNS servers are authoritative. Alternatively, in case of zone delegation, IP addresses of DNS servers located at the lower levels of the hierarchy are returned. The resolution of a hostname is performed by subsequently resolving domains of the hostname from right to left, thereby traversing the hierarchy of the DNS nameservers until the corresponding IP address is obtained.

In practice, not every resolution request has to traverse the whole hierarchy. To reduce the load on the DNS, nameservers use a caching mechanism. For a limited period of time called *time to live* (TTL), DNS resolvers and servers store results of successful DNS queries in a local cache and, when possible, reuse those instead of delegating or issuing queries to other DNS servers. The detailed coverage of DNS mechanism and operations is out of scope of this paper. The interested reader can consult the plethora of existing DNS-related RFCs compiled in [9] and standard literature [5] for more details.

3.2 DNS and Multipolarity

As we outlined above, the DNS is a hierarchy of DNS nameservers, each responsible for resolving hostnames of Internet sites belonging to its zone or pointing to another DNS nameserver if delegation takes place. DNS nameservers authoritative for TLDs (e.g. `.eu`, `.com`) are operated by domain name registries – organizations responsible for managing and technical operation of the TLDs. The root nameservers are operated by governmental agencies, commercial and non-profit organizations. The root zone is maintained by the U.S.-based, non-profit Internet Corporation for Assigned Names and Numbers (ICANN). ICANN was contracted for this purpose by the U.S. Department of Commerce, which thereby holds *de jure* control over the root namespace. Currently the root zone is served by only 13 logical root nameservers, whose number cannot be increased easily due to technical limitations. However, many of those servers are in fact replicated across multiple geographical locations and are reachable via Anycast[2]. As a result, currently most of the physical root nameservers are situated outside of the U.S. [10].

However, the concentration of *de jure* control over the root namespace in hands of a single governmental entity is subject to constant criticism from the Internet community. In theory, this entity has the power to introduce any changes to the root zone file. However, due to the *de facto* dispersal and replication of the root zone, such changes have to be propagated among all the other root nameservers, many of which are beyond the authority of the entity controlling the root zone. In case the entity decides to abuse its power and introduces changes in the root zone by pursuing solely its own benefits, some of the root nameservers may refuse to introduce the changes into their root zone files, which, in the end, may lead to the uncontrolled and permanent fragmentation of the Internet, undermining its basic principles and increasing business risk globally.

These consequences, as well as the fact that such changes have not occurred until now, allow to assume that the Internet is not directly dependant on the entity managing the root namespace, and that it is highly unlikely for this entity to introduce any changes impeding fair and global Internet access. As a consequence, the Blocking Attack is not realistic with DNS without severe risks to the initiating country.

4   MONS – Multipolar ONS

In this section we propose modifications of the current ONS architecture that would allow to distribute the control over the ONS root between several independent parties, thus, solving the issue of unilateral root control.

4.1   Replicated MONS

One of the main reasons why the DNS was chosen for implementing the EPC resolution is, probably, the alleviation of effort required to introduce the ONS on a global scale: The DNS is considered by many practitioners as a mature and time-proven architecture.[3] Its choice allows to deploy the ONS using existing DNS software and rely on best practices accumulated during decades of the DNS being in use. As a result, the deployment of a local ONS nameserver can be relatively easily performed by a system administrator with DNS experience

---

[2] Anycast is a routing scheme that allows to set up one-to-many correspondence between an IP address and several Internet sites so that when an actual communication takes place the optimal destination is chosen (for DNS use cf. RFC 3258).

[3] For dissenting arguments, however, see e.g. [11], [8].

using freely available software. Thus, if we want to modify the existing ONS architecture, it makes sense to stay consistent with the DNS protocol.

The ONS root will run on six locally distributed server constellations, all operated by VeriSign [3] (Fig. 3(a)). This strongly contrasts with the DNS architecture, where the root nameservers are operated also by numerous other entities [10]. A straightforward approach to avoid the unipolarity of the ONS is to replicate the ONS root between a number of servers operated by independent entities, and to synchronize the instances of the root zone file with a master copy published by EPCglobal. To restrict the amounts of incoming queries, each root nameserver could be configured to cover a certain area in the IP topology and respond only to queries originating from there.

Such replicated ONS root nameservers could provide their services in parallel with the global ONS root operated by VeriSign. The resolving ONS servers of organizations and Internet Service Providers (ISP) should be configured on the one hand with the domain name or IP address of the global ONS root (`onsepc.com`), or, more efficiently, the server responsible for SGTIN (`sgtin.id.onsepc.com`), on the other hand also with the corresponding replicated ONS server (e.g. `sgtin.id.onsepc-replication.eu`), potentially avoiding Anycast constructions like those used as later add-ons for DNS.

To evaluate the feasibility of this approach and the amount of data that has to be replicated, we approximately calculate the size of the ONS root zone file by estimating the number of RRs stored there, which define mappings between Company Prefixes and domain names of the corresponding ONS nameservers. Today, there are about one million registered Company Prefixes.[4] We assume that at a certain time in future most of them will have corresponding EPCIS services. The ONS root zone file is a plain text file consisting of a number of NS RRs. As an example, consider an EPC number 400453.1734.108265 that can be resolved into one of two ONS nameservers:

```
1737.400453.sgtin.onsepc.com IN NS ons1.company.com
1737.400453.sgtin.onsepc.com IN NS ons2.company.com
```

IN stands for Internet, and NS indicates that the record defines a nameserver authoritative for the domain. The number of nameservers responsible for the same zone cannot exceed thirteen, and the DNS specification recommends having at least two. In practice, however, their number usually varies from two to five.

Assuming the average number of ONS nameservers per company (N) as four, the average length of an NS record (L) as 60 symbols, and that one symbol takes one byte, and the number of registered Company Prefixes (P) as one million, we can roughly estimate the size of the ONS root zone file containing the RRs for all currently registered EAN.UCC Company Prefixes as $N \times L \times P$, which is slightly above 200 megabytes. By using compression a text file may be reduced to 10-20% of its original size. Thus we conclude that the distribution and regular renewal of the root file presents no technical difficulties. The master root file can be shared between ONS roots by the means a simple file transfer or a

---

[4] `http://www.gs1.org/productssolutions/barcodes/implementation/` (09/2007).

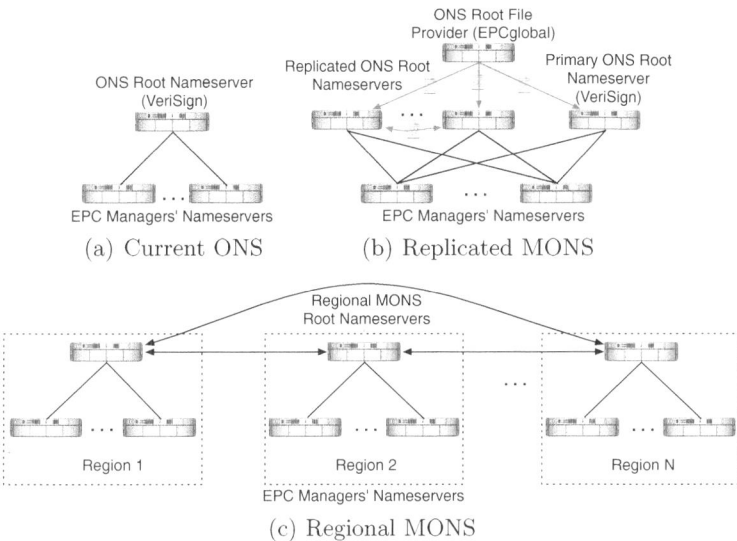

(a) Current ONS   (b) Replicated MONS

(c) Regional MONS

Fig. 3. MONS Architectures

peer-to-peer file sharing protocol. The architecture is illustrated at Fig. 3(b) and will be further referred to as Replicated MONS.

The key requirement of Replicated MONS is the public availability of the ONS root file. As soon as the root file is published and regularly updated, the replicated roots can be deployed independently from each other. In case those new roots will be configured to cover only certain areas, locations beyond their bounds will still be able to use VeriSign's nameservers, remaining vulnerable to the Blocking Attack.

4.2   Regional MONS

The architecture described in the previous section provides a solution which allows any entity to maintain a copy of an ONS root nameserver, enhancing the availability of the ONS. However, due to the necessity to cope with a high load, such nameservers might not be accessible globally, potentially resulting in a (from a global perspective) unstructured patchwork of areas with ONS root redundancy. The high load on the root nameservers will be mainly caused by the size and frequent updates of the root zone file. Compared to the DNS root zone file, which contains RRs on about 1500 TLD nameservers and currently has a size of about 68 kilobytes[5], the ONS root zone file will contain RRs for *all* EPC Managers' ONS nameservers registered at EPCglobal. With RFID becoming ubiquitous, their number is expected to grow rapidly, resulting in millions of RRs. Also, due to a higher volatility of ONS root RRs, their TTL parameters

---

[5] http://www.internic.net/zones/ (09/2007)

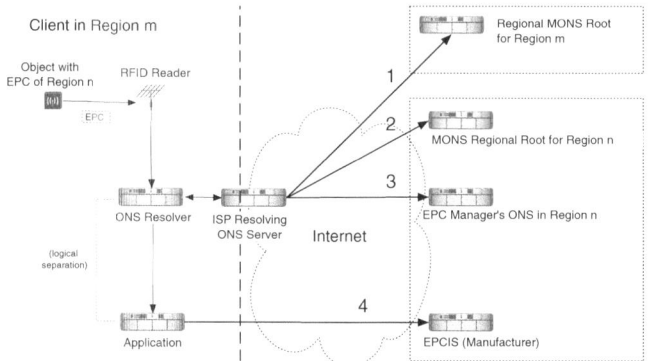

Fig. 4. Regional MONS Resolution

might be assigned lower values as compared to the RRs of the DNS root. As a result, the ONS RRs will be cached for shorter periods of time and a larger number of queries will be reaching the ONS root nameservers.

In this section we suggest a more radical alteration of the existing ONS architecture that will allow to reduce the size of the root zone file and the frequency of its updates by splitting it between a number of *regional root nameservers*, at the same time offering a structured way to achieve area coverage for redundancy. A zone file of each regional nameserver contains RRs that correspond to EPC Managers belonging to a region for which a nameserver is authoritative. The membership to a region might be determined by a company's registration address, regional GS1 department that issued the Company Prefix, or other properties.

The architecture is depicted in Fig. 3(c), while the resolution process is presented in Fig. 4. In case the resolving nameserver and the EPC Manager (who corresponds to the EPC being resolved) belong to the same region ($n = m$), the step 2 is omitted and the resolution process is almost identical to the one depicted in Fig. 2: The regional root nameserver delegates the query to the nameserver of the EPC Manager which returns the address of the EP-CIS. However, if $n \neq m$, the query is redirected to the regional root nameserver authoritative for the Region n (step 2), which in turn delegates it to the nameserver of the EPC Manager. We will refer to this architecture as *Regional MONS*.

Compared to the ONS resolution process described in Section 2.1, the case of the delegation of a query from one regional ONS nameserver to another (step 2) introduces an additional resolution step. Consequently, this requires an extension of the EPC scheme and the introduction of a new prefix that will be resolved at this step. Following the approach for constructing an EPC, a natural choice would be a *regional prefix* pointing to a country or a region of origin for a given product. The introduction of this regional prefix requires an update of the EPC encoding standards, which might result in a lengthy and costly process.

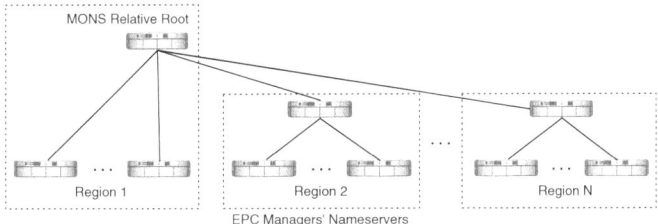

Fig. 5. Relative Hierarchy of Regional MONS Nameservers

However, the EPC encoding schemes defined in [7] already contain enough information to unambiguously associate an EPC with a certain region. The first three digits of the EAN.UCC Company Prefix identify the country of GS1 membership for the company (e.g. codes 400-440 are reserved for Germany). Therefore, an alternative to the introduction of a new regional prefix field would be to use these digits for associating EPC identifiers with corresponding regions. Each regional root nameserver will be responsible for one or several regional prefixes.

Note that a resolver still sees the Regional MONS architecture as a hierarchy: The MONS root of its region is being perceived as the root of the whole hierarchy (Fig. 5). We call such a structure a *relative hierarchy*. A regional nameserver authoritative for a region from which the resolution takes place is called its *relative root*. This allows to implement the Regional MONS within the DNS framework, reflecting the approach described in the ONS specification.

In the following, we assume that the regional prefix is defined as the first three digits of the Company Prefix. To access an EPCIS that could provide data about a given EPC identifier, the identifier is again translated into a DNS-compatible address, but now the first three digits of the Company Prefix have to be explicitly separated by dots and placed to the right of the rest of the inverted EPC identifier (e.g. 1734.453.400.sgtin.id.onsepc.com). Assume that the domain name of the regional nameserver authoritative for zone 400.sgtin.id.onsepc.com is ns1.mons.eu. An ONS client physically located at the same region is configured to sends all its ONS queries to ns1.mons.eu (step 1 at Fig. 4), which it views as the relative root of the Regional MONS. Correspondingly, a resolver that belongs to a different region will be configured with the address of a different regional root, also viewed as relative root. In this example we deliberately choose the domain name of the regional root to have the TLD (.eu) corresponding to the region of its authority. This avoids the dependency on entities administering regional nameservers domains and excludes the possibility of a Blocking Attack from their side. Note, that the resolution process described above does not require an EPC identifier to be translated to the domain name resolvable by the DNS of the Internet. The only domains relevant to the ONS resolution are the dot-separated EPC identifier and the domain pointing out in which format an

EPC number is stored. This makes the three rightmost domains abundant since 1734.453.400.sgtin is already sufficient for unambiguous ONS resolution.

By appointing specific nameservers to regions, Regional MONS naturally shifts the load to nameservers authoritative for economically developed or industrial countries, since regional prefixes of such regions will occur on the majority of the EPC identifiers. Moreover, regions whose export values are too low, or who are not interested in maintaining their own Regional MONS root nameservers could delegate this responsibility to third parties, as it is sometimes done with country code TLDs [10]. Once their situation changes, they can take back their reserved share of the system by a minor change in the table of Regional MONS Roots (MONS Root Zone).

4.3   Regional MONS Prototype

In this section we present a possible fragment of the Regional MONS architecture implemented using BIND DNS Server software. BIND (Berkeley Internet Name Domain) is the most common DNS server in the Internet and the *de facto* standard for Unix-based systems. ONS can be deployed using standard DNS software, so it is very likely that a considerable portion of ONS nameservers will be using BIND. In our sample scenario we consider two regions with regional codes 400 and 450 and two EPCISs, each providing information about one of the following SGTIN formatted EPC identifiers: 400453.1734.108 and 450321.1235.304.

The main configuration file of a BIND server is the named.conf. RRs for namespaces are stored in zone files often named *namespace*.db. Fig. 6 presents a possible configuration of four ONS nameservers that constitute this fragment of the Regional MONS hierarchy. The fragment includes two regional MONS root nameservers authoritative for regional prefixes 400 and 450, correspondingly, and two nameservers of EPC Managers.[6] The regional roots are configured as relative roots of the sgtin zone and as authorities for the respective regional codes (400.sgtin and 450.sgtin, correspondingly). The sgtin.db file describes the relative root zone (sgtin) by declaring the nameserver as the authority for this zone and referring to the content of onsroots.db file, which represents the MONS Root Zone. This file is the same for all regional roots and defines the delegation of the zones (using the regional codes) to the regional roots. The RRs of the 400.sgtin.db and 450.sgtin.db files introduce a further delegation step by pointing to the nameservers of the respective EPC Managers that complete the resolution process by returning the URI of the requested EPCIS via NAPTR RR. To make the zone files less dependent on infrastructure changes in the MONS hierarchy, they may contain only NS records without mentioning the corresponding IP addresses in A records. So, if one or several nameservers has its IP address changed the zone files still remain consistent. However, this can prolong the resolution process, since ONS nameservers will have to query the DNS to resolve domain names to IP addresses.

---

[6] Note that all domain names, IP addresses and URIs in this example are fictional.

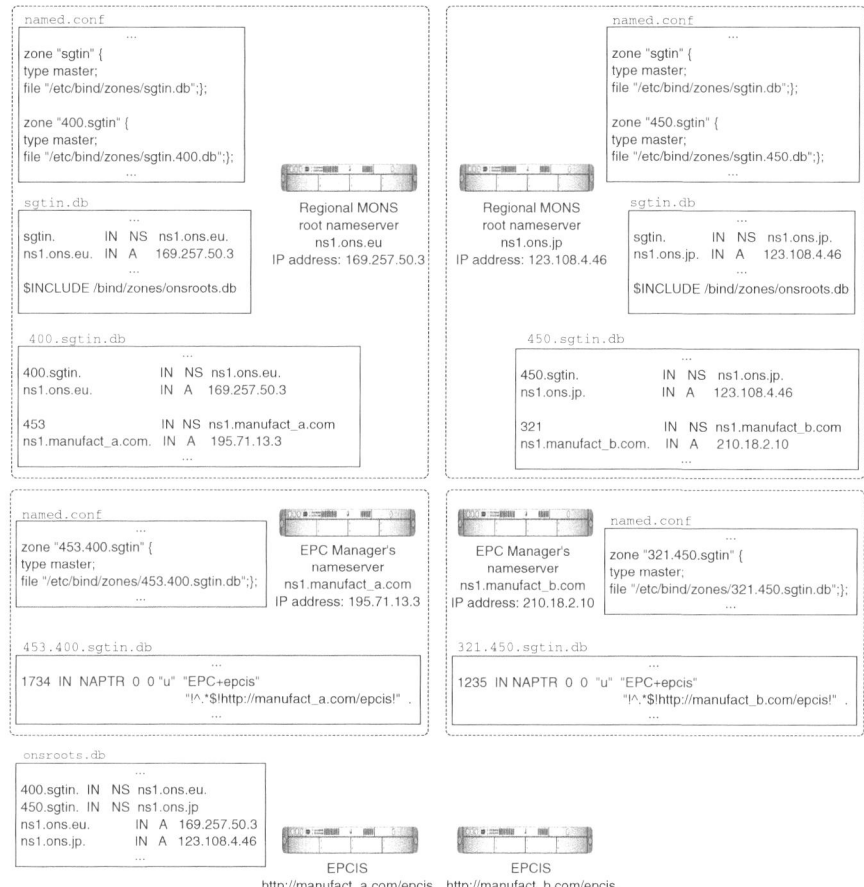

Fig. 6. Fragment of Regional MONS Hierarchy

4.4 Modularity

One further advantage of Regional MONS is that each region could implement different resolution architectures for its own subsystem below the root zone. For example (see Fig. 7), a region r could use the original ONS specification based on the DNS, another region n could use a centralized search system, while yet other regions, like m, could implement subsystems based on Distributed Hash Tables (DHT), e.g. the OIDA system proposed in [4]. Delegation between MONS and heterogeneous subsystems can be established by bridging nodes that are able to use both protocols. In the DHT case for example, a DHT node queried by external DNS clients uses the DNS protocol to answer. However, to communicate with other DHT nodes, the specific overlay network communication is used, for example as defined in Chord [12]. This combination of DNS and DHT has been successfully implemented for general DNS use, for example in CoDoNS [11].

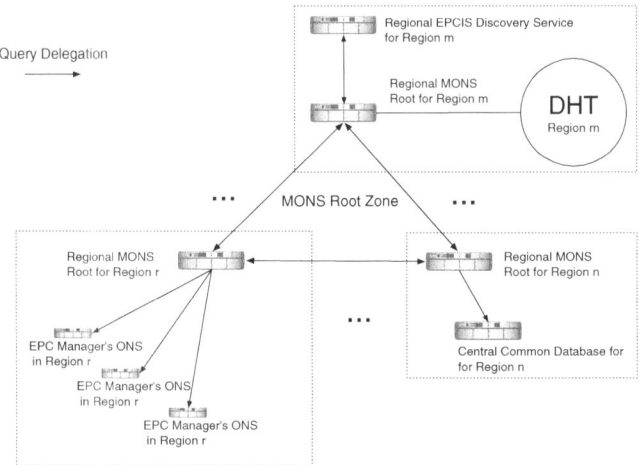

Fig. 7. Modularity of MONS Subsystems

5   MONS Data Authenticity

Today's Internet has to be regarded as a highly insecure environment, a fact that has been acknowledged not only by the security community, but also political institutions [13]. Surprisingly, security measures have not been considered intrinsically from the beginning in the EPCglobal architecture standards [8], but seem to be held as optional and mostly to be added later by its users [1]. Besides availability and confidentiality risks of the EPCglobal Network and the ONS in particular, a major concern is the lack of authentication methods in the current ONS standard [2]. Without additional security measures, global business systems depending on the ONS, as it has been designed in the standard so far, could suffer from cache poisoning and man-in-the-middle attacks [14], leading to spoofed EPCIS address information, and potentially also to forged EPC information, or via additional vulnerabilities, malware infection initiated by malicious servers. Adding countermeasures like DNS Security Extensions (DNSSEC) later, however, will also have an impact on properties of the whole system, like performance, security and privacy, as well as in our case, multipolarity.

In this section we first take a short look at the recent DNSSEC standards, discuss how DNSSEC could be used in a straightforward way to secure ONS data, leading to a substructure of DNSEC we propose to call ONSSEC. Finally we suggest mechanisms to achieve multipolarity for ONSSEC, thereby enabling its use for MONS (short for Regional MONS from now on).

5.1   DNSSEC

To address the lack of authentication in the DNS, a set of mechanisms called DNSSEC (DNS Security Extensions) has been designed, the recent version being presented in [15] and related other RFCs. The DNSSEC provides data integrity and authenticity for the delivered DNS information by using public-key cryptography to sign sets of resource records (RRs). It uses four resource record types: Resource Record Signature (RRSIG), DNS Public Key (DNSKEY), Delegation Signer (DS), and Next Secure (NSEC), the last one is used to provide authenticated denial of existence of a zone entry, for details cf. [15]. Each DNS zone maintainer is also responsible for providing a signature of those zone files. These signatures are stored in an RRSIG record. The server's public key could be transferred out-of-band, or be stored and delivered via DNS itself using an RR of type DNSKEY.

The use of separate zone-signing and key-signing keys enables easy resigning of zone data without involving an administrator of the parent zone [5]. However, having a signature and an apparently corresponding public key does not guarantee authenticity of the data – the public key and identity must be securely linked by a trusted entity, most practically, by the maintainer of the respective parent zone. To be able to verify an arbitrary DNS public key in a scalable way, chains of trust down from the (necessarily trusted) root of the DNS would be necessary, where each parent DNS server signs the keys of its children, after having verified its correspondence to the correct identity by some external means.

Even after a major redesign of the protocol (and its RRs) in 2005, cf. RFC 4033 [15] (which replaces RFC 2535 from 1999 that in turn obsoleted the original RFC 2065 dating from 1997), DNSSEC is not yet widely established throughout the Internet, though recent developments like the signing of some countries' TLD seem to indicate a brighter perspective for DNSSEC [16]. Reasons for the slow DNSSEC adaption include, first of all, reluctance to major changes for critical services like DNS, scalability problems of key management, the administrative problem of building chains of trust between servers of many different organizations. There also is the problem of establishing a critical mass of DNSSEC users with different incentives [17]. Despite these problems, the establishment of a new global business architecture like the EPCglobal Network could be a major opportunity to launch ONSSEC, the adaption and restriction of DNSSEC to ONS use. However, DNSSEC suffers from a major unipolarity problem: Who should control the anchor of trust, the keys for the root zone? This problem must be solved for a multipolar ONS, to avoid unwanted indirect unipolarity for MONS introduced by its security extensions.

5.2   ONSSEC

DNSSEC can be applied to MONS as follows, cf. Fig. 8: Each Regional MONS Root provider signs the key-signing keys of all EPC Managers in its region. This is major administrative task and has to involve the verification of the EPC Manager's identity. This procedure is, however, less cumbersome then signing *all* subdomain keys of a given TLD, rendering ONSSEC introduction more scalable than general DNSSEC where probably also more delegation steps are involved. The EPC Managers then are able to sign their own zone-signing keys and the actual zone data. They can repeat the latter procedure after each change in zone

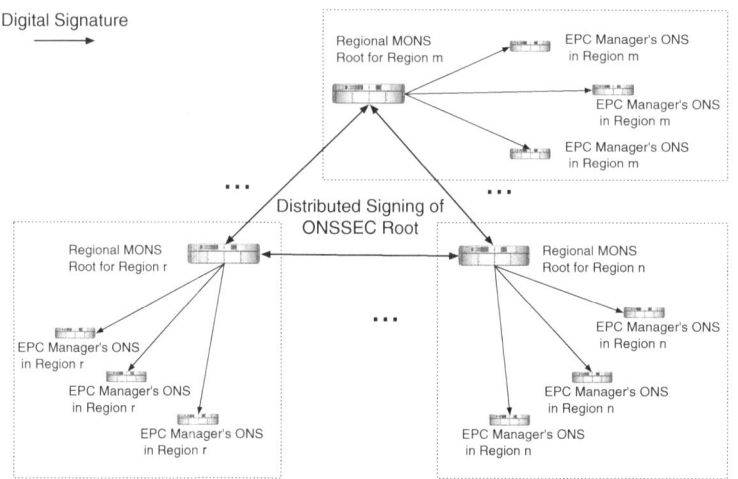

Fig. 8. Multipolar ONSSEC Trust Structure

data without contacting the regional root; they are also able to periodically change their zone-signing keys for better long-term security. The EPC Manager's nameservers can now answer MONS queries by returning the actual zone information in combination with the signature. This signature can be verified by a client by retrieving the public key of the regional MONS root. Here another (cf. section 4.3), bigger problem of using the flexible option of general DNS names in (M)ONS resource records becomes apparent (e.g. in URIs of NAPTR records for EPCIS, see Fig. 6): Without an established global trust structure and ubiquitous use of DNSSEC, arbitrary DNS names and resolution steps would not easily be covered by authentication measures. As long as this situation holds, the tradeoff between flexibility vs. lack of authenticity needs to be constantly evaluated.

With the described Regional MONS architecture, there would be multiple roots of trust. This situation could be impractical, because clients who often resolve EPCs of foreign regions would have to trust multiple public keys, those of the local and all foreign regional MONS roots. With DNSSEC, it is often stated as best practice that a single entity should control the root zone key signing keys. It is, however, subject to current international debate, which organization should represent this entity – for example, interest has been expressed by US authorities like the Department of Homeland Security [18]. A similar problem exists for the MONS root zone (the `onsroots.db` of the prototype in section 4.3). In the following section, we briefly discuss options for a solution.

5.3   Multipolarity for the ONSSEC Root

Multipolarity for the root key control of ONSSEC (that is DNS Security Extensions applied to (M)ONS) could be achieved by multiple signatures (each regional

MONS root would sign the root zone) [19], or more elegantly and scalably, by the use of one virtual ONSSEC root by applying threshold cryptography. An *(n, t)-threshold cryptography scheme* allows n parties to share the ability to perform a cryptographic operation (e.g., applying a digital signature), so that t $(t \leq n)$ parties can perform this operation jointly, but at most $t - 1$ (malicious) parties are not able to do so, even by collusion [20, pp. 525]. Famous threshold secret sharing schemes include [21], using polynomial interpolation, and [22] based on intersection of n-dimensional hyperplanes. Secret sharing could be used to share the private key of the virtual ONSSEC root, but once used the whole private key might become compromised.

More secure are threshold function sharing schemes, extensions of the basic secret sharing, which allow for digital signatures without letting a single party know the complete key during operations, see e.g. [23,24] for schemes with usable performance properties. The signing of the regional root keys and the MONS root zone should be quite a rare operation in comparison to the signing of actual manufacturer zone data. Therefore, these schemes could be implemented without major performance penalties on the whole system. In summary, using threshold cryptography would enable the distributed and multipolar signing of the MONS regional root keys (Fig. 8), as well as the MONS root zone that contains address data of all Regional MONS Roots.

6   Conclusion and Future Research

In this paper we presented MONS, a practical architecture to achieve multipolarity in the ONS. We also showed how multipolarity in corresponding authentication extensions can be achieved. To our knowledge, this is the first extensive discussion and solution proposal of the multipolarity problem for ONS, which in a future "Internet of Things" may have even more detrimental consequences than the analogous problem currently debated for DNS [19]. We focus so far on a technical perspective, where our future work will include a sample implementation of distributed signing of the ONSSEC root zone, which may also become relevant for DNSSEC. On the policy side, analysis of the practical political and administrative challenges of distributing control over the ONS is an important line for future research. Not last, there is urgent need to solve further multilateral security problems of ONS and related systems like MONS, especially their possible impact on corporate and individual privacy.

References

1. EPCglobal: The EPCglobal Architecture Framework – Version 1.2. In: Traub, K. (ed.) (September 2007)
2. Mealling, M.: EPCglobal Object Naming Service (ONS) 1.0 (2005)
3. EPCglobal: Implementation of the EPCglobal Network Root ONS. EPCglobal Position Paper (November 2005)

4. Fabian, B., Günther, O.: Distributed ONS and its Impact on Privacy. In: Proc. IEEE International Conference on Communications (IEEE ICC 2007), Glasgow, pp. 1223–1228 (2007)
5. Liu, C., Albitz, P.: DNS and BIND, 5th edn. O'Reilly & Associates, Sebastopol (2006)
6. EPCglobal: EPC Information Services (EPCIS) Version 1.0 Specification (April 2007)
7. EPCglobal: EPC Tag Data Standards Version 1.3 (2006)
8. Fabian, B., Günther, O., Spiekermann, S.: Security Analysis of the Object Name Service. In: Proceedings of the 1st International Workshop on Security, Privacy and Trust in Pervasive and Ubiquitous Computing (SecPerU 2005), with IEEE ICPS 2005, Santorini, pp. 71–76 (2005)
9. Salamon, A.: DNS related RFCs, http://www.dns.net/dnsrd/rfc/
10. Gibbard, S.: Geographic Implications of DNS Infrastructure Distribution. The Internet Protocol Journal 10(1), 12–24 (2007)
11. Ramasubramanian, V., Sirer, E.G.: The Design and Implementation of a Next Generation Name Service for the Internet. In: SIGCOMM 2004: Proceedings of the 2004 conference on Applications, technologies, architectures, and protocols for computer communications, pp. 331–342. ACM Press, New York (2004)
12. Stoica, I., Morris, R., Liben-Nowell, D., Karger, D.R., Kaashoek, M.F., Dabek, F., Balakrishnan, H.: Chord: A Scalable Peer-to-peer Lookup Protocol for Internet Applications. Networking, IEEE/ACM Transactions on 11(1), 17–32 (2003)
13. PITAC: Cyber security - a crisis of prioritization (February 2005), http://www.nitrd.gov/pitac/reports/2005030_cybersecurity/cybersecurity.pdf
14. Atkins, D., Austein, R.: Threat Analysis of the Domain Name System (DNS). In: Request for Comments - RFC 3833 (2004)
15. Arends, R., Austein, R., Larson, M., Massey, D., Rose, S.: DNS Security Introduction and Requirements. In: Request for Comments - RFC 4033 (March 2005)
16. Friedlander, A., Mankin, A., Maughan, W.D., Crocker, S.D.: DNSSEC: A Protocol toward Securing the Internet Infrastructure. Commun. ACM 50(6), 44–50 (2007)
17. Ozment, A., Schechter, S.E.: Bootstrapping the Adoption of Internet Security Protocols. In: Proc. of the Fifth Workshop on the Economics of Information Security (WEIS 2006) (2006)
18. Leyden, J.: Homeland Security Grabs for Net's Master Keys (2007), http://www.theregister.co.uk/2007/04/03/dns_master_key_controversy/
19. Kuerbis, B., Mueller, M.: Securing the Root: A Proposal for Distributing Signing Authority. Internet Governance Project Paper IGP 2007-2002 (2007)
20. Menezes, A.J., van Oorschot, P.C., Vanstone, S.A.: Handbook of Applied Cryptography. CRC Press, Boca Raton (1997)
21. Shamir, A.: How to Share a Secret. Commun. ACM 22(11), 612–613 (1979)
22. Blakley, G.R.: Safeguarding Cryptographic Keys. Proc. of the National Computer Conference 48, 313–317 (1979)
23. Shoup, V.: Practical Threshold Signatures. In: Preneel, B. (ed.) EUROCRYPT 2000. LNCS, vol. 1807, pp. 207–220. Springer, Heidelberg (2000)
24. Kaya, K., Selçuk, A.A.: Threshold Cryptography based on Asmuth-Bloom Secret Sharing. Inf. Sci. 177(19), 4148–4160 (2007)

Discovery Service Design in the EPCglobal Network
Towards Full Supply Chain Visibility

Chris Kürschner[1], Cosmin Condea[1], Oliver Kasten[1], and Frédéric Thiesse[2]

[1] SAP (Switzerland) Inc., SAP Research CEC St. Gallen, Blumenbergplatz 9,
CH-9000 St. Gallen, Switzerland
{chris.kuerschner,cosmin.condea,oliver.kasten}@sap.com
[2] ITEM-HSG, University of St. Gallen, Dufourstrasse 40a, CH-9000 St. Gallen,
Switzerland
frederic.thiesse@unisg.ch

Abstract. The EPCglobal Network, an emerging standard for RFID,
aims to raise visibility in supply chains by enabling interested parties to
query item-level data. To get there, however, a critical piece is yet missing: a Discovery Service to identify possibly *unknown* supply chain actors
holding relevant data for specific EPC numbers of individual products.
Unfortunately, the Discovery Service architecture as initially conceived
by EPCglobal needs revision as it either infringes the confidentiality of
participating companies or its use is limited to identifying only participants already known. Against this background, this paper first discuusses
the limitations of the architecture under consideration by EPCglobal
and presents an alternative, more adequate Discovery Service design.
Our concept encourages participation in the network while ensuring information provider confidentiality. Secondly, we present a roadmap for
extending the existing EPCglobal Network with two critical services: an
automated contract negotiation service and a billing service.

1 Introduction

Cost pressure in supply chain related processes has steadily been increasing over
the last years. Hence, companies put much effort into reducing inefficiencies in
the supply chain including incorrect demand and sales forecasting, low on-shelf
availability, and inaccurate inventory levels [13]. However, the relevant information which can be used to overcome these problems is distributed among the
partners within a supply chain [19]. Against this background, being effective and
efficient in matching demand with supply requires a tight collaboration between
supply chain parties [12].

Although collaboration promises mutual benefits for the partners, those benefits are rarely realized [20]. There are two critical issues to be overcome. First,
in today's complex and dynamic supply networks, each company within the network has only partial knowledge about the participation of other companies.
Hence, retrieving complete information regarding the flow of goods through the

C. Floerkemeier et al. (Eds.): IOT 2008, LNCS 4952, pp. 19–34, 2008.
© Springer-Verlag Berlin Heidelberg 2008

network requires high effort for locating all actors. Second, even if a company was able to locate all relevant supply network participants, there might be a lack of incentive to share sensitive operational data. Moreover, information sharing is usually based on contracts. Selecting and negotiating contracts is a non-trivial challenge for many companies and one common supply chain problem [25]. As a consequence, gaining sufficient transparency within a supply network coordinated by contracts cannot be established at acceptable costs and supply networks are bound to be inefficient and ineffective.

The EPCglobal Network attempts to solve this issue by enabling an interested party to find data related to a specific EPC number and to request access to these data. This infrastructure uses radio frequency identification (RFID) technology and leverages the Internet to access large amounts of information associated with Electronic Product Codes (EPC) [15]. A central component of this network realizing the abovementioned functionality is the Discovery Service which, according to EPCglobal, is to be mostly employed when it is "impractical to follow the chain" because participants are not known in advance [16]. In their current view, a client first contacts the Discovery Service with an EPC number. Then, the Discovery Service, based on the published records, replies with the owner of the information so that the client can directly contact the information owner for further details. Unless access rights accompany the published data within the Discovery Service, this has a privacy implication for the identity of the information owner is revealed irrespective of its decision to provide the information or not. On the other hand, maintaining access rights on the Discovery Service increases the software and management complexity. In addition, since these access rights can be strictly defined for known partners, this solution prohibits interaction with unknown organizations.

To counter this problem, our proposed new scheme makes the Discovery Service act as a proxy to forward a client request to the information owner. This solution, which we call the "Query Relay", allows for full access control at each organization. The information owner may reply with detailed information directly to the client or simply ignore the request, thus not revealing its identity. Furthermore, since the client request is relayed to possibly previously unknown partners, our complete solution envisages, besides the Discovery Service designed as described above, two more components. The first is an *automated contract negotiation service* whose role is to establish a one-to-one contract between two unknown companies stipulating the terms and conditions under which data can be shared. The second is a *pay-per-information pricing mechanism* that is supposed to give the information owner an incentive to provide its data, given it is financially recompensated. Nevertheless, note that the focus of this paper is on the architectural design of the EPC Discovery Service where we compare the design currently under EPCglobal consideration to our newly introduced "Query Relay". The two additional components mentioned above are only presented on a more conceptual level. In any case, the general emphasis is on the interaction among companies with no a priori knowledge about each other [16].

The remainder of the paper is organized as follows. In Section 2, we give an overview of existing discovery service concepts. We then present in Section 3 the requirement analysis, followed by the actual design solution in Section 4. Here, we compare the two Discovery Service design options – the original design as suggested by the EPCglobal or by the related literature and our own proposed design. Then, in Section 6, we describe how to enhance the EPCglobal infrastructure with an automated contract negotiation service and pay-per-information pricing. Our paper closes with an outlook on further research and a summary of our findings including managerial implications.

2   Related Work

A notable amount of research has been conducted in the area of discovery services. Alternative terms in literature are 'lookup service', 'directory service' and 'naming service'. In general, a discovery service provides a method for establishing contact between a client and a resource. A number of different approaches exist which vary according to the specific requirements they are designed to meet. In the following, we show that the existing solutions are not suitable for the EPC Discovery Service design.

The Domain Naming Service (DNS) is an essential component of the Internet [22]. It performs global discovery of known services / domains by mapping names to addresses. An assumption of this solution is that keys uniquely map to services, and that these keys are the query terms. Furthermore, all resources are publicly discoverable and access control is done at the application level rather than on the discovery service level [8]. But in contrast to Internet where domain addresses are freely available, EPC-related information needs to be protected and only selectively shared. The Domain Name Service is therefore not a good fit to build EPC Discovery Services.

Another example is the Service Location Protocol (SLP) which provides a framework to allow networking applications to discover the existence, location, and configuration of services in enterprise networks [23]. SLP eliminates the need to know the names of network hosts; only the description of the service is needed based on which the URL of the desired service is returned. However, SLP is not a global resolution system for the entire Internet; rather it is intended to serve enterprise networks with shared services, which makes it unsuitable for EPC Discovery Services. Furthermore, no authentication of an entity looking for appropriate services is provided. Finally, because access control is missing, confidentiality of service information cannot be guaranteed [8].

Last, but probably closest, are the first implementations of a Discovery Service in the context of supply chains done by IBM [6] and Afilias [2]. While the IBM demo represents a good start, showcasing the feasibility and usefulness of Discovery Services, there is no indication of a requirements-driven design. The Afilias prototype is more mature and the first to illustrate how security within the Discovery Service can be build. Nonetheless, they both follow the EPCglobal concept whose limitations will be explicitly addressed later in the present work.

3   Requirements Analysis

Based on interviews conducted with experts and end users, literature review and project work, we identified several issues that are highly relevant for the design of the Discovery Service and for the wide adoption of EPCglobal Network.

3.1   Data Ownership

From a technical standpoint, publishing data to the Discovery Service is readily defensible, particularly, when taking into account recent developments in information and telecommunication technologies. Technical considerations, however, represent only part of a more complex story in which less tangible issues play a critical role. That is, from an economic standpoint, data control aspects need to be considered. Our investigations show, that at least some companies are not willing to even share the EPC and the related EPCIS addresses with other companies.

It seems that local control can be optimal even when there are no technical barriers to a more centralized solution, like observed by Markus [21]. Key reason for the importance of data ownership is self-interest, that is, owners have greater interest in system success than non-owners. As a consequence, ignoring ownership aspects for system development might be a possible explanation for failures [26]. Based on this background, Davenport et al. state "No technology has yet been invented to convince unwilling managers to share information..." [9]. Today, some information is just too valuable to give it away. According to van Alstyne et al., we define data ownership as the right to determine data usage privileges to other companies and as the ability to track the actual usage [26].

Based on our findings, there are two requirements for the Discovery Service design in regards to data ownership. These are:

Requirement 1. *Companies shall be in complete control over their data including EPCIS addresses, read events, business data as well as setting of detailed, fine-grained access rights.*

Requirement 2. *Companies shall be able to track the usage or the requests upon their data. Particularly, duplications of data at the Discovery Service level should be avoided.*

3.2   Security

Security has always been a critical issue in distributed systems. In our concrete context, it becomes a decisive factor for the broad adoption of the EPCglobal Network. At the heart of the network, enabling the detection of information sources, the Discovery Service consequently demands for very strict security measures that realistically justify inter-company collaborations and data sharing.

The notion of a secure system is linked in the literature to that of dependability [18]. Simply put, a dependable system is one that we trust to deliver its services. More specifically, dependability comprises availability, reliability, safety, confidentiality, integrity, and maintainability. Availability is defined as a system's immediate readiness for usage whereas reliability refers to the continuity of service over a prolonged period of time. Confidentiality expresses the property of a system to disclose information only to authorized parties. Integrity is the characteristic that prevents improper alterations to a systems assets. To ensure security, one needs to associate integrity and availability with respect to authorized access together with confidentiality [18]. While all of the above security features are essential, there are two requirements that stand out for the architecture of the Discovery Service. These are:

Requirement 3. *The confidentiality of both the publisher data and client query shall be ensured by the Discovery Service design.*

Requirement 4. *The Discovery Service architecture shall ensure a high overall system availability and reliability.*

3.3 Business Relationship Independent Design

Nowadays, the business environment is dynamic. Different customer demands, globalization, discovery of uncharted market opportunities, outsourcing, innovation and competition, the continuously varying technological landscape are just some of the major factors that determine significant partnering changes in supply chains. To that end, in order to increase their own strategic flexibility, many companies find changing supply chain partners simpler and also cheaper than changing internal processes [7].

To protect its data from unauthorized access, a company must define and maintain fine-grained, customized permissions for each of its partners. As soon as a company modifies its set of trading partners or simply its existing collaborations, it also needs to define access rights reflecting the new business relationships. We learnt from our investigations that companies want to minimize the access control maintenance effort. This leads to the establishment of an additional important requirement for the Discovery Service, that is:

Requirement 5. *Changes in business relationships shall not affect the way in which a company interacts with the Discovery Service.*

3.4 Organic Growth

One key issue of the EPCglobal Network is organic growth, that is, over time more and more companies will participate in the network and the data volume will increase. In this context, Rogers [24] stated that adopters of any new idea or innovation can be categorized as innovators (2.5%), early adopters (13.5%), early majority (34%), late majority (34%) and laggards (16%). Hereby, an adopter's willingness and ability to adopt an innovation depends mainly on their awareness, interest, evaluation results, and successfully performed trials.

Based on these findings, we have to be aware of the fact that initially only a few companies will be willing and able to adopt the EPCglobal Network. Over time, more and more companies will join, leading to a massive increase in data volume. Thus, supporting an organic growth is crucial for the design of the Discovery Service. Furthermore, as we stated before, the more companies of a supply chain or supply network take part in the EPCglobal Network, the more valuable it is. By lowering the threshold for joining the network for less innovative companies, we will be able to foster the adoption of the EPCglobal Network. Given the aforementioned aspects, we draw another important requirement for the overall Discovery Service solution.

Requirement 6. *The Discovery Service architecture shall encourage participation in the EPCglobal Network.*

3.5 Scalability

Another requirement is the scalability of the service. Scalability refers to the ability to handle large amounts of requests and data. Concretely, Discovery Services need to be able to serve queries of potentially millions of clients. At the same, time Discovery Services need to be able to store EPC numbers and links form EPCISs at the rate of new instances of products being produced. We expect the volume of EPC number updates to be several times the volume of client queries.

Requirement 7. *The Discovery Service architecture shall be highly scalable, able to handle the network traffic both in terms of data volume and number of participants.*

3.6 Quality of Service

The concept of fitness for use is widely adopted in quality literature. This concept emphasizes the importance of taking a consumers viewpoint in regards of quality [11]. When designing the Discovery Service, we need to consider the querying party's viewpoint of fitness for use, that is, return data that are suitable for decision making. Following information systems literature, user satisfaction and information quality are two major issues for the success of information systems [10]. In more detail, these two dimensions include specific concepts such as accuracy, timeliness, completeness, precision and relevance, as well as accessibility and interpretability [4] [17] [28].

Based on the above findings, the Discovery Service needs to be designed in a way that the companies that are requesting information about a certain EPC receive data of an appropriate quality. Our findings indicate that a complete and correct response are always relevant. This conduces to the following requirement:

Requirement 8. *The query result shall be complete and correct, respecting the client's access rights defined separately by each information provider.*

4   EPC Discovery Service Solution Concept

The EPCglobal Network is a collection of interrelated standards for hardware, software and data interfaces that aims at enhancing the supply chain through the use of Electronic Product Code (EPC) - a globally unique and standardized identifier contained in each RFID tag labeling a product. More specifically, the components of the EPCglobal Network provide the functionality to capture and share RFID information in today's complex trading networks [16].

The network consists of three main components: the Object Name Service (ONS), the EPC Information Services (EPCIS) and the EPC Discovery Service. ONS is the authoritative directory of information sources available to describe an EPC associated to a product - typically this is the manufacturer. EPC Information Services are the actual repositories at each company's site that store data about unique items in the supply chain. The EPC Discovery Service component is essentially a chain-of-custody registration service that enables companies to find detailed, up-to-date data related to a specific EPC and to request access to those data.

Next, we describe two different Discovery Service designs – the one considered by EPCglobal in Section 4.1 and our alternative design, the Query Relay, in Section 4.2. For the description of both designs, the following *assumptions* are made:

- The client is pre-configured with the service location of the Discovery Service to avoid bootstrapping related issues.
- Each information provider publishes EPC-related information to the Discovery Service in advance.
- The actual access to data is managed locally at the EPCIS where the company itself determines which trading partners have access to which information [15].
- The Discovery Service provider is a trustworthy entity that provides the service as described and does not misuse the data.

4.1   Directory Look-Up Design

We present here the Discovery Service design currently under consideration at EPCglobal. The description is based on the review of existing literature [2] [6] [14] [15] [27]. Figure 1 illustrates in detail the process of retrieving information related to a specific EPC. It is organized along the traditional client-server architecture. The concrete steps are:

(1) A company – hereafter referred to as *client* or *information requester* – asks for information about a specific EPC-equipped product. It sends its query request to the Discovery Service;
(2) The Discovery Service looks up all EPCIS addresses that hold data about the provided EPC number;
(3) The Discovery Service sends the related addresses back to the requesting company;

(4a) and (4b) The client queries the given EPCISs directly.

(5a) and (5b) The queried information providers A and B check the client's access rights;

(6) Given the client is authorized to get information, the query result is sent by the information providers. In our example, only Company B replies.

Some key observations: unless there are access control policies at the Discovery Service set by A, the information requester knows that Company A did not deliver any information and, moreover, the identity of Company A is revealed to the information requester. This might generally not be a desired effect. However, if access rights are set to the Discovery Service, we run into the duplication problem. Concretely, this means that control policies are published by Company A along with any data to the Discovery Service. Redundancies are thus created and the complexity of the data structures within the Discovery Service significantly increases.

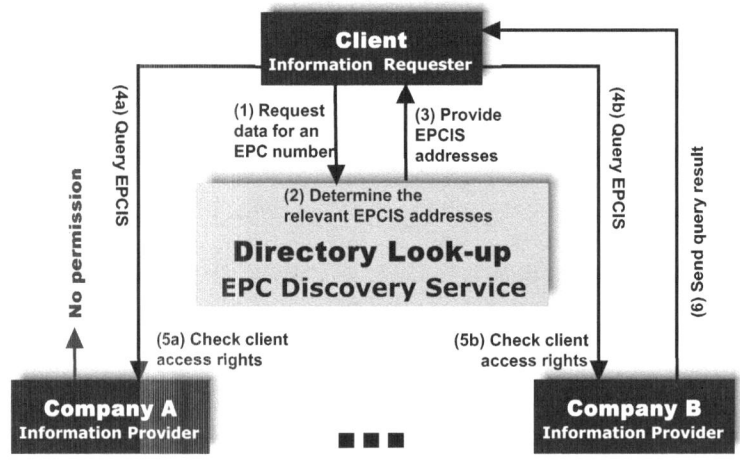

Fig. 1. Directory Look-up Design of the EPC Discovery Service. Note: initial publishing of records omitted.

4.2 Query-Relay Design

Based on the requirements presented in the Section 3, we suggest the following Discovery Service solution, as depicted in Figure 2. To the best of our knowledge, this design is innovative in the context of EPCglobal Network. We proceed by explaining every step of the querying process in detail:

(1) An information requester is asking for detailed information associated with an EPC;

(2) The Discovery Service looks up the relevant EPCIS addresses;

(3a and 3b) The Discovery Service forwards the complete query to the identified information providers;

(4a and 4b) The queried information providers A and B check finegrained access rights defined for the client;

(5) If access is granted, the query is answered directly to the querying company.

Some key observations: each company defines its access rights at the EPCIS level but does not duplicate these to the Discovery Service level. The Discovery Service only holds a list of EPCs and related EPCIS addresses. The main purpose of the service becomes now the routing of queries. Notice additionally, that the information requester does not become aware that Company A denied access to its information.

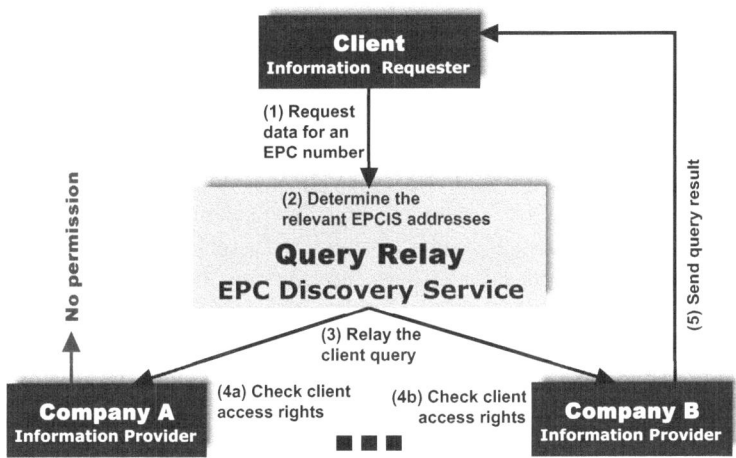

Fig. 2. Query Relay Design of the EPC Discovery Service. Note: initial publishing of records omitted.

5   Discussion and Evaluation

In this section we comprehensively compare the two suggested solutions. As evaluation criteria we use the identified requirements from Section 3.

Data Ownership. In the Directory Look-up solution, any published data must be accompanied by access rights at the Discovery Service. If companies fail to do so, they would leave their own published data open to any random client and, hence, lose control over it – a violation of Requirement 1. But even with access rights maintained and stored at the Discovery Service together with the published data, control is merely delegated to the Discovery Service provider. In this case, companies might still find difficult to track the requests upon their data within the Discovery Service. Besides, the access rights duplication problem remains unresolved and thus Requirement 2 is not met.

The Query Relay design, on the other hand, abides both Requirement 1 and 2. It is very lightweight, avoids heavy duplication and the data published within the Discovery Service is used for routing only. Nothing is ever returned to the client, whose complete access to the data is controlled directly and locally at the EPCIS of the information provider.

Security. First and foremost, the security of the Discovery Service must address the interaction between a publisher and a client and their prospective information exchange. For both of them, confidentiality correlated with availability and reliability of the system are the supreme necessities.

We first address publisher confidentiality. The Directory Look-up design ensures publisher confidentiality if and only if access controls are enforced within the Discovery Service. As already stated, these access rights must be adjunct to the published data; otherwise, the published identity / EPCIS address, together with any additional data contained in the published records for a particular EPC, could be freely revealed. In the Query Relay design however, the condition guaranteeing publisher confidentiality is less strict. Concretely, it is only required that the Discovery Service provider is trustworthy in the sense that it would not further disclose the published records. Hence, the part referring to publisher confidentiality of Requirement 3 is better satisfied by the Query Relay.

Second, we analyze client confidentiality. This mainly refers to unveiling the query itself. The information held within a query reflects the strategic intents of a client. This is considered sensitive information. In the Directory Look-up design, the client query is treated confidentially by the Discovery Service, in the sense that it is not passed further to EPCISs. This is opposed to the Query Relay design, where the client query is straightly forwarded to EPCISs, without the control of a client. These EPCISs might be malicious and thus the client is put at risk. However, a similar risk could be pertinent to the Directory Look-up design as well: once the client has received the EPCIS list, it may query potentially malicious information services not identifiable as such by the Discovery Service. Hence, the part referring to client confidentiality of Requirement 3 is satisfied in approximately the same manner by both designs.

Finally, we examine availability and reliability. These two properties, forming Requirement 4, are similarly fulfilled by both designs. More precisely, if the Discovery Service is down, no response at all is given to the client. We particularly consider the case where the client as well as the information providers have a transient connection to the network. The problem can be overcome in the Directory Look-up design by duplicating information providers' data to the Discovery Service level during uptime. However, this is rather unlikely as companies want to be in control of their data and, thus, are reluctant to duplicating data to a third party system.

Business Relationship Independent Design. As described previously, this requirement solicits that business relationships do not influence the way in which companies interact with the Discovery Service. In the center of attention are here the access rights again.

In the Directory Look-up design, there are two levels of access control: one on the EPCIS and another one on the Discovery Service. As the access rights encode a company's business relationship with its trading partners, any partnering changes directly trigger a modification in the access rights. Following the Directory Look-up design, this modification must be reproduced without delay both in the access rights on the EPCIS and on the Discovery Service. Maintaining permissions in this dual manner creates an additional effort, increases software complexity and even introduces a potential attack point as compared to the Query Relay design, where there is a single point of access control – at the EPC Information Service. Requirement 5 is consequently fulfilled only by the Query Relay design.

Organic Growth. To launch the EPCglobal Network in terms of partakers, the concept of *clusters* has been introduced. A cluster is an environment comprising only *trusted* business partners of a supply chain that are serviced by their own Discovery Service. With more and more companies interested to join the network over time, it is natural to assume that two or more clusters will merge. We now analyze how the two discussed Discovery Service models facilitate cluster merging. In the Directory Look-up, when opening up a cluster, at least some information, like published references to sources of detailed EPC data, is revealed to newcomers. As a consequence, before merging, there must be a unanimous and mutual consent on granting newcomers' permission to join from all the members of both clusters. Yet, this might be very hard to achieve because one or more companies within a cluster might identify competitors in the other cluster. Their data ownership concerns are thus significantly augmented. As a result, the evolution of EPCglobal Network is in fact hindered with the Directory Lookup approach. Contrarily, the Query Relay solution makes it easier for companies to join the EPCglobal Network anytime and, most important, companies already taking part in the network are not affected at all. If the established companies want to extend their network, they simply have to provide appropriate access rights locally on the EPCIS level. Any modification to these access rights will be immediately functional network-wide while allowing publishers to be in agreement exclusively with the Discovery Service provider. Requirement 6 is, as a consequence, much better fulfilled by the Query Relay design.

Scalability. Both designs do not address scalability directly but need to rely on additional mechanisms. The issue of client queries is slightly more accentuated in the Query Relay approach because queries need to be forwarded. On the other hand, the much more significant EPC number updates are handled equally in both design approaches and thus they are susceptible to the same scalability issues. Forwarding client queries can be seen as routing data (i.e., the query) based on EPC numbers contained in the queries. There exists a large body of work on routing high volumes of structured data (e.g., queries) in internet-scale overlay networks, particularly in the publish-subscribe and content-based routing research communities [1] [3] [5]. We believe that existing technologies suffice to cope with forwarding queries. Likewise, for handling and storing mass updates of EPC numbers, suitable mechanisms exist and need to be identified. This is a top priority for future work.

Quality of Service. As mentioned before, data quality in the context of EPC-global Network primarily refers to the completeness and correctness of query results. A query result might be incomplete or incorrect for either of the following three reasons: first, not every read EPC event has been internally stored in the information provider's EPCIS repository; second, the information provider has fabricated EPC events; and third, an information provider consciously denies access because the requested information is too sensitive.

The first two cases are nothing that a Discovery Service, independent from the chosen architecture, could combat. Achieving full supply chain visibility in these cases is compromised anyway. In the last case however, without exposing the information provider, the Discovery Service might assist the client with information whether the query result is complete or correct. For instance, it can inform the client of the length of the trace, i.e. the number of nodes containing detailed EPC information. Consider now the Directory Look-up design and, in addition, realistically assume that the information provider did not publish its sensitive EPC data to the Discovery Service. Unlike in Query Relay design, the Directory Look-up might allow a client to identify the information provider who did not answer. While this constitutes a slight advantage in regards to adjuvant information on completeness, on the negative side, from a security standpoint, this property breaches the confidentiality of the information provider. We can thus conclude that Requirement 8 is better satisfied by Directory Look-up at the expense of infringing information provider confidentiality.

6   The Complete EPCglobal Network Vision

In a competitive environment with numerous parties like the supply chain networks of today, cross-organizational collaboration and information sharing become fundamental. To appropriately support them, very strong security must be enforced at the level of the EPCglobal Network. Until now, the discussion within the EPCglobal working group or any related literature along security measures only refers to the definition of powerful access rights. The pure access rights are, however, not sufficient for enabling companies to interact with strangers – parties that are neither trusted nor distrusted, but just not recognized. This is because any company would define its access rights permitting another company to retrieve its operational data no sooner than the two of them have established a first contact. This is counter-intuitive to the general concept of EPCglobal Network, whose main purpose is to foster collaboration, independent of previous business relationships. Since interaction with alien partners is often unavoidable, sometimes even necessary, in today's increasingly dynamic supply chain networks, it should by no means be obstructed.

The arguments given above clearly point to several challenges that need to be overcome in the context of EPCglobal Network. First, all actors in the supply chain who were custodians of a product must be located. This includes both known and unknown parties and lies under the responsibility of the EPC Discovery Service. Second, there must be a mechanism to mediate trust between two a priori unknown parties. For the known partners, the exchange of information is usually mandated by a one-to-one contract. Complementarily, for the unknown ones there should be a mechanism which negotiates the one-to-one contract automatically. This is a truly daring and intricate pursuit. Therefore, there must exist an incentive for this kind of dicey collaboration. In most cases, money constitutes a reasonable argument. This brings us to the third and last challenge, namely a system able to automatically bill the companies retrieving information. Most naturally, the billing takes the form of pay-per-information. To summarize, we envision the complete solution comprising the following three components. These three building blocks together with interaction between two companies – a client and a server – can be modeled as depicted in Figure 3.

1. **EPC Discovery Service.** The main purpose of the Discovery service is to locate all actors in the trading network who own information about an EPC of interest to the client.
2. **Automated Contract Negotiation Service.** The main purpose of this service is to replace or support the conventional paper contracts and human negotiation leading to the establishment of those contracts among the supply chain actors.
3. **Pay-per-information Pricing Model.** The main purpose of this service is to incentivize inter-organizational information sharing through financial recompensation.

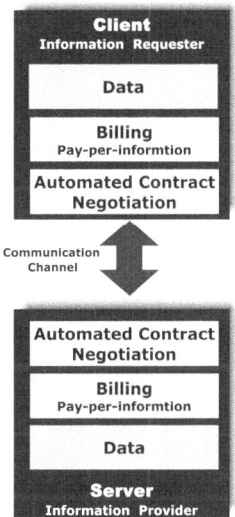

Fig. 3. Enhancements to the EPCgloblal Network at the EPCIS level: automated contract negotiation and pay-per-information pricing

The EPC Discovery Service acts as an enabler of the entire solution scope, including automated contract negotiation and pay-per-information pricing. There is a subtle yet significant difference between the two designs in regards to the extent a client query reaches the relevant information services holding information for an EPC of interest. In the Directory Look-up design, due to access rights which companies could justifiably store within the Discovery Service to avoid openness, a client query will never reach industrial parties not known in advance. In other words, only publishers that have explicitly granted access to the client can answer. This represents an impediment towards achieving full supply chain visibility, as it prevents interaction with strangers. Note that storing no access rights within the Discovery Service is *not* an option as it ultimately leads to the infringement of publisher confidentiality. On the other hand, with the Query Relay design, a client query reaches all relevant information services for an EPC. Thus, besides known partners, companies that are nothing but merely strangers to the client are also given the chance to provide information.

7   Conclusion and Future Work

In this paper we presented an alternative design for the EPC Discovery Service. This design was driven by the requirements we elicited from interviews and literature research. We identified five major requirements – data ownership, security, business relationship independence, organic growth, and quality of service. The discussion along these requirements showed that the originally suggested Directory Look-up Discovery Service design only partially complies. Most prominently, this design does not encourage participation in the EPCglobal Network even when clusters of trusted parties are formed and poses serious problems to publisher confidentiality. Against this background, we proposed a lightweight Discovery Service, the so-called Query Relay, that only stores the EPCs and related EPCIS addresses. When querying for more information about an EPC, the Discovery Service simply forwards the query to the relevant EPCISs. Access rights are checked only locally at the EPCIS level and the initial query is answered directly to the querying company. As opposed to the Directory Look-up, no data – including access rights – need to be duplicated to the Discovery Service level. By having such a design, companies will be in complete control of their data. In effect, we create a significantly more secure solution, lower the threshold for companies that want to take part in the network and, thus, support an organic growth. Above all, the Query Relay design makes it easier for a client to deal with parties unknown a priori, which increases the chance to attain full supply chain visibility. In this respect, we envision an automated contract negotiation service incorporated in the EPCglobal infrastructure. Additionally, a pay-per-information pricing model might be supported as an incentive for the information provider to share its operational data.

We see potential for further research in several directions. First of all, in the Query Relay model, the development of a more selective and intelligent routing of client requests is desired. This can also have a security benefit since client requests are not disseminated further than necessary, compromising the confidentiality of the client. Mechanisms to enforce the registration of accurate routing data will be an essential research area, especially since it will be hard to detect if an EPCIS has published correct routing data to the Discovery Service and simply refuses to respond, or if it cannot respond because it has published inaccurate data. Second, it is crucial to precisely assess the scalability of both Discovery Service designs. Third, the challenges related to incompleteness of the query result need to be examined. Finally, an equally important matter is the standardization of EPCglobal interfaces which may need to be tailored to suit our proposed solution, particularly addressing the automated contract negotiation. As automated negotiations typically follow an offer-counteroffer workflow, it is interesting to inspect what could be simple yet reliable offers / counteroffers in our environment.

References

1. Achieving Scalability and Throughput in a Publish/Subscribe System. Technical Report RC23103, IBM Research Division Thomas J. Watson Research Center, P.O. Box 704, Yorktown Heights, NY 10598 (February 2004)
2. Afilias. RFID Discovery Services, information request on (December 19th, 2006), http://www.afilias.info/rfid
3. Marcos, K., Aguilera, R.E., Strom, D.C.: Matching Events in a Content-based Subscription System. In: PODC 1999: Proceedings of the Eighteenth Annual ACM Symposium on Principles of Distributed Computing, pp. 53–61. ACM Press, New York (1999)
4. Bailey, J.E., Pearson, S.W.: Development of a Tool for Measuring and Analyzing Computer User Satisfaction. Management Science 29(5), 530–545 (1985)
5. Baldoni, R., Virgillito, A.: Distributed Event Routing in Publish/Subscribe Communication Systems: A Survey (revised version). Technical report, MIDLAB 1/2006 - Dipartimento di Informatica e Sistemistica (2006)
6. Beier, S., Grandison, T., Kailing, K., Rantzau, R.: Discovery Services – Enabling RFID Traceability in EPCglobal Networks. In: 13th International Conference on Management of Data (2006)
7. Bozarth, C., Handfield, R.B.: Introduction to Operations and Supply Chain Management. Prentice-Hall, Englewood Cliffs (2006)
8. Czerwinski, S.E., Zhao, B.Y., Hodes, T.D., Joseph, A.D., Katz, R.H.: An Architecture for a Secure Service Discovery Service. In: Proceedings of the 5th Annual ACM/IEEE International Conference on Mobile Computing and Networking, pp. 24–35 (1999)
9. Davenport, T.H., Eccles, R.G., Prusak, L.: Information Politics. Sloan Management Review, Fall, 53–65 (1992)
10. Delone, W.H., McLesan, E.R.: Information Systems Success: The Quest for the Dependent Variable. Information Systems Research 3(1), 60–95 (1992)
11. Dobyns, L., Crawford-Mason, C.: Quality or Else: The Revolution in World Business. Houghton Mifflin, Boston (1991)
12. Fisher, M.L., Hammond, J.H., Obermeyer, W.R., Raman, A.: Making Supply Meet Demand in An Uncertain World. Harvard Business Review 72(3), 83–93 (1994)

13. Roussos, G., Tuominen, J., Koukara, L., Seppala, O., Kourouthanasis, P., Giaglis, G., Frissaer, J.: A Case Study in Pervasive Retail. In: Proceedings of the 2nd International Workshop on Mobile Commerce, Atlanta, pp. 90–94 (2002)
14. GS1 Germany. Internet der Dinge. Management Information. Das EPCglobal Netzwerk. Technical report (March 2005)
15. EPCglobal Inc. The EPCglobal Network: Overview of Design, Benefits, & Security (September 2004)
16. EPCglobal Inc. The EPCglobal Architecture Framework (July 2005)
17. Ives, B., Olson, M.H., Baroudi, J.J.: The Measurement of User Information Satisfaction. Communications of the ACM 26(10), 785–793 (1983)
18. Laprie, J.C.: Dependabable Computing: Concepts, Limits, Challenges. In: Proc. 25th IEEE Int. Symp on Fault Tolerant Computing (FTCS-25), Special Issue, Pasadena, California, pp. 42–54 (June 1995)
19. Lee, H.L., Whang, S.: Information Sharing in Supply Chain. International Journal of Technology Management 20(3/4), 373–387 (2000)
20. Lee, V., Padmanabhan, H.L., Whang, S.: The Bullwhip Effect in Supply Chains. Sloan Management Review 38(3), 93–102 (1997)
21. Markus, M.L.: Power, Politics, and MIS Implementation. Communications of the ACM 26(6), 430–444 (1983)
22. Mockapetris, P.V., Dunlap, K.: Development of the Domain Name System. In: Proceedings of SIGCOMM (August 1988)
23. OpenSLP. An Introduction to the Service Location Protocol (SLP) (information request on January 5th, 2007),
    http://www.openslp.org/doc/html/IntroductionToSLP/index.html
24. Rogers, E.M.: Diffusion of Innovations. The Free Press, New York (1995)
25. Shah, N.: Process Industry Supply Chains: Advances and Challenges. Computers and Chemical Engineering 29, 1225–1235 (2005)
26. Van Alstyne, M., Brynjolfsson, E., Madnick, S.: Why Not One Big Database? Principles for Data Ownership. Decision Support Systems 15, 267–284 (1995)
27. VeriSign. The EPC Network: Enhancing the Supply Chain (January 2004)
28. Wang, R.Y., Kon, H.B., Madnick, S.E.: Data Quality Requirements Analysis and Modeling. In: Proceedings of the 9th International Conference on Data Engineering, Vienna, pp. 670–677 (1993)

Fine-Grained Access Control
for EPC Information Services

Eberhard Grummt[1,2] and Markus Müller[2]

[1] SAP Research CEC Dresden
eberhard.oliver.grummt@sap.com
[2] Technische Universität Dresden
markus.mueller@mailbox.tu-dresden.de

Abstract. Inter-organizational exchange of information about physical
objects that is automatically gathered using RFID can increase the traceability of goods in complex supply chains. With the EPCIS specification,
a standard for RFID-based events and respective information system interfaces is available. However, it does not address access control in detail, which is a prerequisite for secure information exchange. We propose
a novel rule-based, context-aware policy language for describing access
rights on large sets of EPCIS Events. Furthermore, we discuss approaches
to enforce these policies and introduce an efficient enforcement mechanism based on query recomposition and its prototypical implementation.

1 Introduction

RFID is quickly becoming a key technology for novel supply chain management
applications. It enables the automatic identification (*Auto-ID*) of physical objects equipped with small transponder tags. In intra-organizational scenarios,
RFID's advantages over the established bar code regarding efficiency and data
granularity have been used for several years. In inter-organizational settings, the
technology's main potential is to increase the visibility of goods along the whole
supply chain. While gathering and exchanging object-related information does
not pose a general challenge using technology available today, several security
and incompatibility issues remain unsolved.

With the industry organization *EPCglobal Inc.*[1], there is a strong initiative towards overcoming incompatibilities between companies' RFID-related IT infrastructures. EPCglobal fosters open data format and interface standards. Besides
tag data specifications, the most important standard for inter-organizational
data exchange is the *EPC Information Services* (EPCIS) specification [8]. Software systems implementing this specification, called *EPCIS Repositories*, feature
standardized interfaces for capturing and querying EPC-related event and meta
data. Since EPCIS Repositories hold mission-critical, potentially confidential information, access to these interfaces needs to be limited. The EPCIS standard

---

[1] http://www.epcglobalinc.org/, EPC stands for Electronic Product Code.

C. Floerkemeier et al. (Eds.): IOT 2008, LNCS 4952, pp. 35–49, 2008.

explicitly leaves the details how access control is performed to the individual EPCIS implementations [8, pp. 57–58].

In this paper, we present an approach to specify and enforce fine-grained access rights to large quantities of EPCIS event information. Our contribution is twofold. First, we introduce a novel access control policy language called AAL leveraging the structure of EPCIS data. Second, we present an efficient policy enforcement mechanism and its implementation that is based on the concept of SQL query rewriting for relational databases. To our best knowledge, access control for EPCIS data has not been addressed by scientific literature so far. The remainder of this paper is structured as follows: We present the problem statement, assumptions and challenges in Section 2. We give an overview of related work in Section 3. In Section 4, we introduce requirements and our concepts for a policy definition language and an efficient enforcement mechanism. We evaluate our results in Section 5 and conclude in Section 6, giving directions for future research.

2   Problem Statement

We investigate how access control policies for EPCIS Repositories can be defined and enforced efficiently in order to facilitate fine-grained disclosure control for RFID-based events.

2.1   Definitions and Assumptions

Let $C = \{c_1, \ldots, c_n\}$ be a set of companies, each of which operates an EPCIS Repository or an EPCIS *Accessing Application* [27, p. 41]. We assume that every $c_k \in C$ can be reliably identified and authenticated by every other company $c_l \in C$. An EPCIS Repository operated by any company c_m stores only EPCIS Events generated by this company, i.e. information gathered from remote sources is *not* integrated into c_m's repository. A *principal* can be any participating user, administrator, role, company, or system [2, p. 9]. In our context, a *user* is a principal that belongs to a company c_a and tries to access a remote EPCIS operated by a company c_b using an Accessing Application. An *administrator* is a principal that is allowed to grant and revoke access rights to an EPCIS Repository. We refer to *Access Control* (AC) as the process of enforcing an applicable Access Policy. An *Access Policy* (policy for short) is a formal specification that states which user or role has the right to access which EPCIS Events. An *Access Control Mechanism* or *Enforcement Mechanism* (mechanism for short) is a software component ensuring that relevant policies are applied to all access operations carried out by users, prohibiting or limiting disclosure if necessary.

2.2   Introduction to EPCIS

EPCIS Repositories store information about to physical objects. This information is logically represented in the form of *EPCIS Events*. Generally, an event represents a change in state that occurs at a certain point in time. The EPCIS specification defines four event types, namely *ObjectEvent*, *AggregationEvent*, *QuantityEvent*, and *TransactionEvent*. They are used to express object observations, object aggregations, object quantity observations, and connections of objects and business transactions, respectively [8, pp. 39–53]. While the internal processing and storage details may vary from implementation to implementation, an EPCIS Repository needs to provide two interfaces: The *EPCIS Capture Interface* (CI) and the *EPCIS Query Interface* (QI). The CI is used to submit new events to be stored in the repository, while the QI is used to retrieve events of interest. Both interfaces can be implemented in the form of web services. To that end, EPCglobal specifies HTTP and SOAP bindings [8, pp. 108–126].

2.3   Use Case and Challenges

To enable certain applications such as Tracking and Tracing (determining the current and all previous locations of an object), companies need to access events stored in EPCIS Repositories operated by other companies. EPCIS Events are confidential, because they can be used to infer production capacities, inventory levels, sales figures, and business relationships, among others. This is why access control is a prerequisite for the inter-organizational EPCIS deployment.

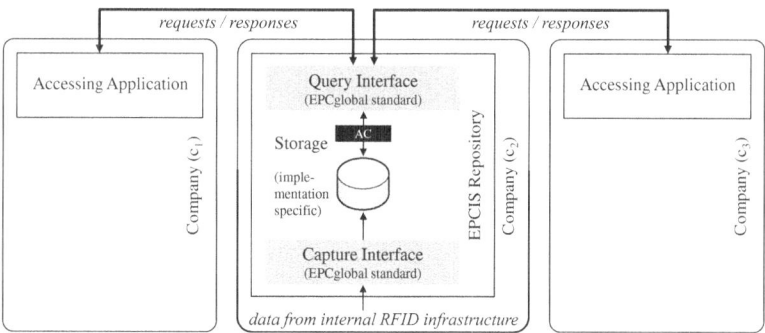

Fig. 1. EPCIS Interfaces and Interactions

Fig. 1 depicts a company c_2 that operates an EPCIS Repository. Via its Capture Interface, RFID-based events gathered locally by c_2 are transferred to a component responsible for their persistent storage. This storage component is not specified by EPCglobal and can be implemented arbitrarily, for example using relational or XML databases. The repository's Query Interface is exposed to external companies. Using Accessing Applications, companies such as c_1 and c_3 can use this interface. An Access Control mechanism is depicted as a logical component between the storage component and the Query Interface. Note that c_1 and c_3 might operate EPCIS Repositories and c_2 might employ an Accessing Application, too. This is not depicted for reasons of clarity.

As different business partners need different subsets of events to get their work done and companies usually try to disclose only a minimum amount of data, administrators need fine-grained means to define access rights. These access rights depend on the content of the events themselves and may also refer to data that is not stored yet, as new events are generated continuously. Instead of static lists defining which event can be accessed by whom, such as ACLs used in traditional AC systems, rules that are dynamically evaluated by the enforcement mechanism are required. Rules can refer to the content of the events and to contextual information such as business relationships and the current time. We introduce further requirements in Section 4.1.

3   Related Work

While security research in the context of RFID has mainly focused on privacy aspects, authentication, and secure reader-tag-communication [17,11,21], confidentiality of RFID-based data after it has been captured and stored has not received much attention so far [1]. At a first glance, sets of EPCIS Repositories can be considered distributed or federated databases, so respective access control models and mechanisms [6] seem to be viable starting points. However, the characteristics of the stored data and its volume pose new challenges [4,23], especially regarding access control [14].

A recent NIST study [15] gives an assessment of established access control approaches. Popular models for non-military systems include Discretionary Access Control (DAC) (as implemented in the form of Access Control Lists (ACLs) [18] or Capabilities by popular operating systems), and Role-based Access Control [9]. The counterpart of DAC is formed by Non-Discretionary Access Control models (NDAC) such as Mandatory Access Control (MAC) whose best-known representative is the Bell-LaPadula model. Besides these established approaches, that do not form a strict taxonomy and can actually be combined in a number of ways, several application-specific security models and mechanisms have been developed. Temporal aspects have been addressed in TRBAC [3], which focuses on temporal availability and dependency of roles. Rule-Based Access Control (RuBAC) is a general term to describe AC systems that grant or deny access to resources based on rules defined by an administrator [15]. Research in the area of context-sensitive access control, e.g. [16,12], strives to simplify access decisions by using environmental sensor data or other information not intrinsic to the system performing the AC. It is mainly focused on pervasive computing for personal use and has not been applied extensively to supply chain management before, even though approaches do exist [13].

Given the vast amount of data expected in future EPCIS Repositories [7,23,28], efficiency of the enforcement mechanism is an important issue. Since for a long time, relational databases will probably remain the dominant storage technology for such repositories, existing approaches to performing AC at the database level are relevant. Oracle's Virtual Private Databases [5] and Sybase's Row Level Security Model [25] use techniques similar to query rewriting [24], basically adding predicates to each query to restrict the accessible rows and columns. Hippocratic Databases (HDB) [19] primarily aim at protecting patients' privacy in medical databases, but the concept can be applied to arbitrary relational databases. Instead of query rewriting, HDB replace the tables a query affects by prepared views. These views reflect the desired access policies in that protected rows, columns, and cells are removed compared to the original table. Our approach differs from all of the above in that we use an enhanced query rewriting technique that not only extends the original query, but recomposes it into several sub-queries in order to achieve flexible row, column, and cell based restrictions specified by rule-based policy definitions.

4   Efficient Access Control for EPCIS Repositories

In this section, we introduce a rule-based, content and context aware Policy Language for describing access rights to large amounts of RFID-based events. Based on specific requirements, the language design and its semantics are described. Furthermore, an enforcement mechanism based on query rewriting for relational databases is introduced.

4.1   Requirements

Based on a number of case studies [4,23,28] and previous work [14], we identified the following access control requirements:

Fine-grained disclosure control. Besides the ability to restrict access to certain events and event types, attribute-level restrictions need to be supported.

Content and context awareness. Access rights to events may depend on their respective content, as well as on contextual (external) information such as business relationships and temporal constructs.

Rules, range, and condition support. Because access rights are usually not assigned to individual events but to (continuously growing) sets of events, rules that may refer to ranges of events fulfilling certain conditions need to be supported.

Automatic reduction of result sets. If a user queries more information than he is authorized to access, the EPCIS repository has to return the respective allowed subset of events, instead of denying the whole query.

Query power restriction. To prevent information leakage due to inference, the query interface's flexibility needs to be restrictable per user or role.

Rapid execution. Due to the expected amount of events, efficiency of the enforcement mechanism in terms of memory consumption and execution time needs to be addressed.

4.2   AAL: A Rule-Based Policy Language for Auto-ID Events

Introductory Considerations. Traditionally, each application or operating system employs proprietary policy representations. Recently, there is a trend towards expressing policies using XML [29]. Besides their inherent extensibility, XML-based policy languages promise to be easier to read and edit and to offer better interoperability. With OASIS *XACML* [26,20], an open industry standard for describing access policies using an XML-based language is available. Nonetheless, we decided to develop an own policy language for the following reasons. First, XACML's general purpose approach trades flexibility against simplicity. In our specific context, this clearly violates the principle of the "economy of mechanism" [22]. Second, despite its name, XACML not only specifies a language, but also an architecture how policies shall be evaluated and enforced. It recommends the separation of the AC mechanism by using a *Policy Decision Point* (PDP) and a *Policy Enforcement Point* (PEP). A PDP receives access requests, evaluates them using applicable policies and returns one of four predefined messages (*Permit, Deny, NotApplicable,* or *Indeterminate*). Based on such messages, a PEP permits or denies user's requests, issuing error messages if necessary. There are two problems with this architecture. Because in our scenario, access rights can be directly dependent on EPCIS Events' attributes, the PDP would have to access them in order to decide about requests. According to the XACML specification, a *Policy Information Point* (PIP) provides access to such external information. However, having to transfer large sets of events from the storage engine via a PIP to the PDP would introduce significant overhead. Furthermore, the property of XACML PDPs to only return one of four messages would make the "automatic reduction of result sets" (cmp. 4.1) impossible.

Language Design. The main concept of our policy language called AAL (Auto-ID Authorization Language) is the notion of *Shares*. Each principal is assigned a set of Shares he is authorized to access. A Share defines a subset of EPCIS Events of a specific event type. Within a Share, all events disclosed to a user will contain only the attributes enumerated for this particular Share. The events of a Share are specified using a set of *Conditions*. Each Condition refers to exactly one attribute. All Conditions have to be fulfilled for an event to appear in a Share. Conditions are specified using *Values* that the respective attribute may hold. Values can be defined by enumeration (e.g. single EPCs) or using ranges and wildcards (such as EPC ranges and EPC patterns). These concepts can be formulated as follows:

$$Authorization(Principal) := Share_1 \cup \ldots \cup Share_k$$
$$Share(EventType, \{Attr_1, \ldots, Attr_l\}) := Condition_1 \cap \ldots \cap Condition_m$$
$$Condition(Attribute) := Value_1 \cup \ldots \cup Value_n$$

The following example defines two Shares. They are depicted as dark cells in Fig. 2. The table illustrates a subset of all ObjectEvents in a system, with each row representing an event (more descriptive attributes such as EPC and eventTime as well as content for all cells were omitted for reasons of clarity).

$$Share_1(ObjectEvent, \{a, b, e, f\}) = ((id \in \{1..4\}) \cap (a \in \{12..16\}))$$
$$Share_2(ObjectEvent, \{b, c, d, e\}) = ((id > 5) \cap (a < 20))$$

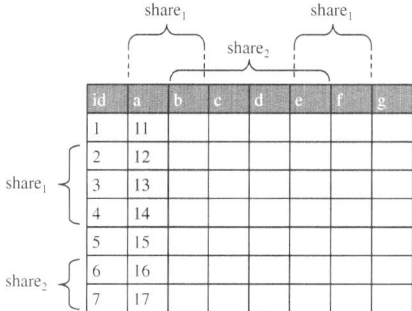

Fig. 2. Visualization of Shares

Language Semantics and Properties. Our policy language is based on whitelisting, i.e. everything that is not explicitly allowed is prohibited. This follows the principle of "Fail-safe defaults" [22]. While not preventing unwanted information leakage due to misconfigurations, this approach makes determining who is allowed to access which events much easier. If Shares containing different sets of attributes overlap (i.e. certain events are in more than one Share), the affected events will contain the union of the respective attributes. A policy can apply to any number of users or roles. Furthermore, policies can extend other policies additively. The structure and semantics of the policy language are illustrated in Fig. 3. Policy instances can be serialized in an XML dialect we defined using XML Schema.

```
Policy
   hasName name
   appliesTo (user|role)+
   extendsPolicies (policyName)*
   contains (Share
      refersTo eventType
      contains (visibleAttribute)+
      fullfillsAll (Condition
         refersTo eventAttribute|contextAttribute
         matchesAny (Value)+
      )*
   )+
```

Fig. 3. Structure and semantics of the policy language

Support for Contextual Information. Using event's attributes such as EPC, eventTime, readPoint, or businessStep together with authorization rules allows for the specification of flexible policies. However, in certain situations contextual information extrinsic to the events is needed for authorative decisions. A built-in function to refer to the current time is now(). Using this function, a user can be granted temporally limited access to certain events by specifying *relative* time intervals. For example, he might be allowed to access all events from a certain business location that are not older than two days. Furthermore, a Condition may refer to a *contextAttribute*. Such attributes can be provided by external *Context Providers* and can be referenced in the form `contextProvider.contextAttribute`. For example, Context Providers can be used to retrieve the current user's identity and transactional information in order to base access control on business relationships.

4.3   An Efficient Enforcement Mechanism Using Query Rewriting

Our enforcement mechanism is based on the assumption that EPCIS Events are stored inside relational databases. This assumption is valid because currently there is no alternative capable of inserting, storing and querying large amounts of structured data with acceptable performance.

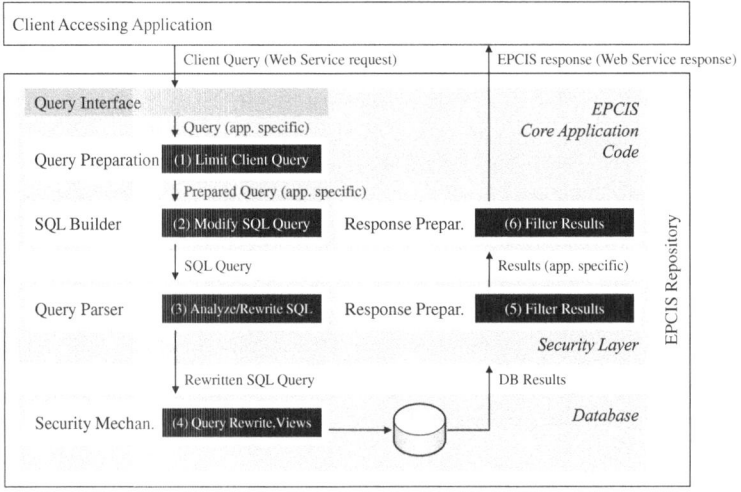

Fig. 4. Possible locations for AC mechanisms

Architectural Considerations. Policy enforcement can be done at a number of logical locations.

First, the user's query submitted using the EPCIS Query Interface can be restricted to affect only the results he is allowed to access. Due to the limited expressiveness of the EPCIS QI, this would not work in many cases. For example, the QI does not allow addressing multiple ranges of EPC at once.

Second, when an EPCIS Application builds an SQL query based on the user's request, it can enrich it with additional predicates that reflect the applicable Policy. The advantage of performing AC using modified SQL queries is that the performance and internal optimization mechanisms of existing databases can be leveraged.

Third, an EPCIS Application can construct an unrestricted SQL query, and instead of directly submitting it to the database driver, pass it through an additional security layer. This layer would analyze the query and rewrite it according to the respective policy. This has the advantage of encapsulating the security-relevant application code instead of mixing it into the query generation algorithm.

Fourth, built-in mechanisms of specific, proprietary database management systems can be used. We discussed some candidates in Section 3.

Fifth, a security layer can filter all result sets returned by the database before passing it on to the EPCIS application.

Sixth, this filtering can also be done by the EPCIS application itself.

Fig. 4 illustrates the six possibilities. Note that it does not depict an actual architecture. The black boxes represent potential locations for the AC mechanisms introduced above.

Translating Policies into Queries. Our approach is based on rewriting an original SQL query to only refer to the events in the intersection of the set the user queried and the set he is authorized to access. As pointed out above, it can either be implemented inside the EPCIS, extending the SQL query building code, or in an external query analyzer/rewriter.

A Share definition can be translated into a relational algebra term as follows (a denotes attributes, v values):

$$Share_k = \pi_{a_1,...,a_l}(\sigma_{(a_1=v_1 \vee ... \vee a_1=v_m) \wedge (...) \wedge (a_n=v_o \vee ... \vee a_n=v_p)}(EventType))$$

Similarly, it can be expressed in SQL:

```
SELECT a₁,...,aₗ FROM EventType WHERE
    (a₁=v₁ OR ... OR a₁=vₘ) AND (aₙ=vₒ OR ... OR aₙ=vₚ)
```

The union of all Shares $Authorization = \bigcup_{k=1}^{n}(Share_k)$ defines the event subset a user is authorized to access. The intersection of $QueriedSet$ (the set he is trying to retrieve using his query) and $Authorization$ is the resulting event set to be returned: $ResultSet = Authorization \cap QueriedSet$. This means that both the Conditions and the attribute enumerations used to define $Authorization$ and $QueriedSet$ need to be intersected.

To build actual SQL statements, separate SELECT statements reflecting these intersections are constructed for each Share and combined using the UNION operator. To keep the semantics of the events, the attributes also need to have the same order in each SELECT statements. Upon executing the query, the database will return the same events several times if they are contained in more than one share. They are aggregated into one event afterwards by the application. Fig. 5 illustrates the necessary steps using pseudocode.

```
query = "";
for each Share {
    Attributes = intersection(UserQuery.Attributes, Share.Attributes);
    Attributes = addNullColumns(Attributes);
    Conditions = removeUnauthorizedConditions
        (UserQuery.Conditions, Attributes);
    Conditions = insertContextValues(Conditions);
    Conditions = "((" + Conditions + ") AND (" + Share.Conditions + "))";
    if (!isFirst(Share)) { query += " UNION ALL "; }
    query += "SELECT " + Attributes + " FROM " + tableName
        + " WHERE " + Conditions;
}
resultSet = execute(query);
return filterAndMergeDuplicates(resultSet);
```

Fig. 5. Pseudocode for Query Construction

For example, assume the following user query and two Share definitions (also depicted in Fig. 6:

$$UserQuery_1(ObjectEvent, \{b, c, d\}) = ((id > 1) \cap (id < 7))$$
$$Share_1(ObjectEvent, \{a, b, e, f\}) = ((id \in \{1..4\}) \cap (a \in \{12..16\}))$$
$$Share_2(ObjectEvent, \{b, c, d, e\}) = ((id > 5) \cap (a < 20))$$

They will result in the following SQL statement:

```
SELECT b, NULL AS c, NULL AS d FROM ObjectEvent WHERE
    (id>1 AND id<7) AND ((id BETWEEN 1 AND 4)
    AND (a BETWEEN 12 AND 16))
UNION ALL
SELECT b, c, d FROM ObjectEvent WHERE
    (id>1 AND id<7) AND (id>5 AND a<20)
```

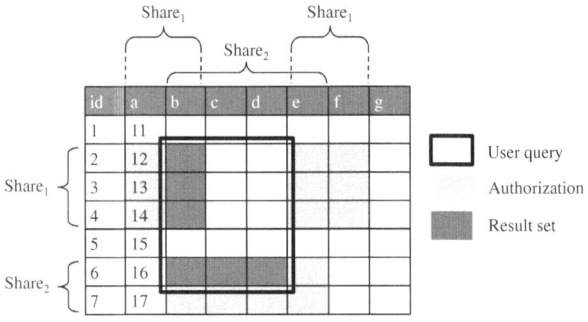

Fig. 6. Visualization of Shares and a user query

5 Evaluation

To evaluate our proposed policy language and mechanism, we implemented a prototype showcasing its applicability. Furthermore, we discuss how the requirements formulated in 4.1 are met and how the design principles for secure systems of Saltzer and Schroeder are reflected.

5.1 Practical Implementation

Our prototypical implementation is based on the *Accada RFID prototyping platform* [10] written in the Java programming language. The Accada EPCIS implementation runs on an *Apache Tomcat* 5.5 server, exposing its interfaces as Web Services based on the *Apache Axis* 1.4 library. For persistent storage, a *MySQL* 5.0 database and the *JDBC* library are employed. Accada's *Query Client* allows for querying the repository and displaying the returned events. We enhanced the Query Client's graphical interface by a user identity selection menu. Because authentication was out of the scope of our work, this replacement for a password or certificate mechanism is a viable simplification. For each user in the system, the policy definition is stored in a separate XML file named <username>.pol. To map the policy's XML structures to Java objects, the library *Simple*[2] 1.4 was used as a light-weighted alternative to *JAXB*. This way, extensions of the policy language do not entail significant updates of the program code. We modified the query generation code in Accada's QueryOperationsModule.createEventQuery method in the package org.accada.epcis.repository, constructing modified SQL queries based on the general idea presented in Section 4.3. Fig. 7 shows a screenshot of the Query Client, a policy definition and a part of a corresponding result set.

Our experience with the implementation of access control into Accada is twofold. First, it showed the technical feasibility of our approach. Reading XML-based policy files and transforming the rules into SQL to restrict a user's query therefore works in practice. Accada proved to be a solid basis for EPCglobal related prototyping activities. Second, the extension of Accada's complex query generation code turned out to be intricate. This is why architecturally, we consider placing the query rewriting code into a separate module a better approach (cmp. 4.3).

5.2 Addressing the Challenges

Our policy language ALL and the respective enforcement mechanism supports fine-grained Using a Share with one Condition, an individual event can be addressed (using suitable values for a simple or compound primary key). Because for such a Share, every column can be addressed directly by enumeration, cell-level control can be achieved.

Using our policy language, businesses can define context- and content-based rules who is allowed to access which information. Especially the ability to define

---

[2] http://simple.sourceforge.net/

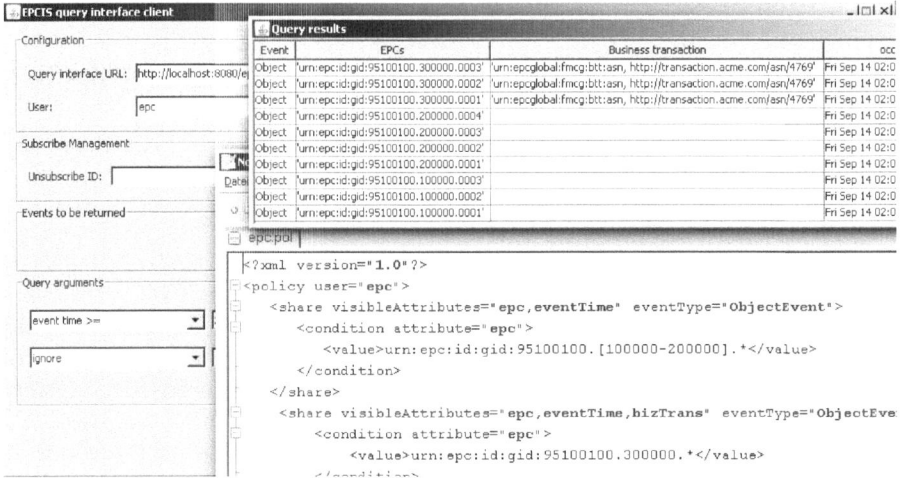

Fig. 7. Screenshot of Query Client, policy definition and result set

relative time ranges and EPC patterns provides flexibility. At policy enforcement time, contextual attributes can be inserted, so information that is not known at policy design time can be included in the access decisions.

By intersecting the queried set and the authorized set, an automatic reduction of the result sets is performed. This is very important, because using an EPCIS compliant query interface, it is not possible for a user to tell the service that he is *not* interested in certain attributes. This, if an AC mechanism rejected all requests that refer to too many attributes, the user would not receive any information in case of a single violated server-side attribute restriction.

Our mechanism restricts the query interface's power by omitting user query restrictions that refer to attributes that are not accessible for him. This is done for each Share individually. However, this is just a basic restriction feature. To prevent undesired information leakage in the face of malicious queries, further investigations in the area of Inference Control would be needed.

Our query rewriting technique involves parsing the user query, resolving context attributes and constructing the final SQL statement. These steps are done once for each user request. The parsing of XML files only need to be performed once after a policy has changed. The overhead for constructing the modified SQL query therefore only consists of several string operations. The actual access control is done by the database that executes the query. It can apply further query optimizations, for example removing or combining redundant predicates.

To further evaluate our approach, we discuss the fulfillment of the seven principles for secure systems of Saltzer and Schroeder [22]:

Economy of mechanism. Our design is simple, so this principle is fulfilled. The policy language's XML Schema definition comprises only 26 lines, compared to 380 lines of the XACML 2.0 policy schema.

Fail-safe defaults. We base access decisions on permission rather than exclusion, so this principle is fulfilled.

Complete mediation. We check every access for authority, so this principle is fulfilled.

Open design. Our design is not secret, so this principle is fulfilled.

Separation of privilege. We do not provide a two key mechanism, so this principle is not fulfilled.

Least privilege. The database runs at a higher security level than necessary in some situations, so this principle is not fulfilled.

Least common mechanism. This principle is not applicable.

Psychological acceptability. Since we do not provide a human interface, this principle is not applicable.

6   Conclusion and Future Work

Based on the access control requirements specific to Auto-ID based collaboration scenarios, we have presented a novel policy language and an efficient enforcement mechanism as well as their implementation. The policy language is expressed in XML and reflects the notion that companies will most probably prefer defining subsets of events to be shared using dynamic rules instead of static access control lists.

We have shown that an enforcement mechanism could be implemented either by restricting a user's query or by filtering a result set. In order to leverage relational databases' optimized query execution and to avoid high memory consumption, we argued that query rewriting is a viable approach. The techniques we presented to modify queries can also be applied to the generation of database views. These could increase query execution performance, possibly at the cost of storage space (in the case of materialized views). We discussed that most of our requirements can be met and that the system design reflects some "golden rules" for secure systems. By providing a prototypical implementation, we proved the plausibility of our approaches.

The outcome of our work are feasible means for administrators to restrict access to single EPCIS Repositories based on the known identities or roles of business partners. So far, this reflects the traditional paradigm of manually assigning rights to users of the system. However, in future dynamic supply chain scenarios, companies who do not know each other beforehand might need to share certain data, with restrictions such as temporal constraints. This is especially true for *traceability queries*, which a certain stakeholder uses to determine all past locations and states of a given object or a class of objects. While we have shown how the enforcement of concrete access policies can be realized, we consider the management of such policies, including their generation, assignment, revokation, and maybe delegation, a challenging and open research issue. Considering the large amounts of both the information and the potential participants, overcoming the need to manually define every single access policy is highly desirable.

In our future work, we will target these issues as well as other access control challenges in global traceability networks such as access control for discovery services, inter-organizational role concepts and concepts for proving and delegating attributes and permissions.

References

1. Agrawal, R., Cheung, A., Kailing, K., Schönauer, S.: Towards Traceability across Sovereign, Distributed RFID Databases. In: IDEAS 2006: Proceedings of the 10th International Database Engineering and Applications Symposium, Washington, DC, USA, pp. 174–184. IEEE Computer Society, Los Alamitos (2006)
2. Anderson, R.J.: Security Engineering: A Guide to Building Dependable Distributed Systems. John Wiley & Sons, Inc., New York (2001)
3. Bertino, E., Bonatti, P.A., Ferrari, E.: TRBAC: A Temporal Role-Based Access Control Model. In: RBAC 2000: Proceedings of the fifth ACM workshop on Role-based access control, pp. 21–30. ACM Press, New York (2000)
4. Bornh"ovd, C., Lin, T., Haller, S., Schaper, J.: Integrating Automatic Data Acquisition with Business Processes – Experiences with SAP's Auto-ID Infrastructure. In: Proceedings of the 30th International Conference on Very Large Data Bases, pp. 1182–1188 (2004)
5. Browder, K., Davidson, M.A.: The Virtual Private Database in Oracle9iR2. In: Oracle Technical White Paper, Oracle Corporation, 500 Oracle Parkway, Redwood Shores, CA 94065, U.S.A (January 2002)
6. De Capitani di Vimercati, S., Samarati, P.: Access Control in Federated Systems. In: NSPW 1996: Proceedings of the 1996 workshop on New security paradigms, pp. 87–99. ACM Press, New York (1996)
7. Do, H.-H., Anke, J., Hackenbroich, G.: Architecture Evaluation for Distributed Auto-ID Systems. In: Bressan, S., Küng, J., Wagner, R. (eds.) DEXA 2006. LNCS, vol. 4080, pp. 30–34. Springer, Heidelberg (2006)
8. EPCglobal Inc. EPC Information Services (EPCIS) Version 1.0 Specification (April 2007),
   http://www.epcglobalinc.org/standards/EPCglobal_EPCIS_Ratified_Standard_12April_2007_V1.0.pdf
9. Ferraiolo, D.F., Kuhn, D.R.: Role-based access controls. In: 15th National Computer Security Conference, Baltimore, MD, pp. 554–563 (October 1992)
10. Floerkemeier, C., Lampe, M., Roduner, C.: Facilitating RFID Development with the Accada Prototyping Platform. In: Proceedings of PerWare Workshop 2007 at IEEE International Conference on Pervasive Computing and Communications, New York, USA (March 2007)
11. Garfinkel, S., Juels, A., Pappu, R.: RFID Privacy: An Overview of Problems and Proposed Solutions. IEEE Security and Privacy 3(3), 34–43 (2005)
12. Groba, C., Groß, S., Springer, T.: Context-Dependent Access Control for Contextual Information. In: ARES 2007: Proceedings of the The Second International Conference on Availability, Reliability and Security, Washington, DC, USA, pp. 155–161. IEEE Computer Society, Los Alamitos (2007)
13. Grummt, E., Ackermann, R.: Proof of Possession: Using RFID for large-scale Authorization Management. In: Proceedings of the European Conference on Ambient Intelligence, Darmstadt, Germany (November 2007)

14. Grummt, E., Müller, M., Ackermann, R.: Access Control: Challenges and Approaches in the Internet of Things. In: Proceedings of the IADIS International Conference WWW/Internet 2007, Vila Real, Portugal, vol. 2, pp. 89–93 (October 2007)

15. Vincent, C., Hu, D.F.: Ferraiolo, and D. Rick Kuhn. Assessment of Access Control Systems. Interagency Report 7316, National Institute of Standards and Technology, Gaithersburg, MD 20899-8930 (September 2006)

16. Hulsebosch, R.J., Salden, A.H., Bargh, M.S., Ebben, P.W.G., Reitsma, J.: Context sensitive access control. In: SACMAT 2005: Proceedings of the tenth ACM symposium on Access control models and technologies, pp. 111–119. ACM Press, New York (2005)

17. Juels, A.: RFID Security and Privacy: A Research Survey. IEEE Journal on Selected Areas in Communication 24(2), 381–394 (2006)

18. Lampson, B.: Protection. In: Proceedings of the 5th Annual Princeton Conference on Information Sciences and Systems, pp. 437–443. Princeton University, Princeton (1971)

19. LeFevre, K., Agrawal, R., Ercegovac, V., Ramakrishnan, R., Xu, Y., DeWitt, D.: Limiting Disclosure in Hippocratic Databases. In: Proceedings of the 30th International Conference on Very Large Data Bases, Toronto, Canada, pp. 108–119 (August 2004)

20. Lorch, M., Proctor, S., Lepro, R., Kafura, D., Shah, S.: First Experiences Using XACML for Access Control in Distributed Systems. In: XMLSEC 2003: Proceedings of the 2003 ACM workshop on XML security, pp. 25–37. ACM Press, New York (2003)

21. Peris-Lopez, P., Hernandez-Castro, J.C., Estevez-Tapiador, J., Ribagorda, A.: RFID Systems: A Survey on Security Threats and Proposed Solutions. In: Cuenca, P., Orozco-Barbosa, L. (eds.) PWC 2006. LNCS, vol. 4217, pp. 159–170. Springer, Heidelberg (2006)

22. Saltzer, J.H., Schroeder, M.D.: The Protection of Information in Computer Systems. Proceedings of the IEEE 63, 1278–1308 (1975)

23. Sarma, S.: Integrating RFID. ACM Queue 2, 50–57 (2004)

24. Stonebraker, M., Wong, E.: Access control in a relational data base management system by query modification. In: ACM 1974: Proceedings of the 1974 annual conference, pp. 180–186. ACM Press, New York (1974)

25. Sybase, Inc. New Security Features in Sybase Adaptive Server Enterprise. Technical Whitepaper (2003)

26. Tim Moses (Editor). eXtensible Access Control Markup Language (XACML) Version 2.0 (February 2005),
http://docs.oasis-open.org/xacml/2.0/access_control-xacml-2.0-core-spec-os.pdf

27. Traub, K., Allgair, G., Barthel, H., Burstein, L., Garrett, J., Hogan, B., Rodrigues, B., Sarma, S., Schmidt, J., Schramek, C., Stewart, R., Suen, K.: The EPCglobal Architecture Framework – EPCglobal Final Version of 1 July 2005 (July 2005),
http://www.epcglobalinc.org/standards/Final-epcglobal-arch-20050701.pdf

28. Wang, F., Liu, P.: Temporal Management of RFID data. In: VLDB 2005: Proceedings of the 31st international conference on Very large data bases, VLDB Endowment, pp. 1128–1139 (2005)

29. Yagüe, M.I.: Survey on XML-Based Policy Languages for Open Environments. Journal of Information Assurance and Security 1, 11–20 (2006)

SOCRADES: A Web Service Based Shop Floor Integration Infrastructure

Luciana Moreira Sá de Souza, Patrik Spiess,
Dominique Guinard, Moritz Köhler, Stamatis Karnouskos, and Domnic Savio

SAP Research
Vincenz-Priessnitz-Strasse 1, Karlsruhe, Germany
Kreuzplatz, 20 Zürich, Switzerland
{luciana.moreira.sa.de.souza,patrik.spiess,dominique.guinard,
mo.koehler,stamatis.karnouskos,domnic.savio}@sap.com

Abstract. On the one hand, enterprises manufacturing any kinds of goods require agile production technology to be able to fully accommodate their customers' demand for flexibility. On the other hand, Smart Objects, such as networked intelligent machines or tagged raw materials, exhibit ever increasing capabilities, up to the point where they offer their smart behaviour as web services. The two trends towards higher flexibility and more capable objects will lead to a service-oriented infrastructure where complex processes will span over all types of systems — from the backend enterprise system down to the Smart Objects. To fully support this, we present SOCRADES, an integration architecture that can serve the requirements of future manufacturing. SOCRADES provides generic components upon which sophisticated production processes can be modelled. In this paper we in particular give a list of requirements, the design, and the reference implementation of that integration architecture.

1 Introduction

In the manufacturing domain, constant improvements and innovation in the business processes are key factors in order to keep enterprises competitive in the market. Manufacturing businesses are standing on the brink of a new era, one that will considerably transform the way business processes are handled.

With the introduction of ubiquitous computing in the shop floor[1], an entirely new dynamic network of networked devices can be created - an Internet of Things (IoT) for manufacturing. The Internet of Things is a concept which first appeared shortly after 2000. Until now, several approaches to describe the IoT have been undertaken of which most have focused on RFID technologies and their application ([5,13]).

Only recently new technologies such as Smart Embedded Devices and Sensor Networks have entered the scene and can be considered as part of the IoT [11].

---

[1] In manufacturing the shop floor is the location where machines are located and products produced.

C. Floerkemeier et al. (Eds.): IOT 2008, LNCS 4952, pp. 50–67, 2008.

Smart Embedded Devices are embedded electronic systems which can sense their internal state and are able to communicate it through data networks. In contrast to this, Sensor Networks not only can measure internal states of their nodes, but also external states of the environment. We group these three technologies - RFID, Smart Embedded Devices, and Sensor Networks - under the notion *Smart Objects*.

Smart Objects are the nerve cells which are interconnected through the Internet and thus build the IoT. For RFID alone it has been shown that it opens fundamentally new ways of executing business processes, and the technology has already been adopted by several key players in the industry. Therefore the focus of this paper lays on Smart Embedded Devices and Sensor Networks and their effects on automatic business process execution.

Although client-server architectures still play an important role in the field of business software systems, the Service Oriented Architecture (SOA) is on the move and it is foreseeable that this architectural paradigm will be predominant in the future. The integration of devices into the business IT-landscape through SOA is a promising approach to digitalize physical objects and to make them available to IT-systems. This can be achieved by running instances of web services on these devices, which moves the integration of back end applications, such as Enterprise Resource Planning (ERP) systems, with the devices one step forward enabling them to interact and create an Internet of Services that collaborates and empowers the future service-based factory.

Enabling efficient collaboration between device-level SOA on the one hand and on the other hand services and applications that constitute the enterprise back end is a challenging task. While the introduction of web service concepts at a level as low as that of the production device or facility automation makes this integration significantly less complex, there are still differences between device-level SOA and the one that is used in the back end. To name but a few of them, device-level services are much more granular, exhibit a lower reliability (especially if they are connected wirelessly) and higher dynamicity and are more focused on technical issues than on business aspects.

These differences can be overcome by introducing a middleware between the back end applications and the services that are offered by devices, service mediators, and gateways. This middleware adds the required reliability, provides means to deal with services appearing and disappearing, and allows intermediate service composition to raise the technical interfaces of low-level services to business-relevant ones.

In this paper we present the SOCRADES middleware for business integration; an architecture focused on coupling web service enabled devices with enterprise applications such as ERP Systems. Our approach combines existing technologies and proposes new concepts for the management of services running on the devices.

This paper is organized as follows: in section 2 we discuss the current state of the art in coupling technologies for shop floor and enterprise applications. Section 3 presents the requirements for our approach, followed by section 4 where we discuss our approach. We propose a prototype for evaluating our approach in section 5 and perform an analysis in section 6. Section 7 concludes this paper.

2 Related Work

Manufacturing companies need agile production systems that can support reconfigurability and flexibility to economically manufacture products. These systems must be able to inform resource planning systems like SAP ERP in advance, about the upcoming breakdown of whole production processes or parts of them, so that adaptation in workflow can be elaborated.

Currently Manufacturing Execution Systems (MES) are bridging the gap between the shop floor and ERP systems that run in the back end. The International systems and automation society - 95 (ISA-95) derivative from the Instrumentation Systems and Automation Society define the standards for this interface [1]. Although MES systems exist as gateways between the enterprise world and the shop floor, they have to be tailored to the individual group of devices and protocols that exist on this shop floor.

By integrating web services on the shop floor, devices have the possibility of interacting seamlessly with the back end system ([9,8]). Currently products like SIMATIC WinCC Smart Access [2] from Siemens Automation use SOAP for accessing tag based data from devices like display panels to PC's. However, they neither provide mechanisms to discover other web-service enabled devices, nor mechanisms for maintaining a catalogue of discovered devices.

The domain of Holonic Manufacturing Execution Systems (HMS) [6] is also relevant to our work. HMS are used in the context of collaborative computing, and use web service concepts to integrate different sources and destinations inside a production environment. They do, however, not offer support to process orchestration or service composition.

Amongst others, European Commission funded projects like SIRENA [4] showed the feasibility and benefit of embedding web services in production devices. However, since these were only initial efforts for proving the concept, not much attention has been given to issues such as device supervision, device life cycle management, or catalogues for maintaining the status of discovered devices, etc. The consortium of the SOCRADES project has integrated partners, code and concepts from SIRENA, and aims to further design and implement a more sophisticated infrastructure of web-service enabled devices. SODA (www.soda-itea.org) aims at creating a comprehensive, scalable, easy to deploy ecosystem built on top of the foundations laid by the SIRENA project.

The SODA ecosystem will comprise a comprehensive tool suite and will target industry-favourite platforms supported by wired and wireless communications. Although EU projects like SIRENA showed the feasibility and benefit of embedding web services in devices used for production, they do not offer an infrastructure or a framework for device supervision or device life cycle. They neither do provide a catalogue for maintaining the status of discovered devices [4]. Changes due to the current development are moving towards a more promising approach of integrating shop floor devices and ERP systems more strongly [14].

Some authors are criticizing the use of RPC-style interaction in ubiquitous computing [12] (we consider the smart manufacturing devices a special case of that). We believe this does not concern our approach, since web services also allow for interaction with asynchronous, one-way messages and publish-subscribe communication.

SAP xApp Manufacturing Integration and Intelligence (SAP xMII) is a manufacturing intelligence portal that uses a web server to extract data from multiple sources, aggregate it at the server, transform it into business context and personalize the delivered results to the users [7]. The user community can include existing personal computers running internet browsers, wireless PDAs or other UIs. Using database connectivity, any legacy device can expose itself to the enterprise systems using this technology.

The drawback of this product is that every device has to communicate to the system using a driver that is tailored to the database connectivity. In this way, SAP xMII limits itself to devices or gateway solutions that support database connectivity.

In [10], we proposed a service-oriented architecture to bridge between shop floor devices and enterprise applications. In this paper however, building on both our previous work and SAP xMII, we show how the already available functionality of xMII can be leveraged and extended to provide an even richer integration platform. The added functionality comprises integration of web service enabled devices, making them accessible through xMII, and supporting the software life cycle of embedded services. This enables real-world devices to seamlessly participate in business processes that span over several systems from the back end through the middleware right down to the Smart Objects.

3   System Requirements

As embedded technology advances, more functionality that currently is hosted on powerful back end systems and intermediate supervisory devices can now be pushed down to the shop floor level. Although this functionality can be transferred to devices that have only a fraction of the capabilities of more complex systems, their distributed orchestration in conjunction with the fact that they execute very task-specific processing, allows us to realise approaches that can outperform centralised systems in means of functionality. By embedding web services on devices, these can become part of a modern ESOA communication infrastructure.

The first step to come closer to realize this vision, is to create a list of requirements. We have come up with this list through interviews with project partners and customers from the application domain, as well as a series of technical workshops with partners form the solution domain. As usually done in software engineering, we separated it in functional and non-functional requirements.

Functional Requirements

– **WS based direct access to devices:** Back end services must be able to discover and directly communicate with devices, and consume the services they offer. This implies the capability of event notifications from the device side, to which other services can subscribe to.

- **WS based direct access to back end services:** Most efforts in the research domain today focus on how to open the shop floor functionality to the back end systems. The next challenge is to open back end systems to the shop floor. E.g. devices must be able to subscribe to events and use enterprise services. Having achieved that, business logic executing locally on shop floor devices can now take decisions not only based on its local information, but also on information from back end systems.
- **Service discovery:** Having the services on devices will not be of much use if they can not be dynamically discovered by other entities. Automatic service discovery will allow us to access them in a dynamic way without having explicit task knowledge and the need of a priori binding. The last would also prevent the system from scaling and we could not create abstract business process models.
- **Brokered access to events:** Events are a fundamental pillar of a service based infrastructure. Therefore access to these has to be eased. As many devices are expected to be mobile, and their online status often change (including the services they host), buffered service invocation should be in-place to guarantee that any started process will continue when the device becomes available again. Also, since not all applications expose web services, a pull point should be realised that will offer access to infrastructure events by polling.
- **Service life cycle management:** In future factories, various services are expected to be installed, updated, deleted, started, and stopped. Therefore, we need an open ways of managing their life cycle. Therefore the requirement is to provide basic support in the infrastructure itself that can offer an open way of handling these issues.
- **Legacy device integration:** Devices of older generation should be also part of the new infrastructure. Although their role will be mostly providing (and not consuming) information, we have to make sure that this information can be acquired and transformed to fit in the new WS-enabled factory. Therefore the requirement is to implement gateways and service mediators to allow integration of the non-ws enabled devices.
- **Middleware historian:** In an information-rich future factory, logging of data, events, and the history of devices is needed. The middleware historian is needed which offers information to middleware services, especially when an analysis of up-to-now behaviour of devices and services is needed.
- **Middleware device management:** Web service enabled devices, will contain both, static and dynamic data. This data can now be better and more reliably integrated to back end systems offering a more accurate view of the shop floor state. Furthermore by checking device data and enterprise inventory, incompatibilities can be discovered and tackled. Therefore we require approaches that will effectively enable the full integration of device data and their exploitation above the device-layer.

Non-Functional Requirements

- **Security support:** Shop floors are more or less closed environments with limited and controlled communication among their components. However, because of open communication networks, this is fundamentally changing. Issues like confidentiality, integrity, availability must be tackled. In a web service mash-up - as the future factory is expected to be -, devices must be able to a) authenticate themselves to external services and b) authenticate/control access to services they offer.
- **Semantic support:** This requirement facilitates the basic blocks primarily for service composition but also for meaningful data understanding and integration. Support for the usage of ontologies and semantic-web concepts will also enhance collaboration as a formal description of concepts, terms, and relationships within a manufacturing knowledge domain.
- **Service composition:** In a SOA infrastructure, service composition will allow us to build more sophisticated services on top of generic ones, therefore allowing thin add-ons for enhanced functionality. This implies a mixed environment where one could compose services a) at device level b) at back end level and c) in bidirectional cross-level way.

In the above list we have described both, functional and non-functional requirements. In our architecture these requirements will be realized through components, each one offering a unique functionality.

4   Architecture

4.1   Overview

In this chapter, we present a concrete integration architecture focusing on leveraging the benefits of existing technologies and taking them to a next level of integration through the use of DPWS and the SOCRADES middleware. The architecture proposed in Figure 1 is composed of four main layers: Device Layer, SOCRADES middleware (consisting of an application and a device services part), xMII, and Enterprise Applications.

The Device Layer is composed of devices in the shop floor. These devices when enabled with DPWS connect to the SOCRADES middleware for more advanced features. Nevertheless, since they support web services, they provide the means for a direct connection to Enterprise Applications. For the intermediate part of the SOCRADES architecture, bridging between enterprise and device layer, we identified an SAP product that partly covered our requirements: SAP xApp Manufacturing Integration and Intelligence (SAP xMII). The features already available in xMII are:

- Connectivity to non web service enabled devices via various shop floor communication standards
- Graphical modelling of business rules
- Visualization Services
- Connectivity to older SAP software through SAP-specific protocols

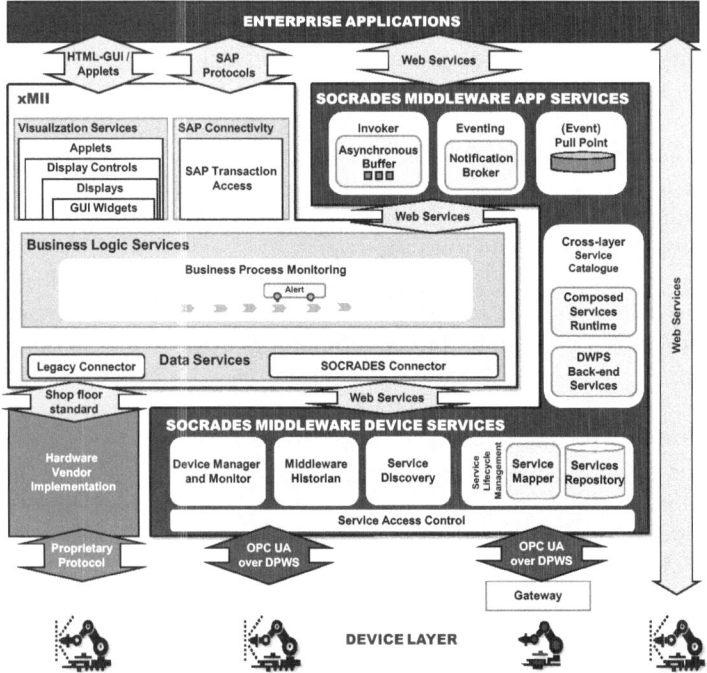

Fig. 1. SOCRADES Integrated Architecture

We decided not to re-implement that functionality but use it as a basis and extend it by what we call the SOCRADES middleware. The SOCRADES middleware and xMII perform together a full integration of devices with ERP systems adding functionalities such as graphical visualization of device data and life cycle management of services running on the devices. In this setting, xMII provides the handling of business logic, process monitoring and visualization of the current status of the devices.

Finally, the connection with Enterprise Applications is realized in three ways. SAP xMII can be used to generate rich web content that can be integrated into the GUI of an enterprise system in mash-up style. Alternatively, it can be used to establish the connection to older SAP systems using SAP-specific protocols.

Current, web service based enterprise software can access devices either via web services of the SOCRADES middleware, benefiting from the additional functionality, or they can directly bind against the web services of DPWS-enabled devices. The data delivered to Enterprise Applications is currently provided by xMII. Nevertheless with the introduction of the SOCRADES middleware and the use of DPWS, this data can be also delivered directly by the regarding devices, leaving to xMII only the task of delivering processed data that requires a global view of the shop floor and of the business process.

4.2   Features and Components of the SOCRADES Middleware

The SOCRADES middleware is the bridging technology that enables the use of features of existing software systems with DPWS enabled devices. Together with SAP xMII, this middleware connects the shop floor to the top floor, providing additional functionality not available in either one of these layers. Although direct access from an ERP system to devices is possible, the SOCRADES middleware simplifies the management of the shop floor devices. In the following, we list this additional functionality and show how the components of the architecture implement them. Brokered Access to Devices. Brokered access means to have a third party mediating the communication between web service clients and servers.

Brokered Access to Devices. Brokered access means to have an intermediate party in the communication between web service clients and servers that adds functionality.

Example are asynchronous invocations, a pull point for handling events, and a publish-subscribe mechanism for events. Asynchronous invocations are useful when dealing with devices that are occasionally connected so that invocations have to be buffered until the device re-appears; they are implemented by the Invoker component. Pull points enable applications to access events without having to expose a web service interface to receive them. The application can instruct the pull point to buffer events and can obtain them by a web service call whenever it is ready. Alternatively, to be notified immediately, the application can expose a web service endpoint and register it at the notification broker for any type of event.

Service Discovery: The service discovery components carries out the actual service discovery on the shop floor level. This component is separated and replicated at each physical site because the DPWS discovery mechanism *WS-Discovery* relies on UDP multicast, a feature that may not be enabled globally across all subsidiaries in a corporate network. All discovered devices from all physically distributed sites and all the services that each device runs are then in a central repository called *Device Manager and Monitor*, which acts as the single access point where ERP systems can find all devices even when they have no direct access to the shop floor network.

Device Supervision: Device Management and Monitor and DPWS Historian provide the necessary static and dynamic information about each DPWS-enabled physical device available in the system. The *device manager* holds any static device data of all on-line and off-line devices while the *device monitor* contains information about the current state of each device. The middleware historian can be configured to log any event occurring at middleware level for later diagnosis and analysis. Many low-level production systems feature historians, but they are concerned with logging low-level data that might be irrelevant for business-level analysis. Only a middleware historian can capture high-level events that are constructed within this architectural layer.

Service Life Cycle Management: Some hardware platforms allow exchanging the embedded software running on them via the network. In a service-enabled shop floor this means that one can update services running on devices. The management of these installed services is handled through the use of the Service Mapper and Services Repository. These components together make a selection of the software that should run in each device and perform the deployment. Cross-Layer Service Catalogue: The cross-layer service catalogue comprises two components. One is the Composed Services Runtime that executes service composition descriptions, therefore realizing service composition at the middleware layer. The second component is the DPWS device for back end services that allows DPWS devices to discover and use a relevant set of services of the ERP system.

Cross-Layer Service Catalogue: The cross-layer service catalogue comprises two components. One is the Composed Services Runtime that executes service composition descriptions, therefore realizing service composition at the middleware layer. The second component is the DPWS device for back end services that allows DPWS devices to discover and use a relevant set of services of the ERP system.

The Composed Services Runtime is used to enrich the services offered by the shop floor devices with business context, such as associating an ID read from an RFID tag with the corresponding order. A compound service can deliver this data by both invoking a service on the RFID reader, and from a warehouse application. A Composed Services Runtime, which is an execution engine for such service composition descriptions, e.g., BPEL [3], is placed in the middleware because only from there, all DPWS services on the shop floor as well as all back end services can be reached. This component can benefit from semantic annotation of the services, both at the shop-floor and the enterprise level. Backed by ontologies, these annotations can be used for semantic discovery of services and for assistance in modelling and verifying processes based on them.

With the sub-component "DPWS device for back-end services" we address the requirement that shop floor devices must be able to access enterprise application services, which can be achieved by making a relevant subset available through the DPWS discovery. This way, devices that run DPWS clients can invoke back end services in exactly the same way they invoke services on their peer devices. Providing only the relevant back end services allows for some access control and reduces overhead during discovery of devices. Co-locating both sub-components in the same component has the advantage that also the composed services that the Composed Services Runtime provides, can be made available to the devices through the virtual DPWS device for back end services.

Security support: The (optional) security features supported by the middleware are role-based access control of devices communication to middleware and back end services and vice versa. Event filtering based on roles is also possible. Both the devices as well as back end and middleware services have to be authorized when they want to communicate. Access control is enforced by the respective component. Additionally, message integrity and confidentiality is provided by the WS-Security standard.

To demonstrate the feasibility of our approach and to make some first evaluations, we implemented a simple manufacturing scenario. We used a first implementation of our architecture to connect two DPWS-enabled real-world devices with an enterprise application.

5   Reference Implementation

In order to prove the feasibility of our concept, we have started realising a reference implementation. From a functional point of view it demonstrates two of the most important incentives for the use of standardized device level web services in manufacturing: flexibility and integration with enterprise software. Indeed, the scenario shows DPWS-enabled devices can be combined easily to create higher-level services and behaviours that can then be integrated into top-floor applications.

The business benefits from adopting such an architecture are numerous:

- lower cost of information delivery
- increases flexibility and thus total cost of ownership (TCO) of machines.
- makes the whole manufacturing process as a whole more visible to the shop floor.
- be able to model at the enterprise layer processes with only abstract view of the underlying layer, therefore easing the creation of new applications and services from non-domain experts.

5.1   Scenario

To support this idea we consider a simple setting with two DPWS devices:

- A robotic arm that can be operated through web service calls. Additionally it offers status information to subscribers through the SOCRADES eventing system.
- A wireless sensor node providing various information about the current environment, posted as events. Furthermore, the sensor nodes provide actuators that are accessible through standard service calls.

The manufacturing process is created on the shop floor using a simple service composition scheme: from the atomic services offered by the arm (such as start/stop, etc.) a simple manufacturing process p is created. The robot is manipulates heat-sensitive chemicals. As a consequence it is identified that the manufacturing process cannot continue if the temperature rises above 45 °C.

The robot may not have a temperature sensor (or this is malfunctioning), but as mentioned before the manufacturing plant is equipped with a network of wireless sensor nodes providing information about the environment. Thus, in order to enforce the business rule, the chief operator uses a visual composition language to combine p with the temperature information published by the service-enabled sensor node: t.

In pseudo code, such a rule looks like:

```
if (t > 45) then p.stopTransportProcess();
```

Furthermore, the operator instantiates a simple gauge fed with the temperature data (provided by t). For this purpose he uses a manufacturing intelligence software and displays the gauge on a screen situated close the robot.

Finally, the sales manager can also leverage the service oriented architecture of this factory. Indeed, the output of the business rule is connected to an ERP system which provides up-to-date information about the execution of the current orders. Whenever the process is stopped because the rule was triggered, an event is sent to the ERP system through its web service interface. The ERP system then updates the orders accordingly and informs the clients of a possible delay in the delivery.

5.2 Components

This section describes the architecture of our prototype from an abstract point of view. Its aim is to understand the functionality whose concrete implementation will be described within the next section.

Functional Components. The system comprises four main components as shown on Figure 2 that we shall briefly describe:

- **Smart Devices:** Manufacturing devices, sensors and Smart Things (i.e. Smart Objects) are the actors forming an Internet of Services in the factory as well as outside of the factory. They all offer web service interfaces, either directly or through the use of gateways or service mediators. Through these interfaces they offer functional services (e.g. start/stop, swap to manual/automatic mode) or status information (e.g. power consumption, mode of operation, usage statistics, etc.).
- **Composed Service:** The component aggregates the services offered by smart objects. Indeed, it is in charge of exposing coarse-grained services to the upper layers. In the case of the robotic arm for instance, it will consume the open(), close() and move(...), methods and use them to offer a doTransportProcess (...) service.
- **Business Logic Services and Visualisation Services:** In our prototype, the business logic services are supported by a service composition engine and visualized using a visualization toolkit. The former component is used to model business rules or higher-level processes, known as business logic services in our architecture. As an example the operator can use it to create the business rules exposed above. The latter component is used to build a plant-floor visualisation of the devices' status and the overall process execution. As an example the operator can instantiate and use a set of widgets such as

gauges and graphs to monitor the status of the machines. The production manager can also use it to obtain real-time graphs of the process execution and status.

– **Enterprise Applications:** This is the place of high-end business software such as ERPs or PLMs. The idea at this level is to visualize processes rather than the machines executing the processes. This layer is connected to the plant-floor devices through the other layers. As such it can report machines failures and plant-floor information on the process visualization and work-flow. Furthermore, business actions (e.g. inform customers about a possible delay) can be executed based on this information.

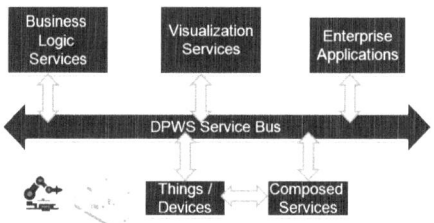

Fig. 2. The DPWS service bus

Cross-Component Communication. In a mash-up, the architecture is not layered but rather flat, enabling any functional component to talk to any other. Such architectures need a common denominator in order for the components to be able to invoke services on one another. In our case the common denominator is the enhanced DPWS we developed. Each component is DPWS-enabled and thus, consumes DPWS services and exposes a DPWS interface to invoke the operations it offers. The service invocations can be done either synchronously or asynchronously via the web service eventing system. For instance the temperature is gathered via a subscription to the temperature service (asynchronous) whereas the transport process is stopped by invoking an operation on the process middleware. Figure 2 depicts the architecture by representing the components connected to a common (DPWS) ESB (Enterprise Service Bus).

5.3 Implementation

The system described in this paper is a reference implementation of concepts described in the architecture rather than a stand-alone concept. Therefore it uses and extends several software and hardware components rather than writing them from scratch. In this section we will briefly describe what these components are and how they interact together, taking a bottom up approach.

Functional Components

– **Smart Devices:** The wireless sensor network providing temperature in-formation is implemented using the Sun Microsystems' SunSPOT sensor

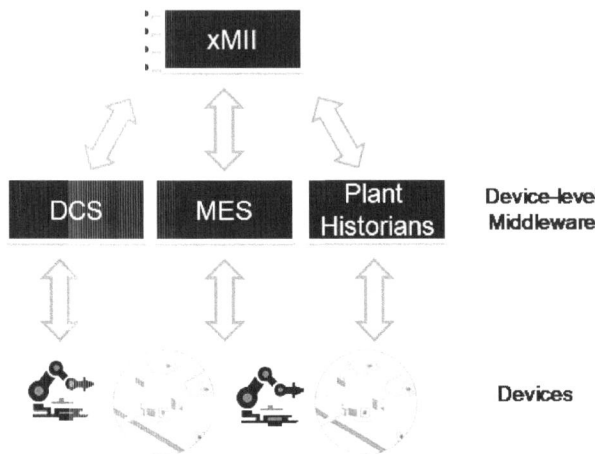

Fig. 3. xMII indirect device connectivity

nodes. Since the nodes are not web services enabled, we had to implement a gateway (as described in our architecture), that would capture the
temperature readings and provide it via DPWS as services one can subscribe to. The gateway component hides the communication protocol between the SunSPOTs and exposes their functionalities as device level web
services (DPWS). More concretely the SunSPOT offer services for sensing
the environment (e.g. `getTemperature()`) or providing output directly on
the nodes (e.g. `turnLightOn(Color)`). The robotic arm was implemented
as a clamp offering DPWS services for both monitoring and control. The
clamp makes these operations available as DPWS SOAP calls on a PLC
(Programmable Logic Controller) over gateway. For monitoring services (e.g.
`getPowerConsumption()`) the calls are issued directly on the gateway standing for the clamp. For control services the idea is slightly different.

– **Composed Service:** Typical operations at the clamp level are `openClamp()`
  and `closeClamp()`. In order to consistently use these operations on the top-
  floor we need to add some business semantics already on the shop floor. This
  is the role of composed services which aggregate an number of coarse-grained
  operations (e.g. `openClamp()`) and turn them into higher level services. This
  way the `start()`, `openClamp()`, `closeClamp()`, `move(x)`, `stop()` operations
  are combined to offer the `startTransportProcess()` service.
– **Business Logic Services and Vizualisation Services:** Services offered
  by both the sensors and the clamp are combined to create a business rule.
  The creation of this business logic service is supported by xMII, SAP's Manufacturing Integration and Intelligence software. As mentioned before, the
  aim of this software is firstly to offer a mean for gathering monitoring data
  from different device aggregators on the shop floor such as MESs (Manufacturing Execution Systems). This functionality is depicted on Figure 3.

Since the SOCRADES infrastructure proposes to DPWS-enable all the devices on the plant-floor, we can enhance the model by directly connecting the devices to xMII. Additionally, xMII offers a business intelligence tool. Using its data visualization services we create a visualization of process-related and monitoring data. Finally, we use the visual composition tool offered by xMII to create the rule. Whenever this rule is triggered the `stopTransportProcess()`operation is invoked on the middleware to stop the clamp.

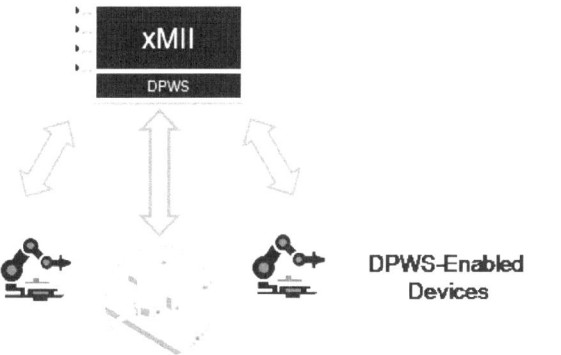

Fig. 4. Direct connectivity to the DPWS devices

– **Enterprise Applications:** Whenever the business rule is triggered, xMII invokes the `updateOrderStatus()`on the ERP. As mentioned before this latter component displays the failure and its consequences (i.e. a delay in the production) in the orders' list. Additionally, if the alert lasts for a while, it informs the customer by email providing him with information about a probable delay.

Cross-Component Communication. Figure 5 presents the communication amongst the components whenever the business rule is triggered. At first the SunSPOT dispatches the temperature change by placing a SOAP message on the DPWS service bus. The xMII is subscribed to this event and thus, receives the message and feeds it to its rules engine. Since the reported temperature is above the threshold xMII fires the rule. As a consequence it invokes the `stopTransportProcess()`operation on the Process Service middleware. This component contacts the clamp and stops it. Furthermore, xMII triggers the `updateOrderStatus()`operation on the ERP. This latter system update the status of the concerned order accordingly and decides whether to contact the customer to inform him by email about the delay.

6   System Analysis

In this section we will discuss the properties of our architecture and give decision makers a framework at hand through which they can assess the concrete value of

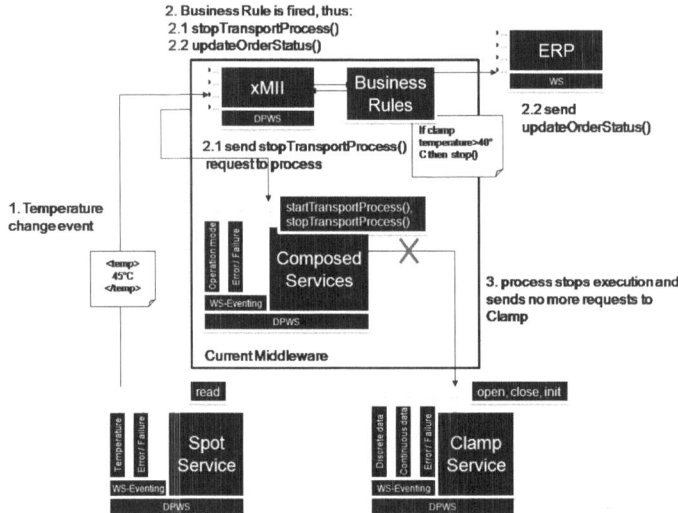

Fig. 5. Interactions when the business rule is triggered

our system for their organisation. Since the work we are presenting in this paper is part of ongoing research, we think it is helpful to have such a framework, in particular to assess future work.

In the field of Systems Management several standards exist [ref. Standards, ITIL, etc.] which aim to support a structured dealing with IT systems. One framework in particular helpful for central corporate functions such as production is the ISO model FCAPS (Fault, Configuration, Administration, Performance, Security). Although being a framework for network management, it is relevant for our architecture because it is enabling low level networked interaction between Smart Objects. Here we will give a first attempt to evaluate the architecture.

- **Fault Management:** Since our system will be part of the manufacturing IT-landscape we need to manage both, faults of particular parts of the manufacturing process and faults in our system. Due to the tight integration these types of faults inherently become the same. In particular the SOA based approach of device integration enables the user to identify faults in his production process, at a level never seen before. It also gives the possibility to build redundancy at system critical stages which ensures fast recovery from local failures. Finally the flexibility given by our SOA approach lets the user decide to what extend he wants to introduce capabilities of quick fault recovery, depending on his individual needs.
- **Configuration Management:** Mainly the two components *Service Lifecycle Management* and *Cross-Layer Service Catalogue* support dynamic configuration management. However, at the current point of view we see code updated to Smart Devices as a major challenge which until today has not

been resolved sufficiently. Configuration also includes the composition of services into higher-level services. In a future version, our Service Discovery module will use semantic annotation of services to find appropriate service instances for online service composition. Using ontologies to specify the behaviour and parameters of web services in their interface descriptions and metadata allows flexible service composition. Especially in the very well defined domain of manufacturing we can make use of existing ontologies that describe production processes.

– **Administrative Management:** The *Device Manager* provides the necessary static and dynamic information about each Smart Device. Through the strict use of web-service interfaces, it will be possible to easily integrate devices into management dash-boards. Through this technically we allow easy and user friendly access to Smart Devices. However, taking the possibly very large number of devices into account, we belief that our middle-ware has deficiencies in offering this user friendly administration. Although this problem is subject to other fields of research such as sensor networks, (e.g, macro programming), we will dedicate our research efforts to the problem.

– **Performance Management:** Already now we can say that local components of our system will scale well in regards to total amount of Smart Objects and their level of interaction. This can be justified since all interaction occurs locally and only a limited amount of Smart Objects is needed to fulfil a particular task. However, it is still an open question, if our system will scale well on a global scale and to what extend it will need to be modularized. For example we will need to investigate whether central components such as device and service registries should operate on a plant level or on a corporate level, which could mean that these parts would have to handle several millions or even billions of devices at the same time.

– **Security Management:** As mentioned in the security support section of the architecture, our system can make use of well established security features which already are part of web-service technologies and their protocols such as DPWS. It is most likely that we will have to take into account industry specific security requirements, and it will be interesting to see, if we can deliver a security specification which satisfies all manufacturing setups.

7   Conclusions

In this paper we have presented SOCRADES, a Web Service based Shop Floor Integration Infrastructure. With SOCRADES we are offering an architecture including a middleware which support connecting Smart Devices, i.e. intelligent production machines from manufacturing shop floors, to high-level back-end systems such as an ERP system. Our integration strategy is to use web services as the main connector technology. This approach is motivated by the emerging importance of Enterprise Service Oriented Architectures, which are enabled through web services.

Our work has three main contributions: First, we elaborated and structured a set of requirements for the integration problem. Second, we are proposing a

concrete architecture containing components which realize the required functionality of the system. Our third contribution is a reference implementation of the
SOCRADES architecture. In this implementation we have demonstrated the full
integration of two Smart Devices into and enterprise system. We showed that it
is possible to connect Smart Devices to an ERP system, and describe how this
is done.

Our next steps include integrating a prototype in a bigger setup and testing
it with live production systems.

Acknowledgements

The author would like to thank the European Commission and the partners
of the European IST FP6 project "Service-Oriented Cross-layer infRAstructure
for Distributed smart Embedded devices" (SOCRADES - www.socrades.eu), for
their support.

References

1. Instrumentation systems and automation society, http://www.isa.org/
2. SIMATIC WinCC flexible, http://www.siemens.com/simatic-wincc-flexible/
3. Web Services Business Process Execution Language Version 2.0 (OASIS Standard),
   (April 2007), http://docs.oasis-open.org/wsbpel/2.0/wsbpel-v2.0.html
4. Bohn, H., Bobek, A., Golatowski, F.: SIRENA - Service Infrastructure for Real-
   time Embedded Networked Devices: A service oriented framework for different
   domains. In: International Conference on Systems and International Conference
   on Mobile Communications and Learning Technologies (ICNICONSMCL 2006),
   Washington, DC, USA, p. 43. IEEE Computer Society, Los Alamitos (2006)
5. Fleisch, E., Mattern, F. (eds.): Das Internet der Dinge: Ubiquitous Computing und
   RFID in der Praxis:Visionen, Technologien, Anwendungen, Handlungsanleitungen.
   Springer, Heidelberg (2005)
6. Gaxiola, L., de Ramírez, M.J., Jimenez, G., Molina, A.: Proposal of Holonic Manufacturing Execution Systems Based on Web Service Technologies for Mexican
   SMEs. In: Mařík, V., McFarlane, D.C., Valckenaers, P. (eds.) HoloMAS 2003. LNCS
   (LNAI), vol. 2744, pp. 156–166. Springer, Heidelberg (2003)
7. Gorbach, G.: Pursuing manufacturing excellence through Real-Time performance
   management and continuous improvement. ARC Whitepaper (April 2006)
8. Jammes, F., Mensch, A., Smit, H.: Service-Oriented Device Communications using
   the Devices Profile for Web Services. In: MPAC 2005: Proceedings of the 3rd international workshop on Middleware for pervasive and ad-hoc computing, pp. 1–8.
   ACM Press, New York (2005)
9. Jammes, F., Smit, H.: Service-oriented paradigms in industrial automation. IEEE
   Transactions on Industrial Informatics 1, 62–70 (2005)
10. Karnouskos, S., Baeker, O., de Souza, L.M.S., Spiess, P.: Integration of SOA-ready
    Networked Embedded Devices in Enterprise Systems via a Cross-Layered Web
    Service Infrastructure. In: 12th IEEE Conference on Emerging Technologies and
    Factory Automation (2007)
11. Reinhardt, A.: A Machine-To-Machine "Internet Of Things". Business Week (April
    2004)

12. Saif, U., Greaves, D.J.: Communication Primitives for Ubiquitous Systems or RPC Considered Harmful. In: 21st International Conference of Distributed Computing Systems (Workshop on Smart Appliances and Wearable Computing), IEEE Computer Society Press, Los Alamitos (2001)
13. Schoenberger, C.R.: RFID: The Internet of Things. Forbes 18 (March 2002)
14. Zeeb, E., Bobek, A., Bohn, H., Golatowski, F.: Service-Oriented Architectures for Embedded Systems Using Devices Profile for Web Services. In: 21st International Conference on Advanced Information Networking and Applications Workshops (2007)

Automation of Facility Management Processes Using Machine-to-Machine Technologies

Sudha Krishnamurthy[2], Omer Anson[1], Lior Sapir[1], Chanan Glezer[1],
Mauro Rois[1], Ilana Shub[1], and Kilian Schloeder[2]

[1] Deutsche Telekom Laboratories at Ben-Gurion University, Beer-Sheva, Israel
{ansono,sapirlio,chanan,roism,shubi}@bgu.ac.il
[2] Deutsche Telekom Laboratories, Berlin, Germany
{sudha.krishnamurthy,kilian.schloeder}@telekom.de

Abstract. The emergence of machine-to-machine (M2M) technologies as a business opportunity is based on the observation that there are many more machines and objects in the world than people and that an everyday object has more value when it is networked. In this paper, we describe an M2M middleware that we have developed for a facility management application. Facility management is a time and labour-intensive service industry, which can greatly benefit from the use of M2M technologies for automating business processes. The need to manage diverse facilities motivates several requirements, such as predictive maintenance, inventory management, access control, location tracking, and remote monitoring, for which an M2M solution would be useful. Our middleware includes software modules for interfacing with intelligent devices that are deployed in customer facilities to sense real-world conditions and control physical devices; communication modules for relaying data from the devices in the customer premises to a centralized data center; and service modules that analyze the data and trigger business events. We also present performance results of our middleware using our testbed and show that our middleware is capable of scalably and reliably handling concurrent events generated by different types of M2M devices, such as RFID tags, Zigbee sensors, and location tracking tags.

1 Introduction

Machine to Machine (M2M) is a term used to describe the technologies that enable smart sensors, actuators, mobile devices, and back-end servers to communicate with each other, for the purposes of remotely controlling and configuring the machines (telematics), remotely monitoring/collecting data from machines (telemetry), and making decisions based on the collected data and sending notifications of unusual situations - often without human intervention. The emergence of M2M technologies as a business opportunity is based on the observation that there are many more machines - defined as things with mechanical, electrical, or electronic properties - than people, and that not only does a machine have more value when it is networked, but that the network becomes more valuable as

C. Floerkemeier et al. (Eds.): IOT 2008, LNCS 4952, pp. 68–86, 2008.

more machines are connected. Harbor Research, a technology consultancy and analysis firm, estimates that by 2010, at least 1.5 billion devices will be connected to the Internet worldwide and this will include devices in the millions of households across the world. Over 45 million new cars are being produced every year and increasingly these are being built with embedded electronic networking capabilities for diagnostics, communications and so on. If we add to this the millions of vending machines, elevators, and security systems, the potential market resulting from M2M solutions is very appealing.

1.1   M2M Opportunities for Telecommunication Operators and Service Providers

Industrial SCADA (supervisory control and data acquisition) systems [1], which were first introduced in the 1980s as a means to remotely operate and collect data from facility systems and equipment, all from a master computer station, may be regarded as the predecessors of the modern day M2M solutions. While M2M and SCADA essentially focus on the same functionality, the fundamental approaches they use to achieve that are quite different. SCADA applications are to a large extent targeted toward specific industries, such as, public utilities including electricity, oil and gas, and water. When SCADA first appeared, few communication options were available and hence, SCADA solutions primarily use proprietary protocols designed for private, licensed radio networks, which makes those solutions more expensive. One of the advantages of M2M-based middleware is that it works with commonly used technologies, such as TCP/IP, IEEE 802.11 wireless LANs, cellular communications technologies, and wired networks such as Ethernet. Using these well-known technologies, some of which are open-source, allows easier device interoperation in M2M systems and facilitates using mass produced, standards-compliant equipment, which makes the implementation less expensive, simpler, and quicker. The recent availability of multiple wireless data transport technologies, such as Bluetooth (for short-range device connectivity), ZigBee (built on top of IEEE 802.15.4 for low-power monitoring, sensing, and control networks), radio frequency identification (RFID) (which uses electromagnetic or electrostatic coupling to transmit status signals from tagged devices) has enabled M2M solutions to be applied to a wide range of applications, such as security and tracking, remote automatic meter reading, vending machines, elevators and escalators, industrial automation, road traffic information, traffic control systems, and telemedicine. These recent, short-range technologies are often used for communicating data in short hops between the wireless devices (in peer-to-peer connections or through mesh networks) and ultimately to a master device that communicates with the Internet through mobile or fixed networks, thereby reducing the power requirements and allowing remote sensors to operate longer between battery changes. M2M nodes can operate autonomously, push information to multiple systems and other nodes, and make some decisions on their own. With SCADA, sensors and controllers are hardwired to the host, which limits the number of sensors the technology can serve and the distance over which data can be transported. In contrast, by utilizing the latest wireless and Internet technologies, M2M is ultimately more flexible, able to adapt to multiple business settings, more efficiently connect a more diverse selection of assets, perform data acquisition and control capabilities from much farther distances.

Given the impact of communication technologies on the emergence of M2M solutions, mobile service providers and telecommunication operators can take advantage of this emerging market in the following ways. As many businesses are becoming more globally based and more reliant on the electronic transfer of data, offering an M2M platform provides an opportunity to expand the cellular customer base to new segments by bundling traditional telecom services with enterprise M2M solutions. Given the high churn rate in mobile voice market, M2M can be useful in retaining existing customers, creating loyalty, and increasing traffic by developing new data services. Furthermore, when the M2M traffic has lower priority than regular voice and data traffic, network operators can improve the network utilization by transmitting the lower priority M2M traffic during off-peak hours.

1.2   M2M for Facility Management

Given that the recent advances in M2M technologies and modules are creating more opportunities for their use in practical scenarios, in this paper, we describe an M2M middleware that we have developed for a facility management application. Facility management is a time and labour-intensive service industry, which can greatly benefit from the use of M2M technologies for automating business processes and predictive maintenance. The M2M middleware that we developed was motivated by the needs of a small and medium-scale real estate service provider that currently manages about 35000 properties, including commercial facilities, utilities such as elevators, fire shutters, ventilators, and technical infrastructure that includes about 23 computer centers, 12000 network nodes and technical buildings, 24000 radio towers, antenna masts and rooftops. The need to manage such diverse facilities poses several requirements for the development of an M2M solution, some of which are listed below:

Remote Monitoring: Round-the-clock fault management requires monitoring of facilities at different frequencies, and triggering of inspection and repair at appropriate times. In some cases, continuous monitoring is required. For example, technical building facilities, including elevators and heating installations need to be working 24x7, as do hot-water systems, power, ventilation, and air-conditioning systems. In certain facilities, the air quality and humidity levels need to be checked continuously. Continuous monitoring of energy consumption using sensors helps in detecting inefficient use of energy and reducing the operating costs of a building. In certain cases, special inspections have to be triggered after adverse weather conditions, such as storm, icing etc. Fire and safety requirements and escape routes have to be monitored regularly, but perhaps not continuously. Regular monitoring is also useful for managing older buildings where harmful substances, such as asbestos, polychlorinated biphenyls (PCB), and heavy metals may be present.

The measurement and inspections help in reporting the presence of harmful substances.

Predictive Maintenance: Problems caused by energy failure in computer centers and technical buildings always involve high costs. Proactive monitoring and predictive maintenance is needed in such environments, in order to minimize faults and maintain high availability of systems.

Security Services: Round the clock security of building and premises is another aspect of facility management that can benefit from M2M technology. This includes access control for critical areas, theft detection and prevention, control, regulation and monitoring of the movement of vehicles, supplies, and goods.

Inventory Management: Inventory taking and data collection are also important for real-estate management, because the quality of information collected impacts the planning accuracy and reliability of decision making.

Prioritized Notification: Since facility management involves data collection from numerous facilities as mentioned above, it is impossible to cater to all of the incoming data simultaneously. Hence, all of the reports from the facility management system need to be linked to the data management center according to priority, so that faults can be cleared according to the necessary response time.

The above requirements illustrate that a facility management application will benefit from an integrated M2M middleware solution that can extract data from diverse physical devices and use that data to automate business processes to ensure the efficient management of different facilities. We have designed and implemented such an M2M solution for facility management and demonstrated and evaluated our solution using a testbed consisting of different RFID technologies, (including passive low-frequency (LF), high-frequency (HF), and ultra-high frequency (UHF) RFID tags, as well as active RFID tags), sensors that communicate using Zigbee radio, and location tracking tags based on Wi-Fi. This paper describes the design, implementation, and evaluation of our M2M middleware. In Section 2, we describe some of the related M2M solutions for facility management. In Section 3, we describe the common services and the design of the middleware. Our middleware includes software modules for interfacing with intelligent devices that are deployed in customer facilities to sense real-world conditions and control physical devices; communication modules for relaying data from the devices in the customer premises to a centralized data center; and service modules that analyze the data and trigger business events. In Section 4, we describe our testbed and provide some details of our implementation. In Section 5, we describe the evaluation of our middleware using our testbed. Finally, in Section 6, we present our conclusions.

2   Related Work

We now briefly discuss some of the related M2M-based solutions for facility management. Frankfurt's Airport (Fraport) recently implemented a new maintenance process replacing the paper-based process with mobile and RFID technology [2]. Fraport's air-conditioning and ventilation system has 22,000 automatic fire shutters and numerous fire doors and smoke detectors. The new system's architecture consists of RFID tags on the fire shutters; mobile devices bundled with an RFID reader and mobile application; and an SAP-based middleware interconnected to a back-end asset management system. While Fraport's solution primarily provides an RFID-based solution for managing airport facilities, our solution is more diverse in that it supports a broader spectrum of M2M technologies, in addition to RFID, and provides common services that are needed for managing facilities that are not limited to airports. Pakanen et al. [3] proposed a system where a microprocessor device interfaced with an Air Handling Unit (AHU) transmits sensor data from the AHU to the Internet through a Public Switch Telephony Network (PSTN) and GSM networks. In addition, a web-based user interface was created to enable remote control of the AHU. Watson et al. [4] applied M2M technology to five commercial buildings in a test of Demand Responsive (DR) energy management. The goal was to reduce electric demand when a remote price signal rose above a predetermined price.

Some of the limitations of existing M2M solutions that we have attempted to address in our solution include the following. Few M2M solutions offer a tight integration with the IT infrastructure. In order to maximize the use of the data collected, we need better integration with enterprise applications (such as, database, ERP, CRM). We need M2M solutions that support end-to-end automation that includes automatic billing, automatic report generation, workflow management, etc. Furthermore, although current technology allows various types of terminals to access information (such as fixed and mobile devices), many M2M solutions only support access through PCs. In order to allow information to be accessed anytime and from anywhere, we need solutions that can easily adapt and extend access to emerging client devices that use newer technologies.

3   Middleware Design

Based on the requirements of the facility management usage scenarios presented in Section 1.2, we have identified some common service modules to form the core services of our M2M middleware. We first enumerate these services and then describe the design of the middleware to realize these services.

3.1   Common Services

- **Access control:** The facility management industry employs several temporary maintenance workers that need to be provided access to specific facilities for the duration of their work period. Hence, access control is an essential functionality that can benefit from M2M technology. Workers can be equipped with tags that describe their work order. The access control system needs to automatically authenticate their entry for the duration of their work order. When the work order is completed, the access should be invalidated and the duration of the work needs to be recorded to generate the billing and payment for the worker.

– **Real-time location tracking:** In certain facilities, it may be necessary to keep track of the location of workers and critical assets within an indoor environment. M2M technologies can be employed to enable real-time location tracking.

– **Decentralized data storage and logging:** Workers may need to work in harsh environments where long-range communication may not be possible. Hence, the M2M solution should allow some of the information, such as operating instructions, maintenance history, and troubleshooting information, to be stored in a decentralized manner. This can be done by recording brief information on read-write tags attached to physical devices. Equipping the workers with portable readers will allow them to quickly read the information off the tags when they are in the proximity of the devices. Upon completion, workers can briefly update the information on the tags with their own work history, in addition to providing a more detailed update when they have access to a backend system.

– **Device inventory:** M2M solutions for facility management should enable knowledge of the connected devices in a facility, their functionalities, and attributes. Instead of scheduling inventory management as a separate task, new inventory can be recorded when goods are brought into a facility and management of existing inventory can be done when workers are within a facility to perform maintenance.

– **Remote monitoring:** The main functionality of an M2M solution is to sense and monitor a remote environment that houses a set of remote devices.

– **Predictive maintenance:** The ability to use the data collected from sensors to perform data analytics and make proactive decisions, such as whether to repair or buy new equipment and when it is time to schedule a maintenance check, is useful for providing high availability of facilities.

– **Device management and control:** The M2M solution should support bidirectional communication. In addition to allowing devices to transmit the data they monitor from the facility premises to a remote data center, the M2M solution should allow customers to remotely configure the devices in their facilities. This remote control is necessary for centrally updating the device software, changing the device settings, scheduling the exchange of information, controlling the periodicity of automated tasks and alarm thresholds. Such device interaction requires messaging protocols that support point-to-point communication with individual devices as well as multicast communication with a group of similar devices.

– **Administration:** An M2M solution needs to offer customized interfaces to a hierarchy of system administrators, programmers, and general users that not only allows them to administer devices in remote facilities, but also administer the machines hosting the M2M modules themselves.

– **Notification:** From the data collected from the remote devices, rules should be triggered to automatically generate reports and send notifications of events and exceptions with varying levels of significance to recipients or applications, and in forms that can be customized for each situation. The M2M solution should accommodate different notification methods, such as email and SMS to a mobile device, and allow the recipients to choose their preferred method of notification.

– **Billing:** Upon completion of a work order, the M2M solution should automatically trigger the billing module to generate an invoice for paying the worker and charging the appropriate customer for the services.

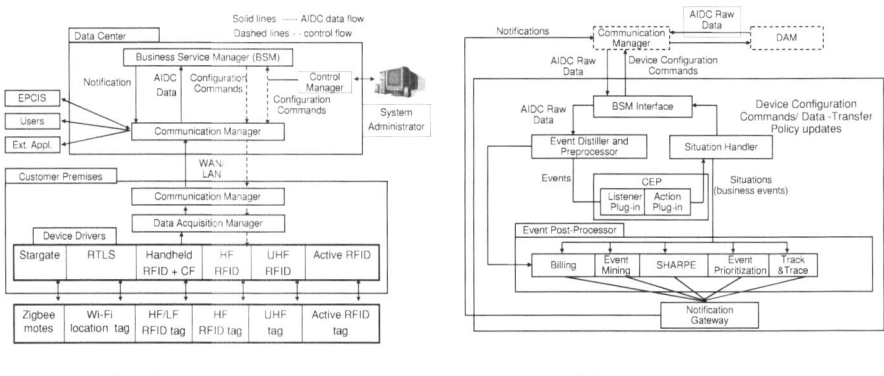

(a) Overall architecture   (b) BSM architecture

Fig. 1. Middleware design

3.2 Description of the Architecture

Figure 1(a) illustrates the M2M software architecture we have implemented for providing the services described above for a facility management application. The architecture consists of a data acquisition manager (DAM) that is responsible for interfacing with the physical devices in the facilities; a communication manager (COM) that transfers data from the devices distributed across the facilities to a remote data center and relays control messages to the devices; a business service manager (BSM) that analyzes the data and automates the business processes; and a control manager (CON) that provides an interface for application hosting and administering the M2M system. These architectural modules are distributed between the facility (customer) premises and the data center. In designing these modules, reliability and scalability are important, so that customers can receive the service within a specified time period. We have also attempted to address some of the limitations of current M2M-based facility management solutions that were mentioned in Section 2, by aiming for a flexible design across the different M2M modules and supporting a diverse spectrum of M2M devices that use different sensing as well as communication technologies, closely integrating the data from remote devices with enterprise resources (databases, business applications, ERP), and supporting multiple forms of access that includes mobile devices. We now provide a detailed description of each of the architectural modules and the challenges involved in their design.

Data Acquisition Manager(DAM): Data acquisition is responsible for interfacing with and acquiring data from a variety of sensors that perform automatic identification and data collection (AIDC). These AIDC devices are programmed to collect and aggregate data, and sometimes react to actions and conditions such as motion, voltage, pressure, flow, or temperature. They can serve as controllers and actuators for machines, such as air conditioners, elevator pumps, and traffic lights. Some of the AIDC technologies that we have in our testbed are listed in Table 1 and this consists of different RFID readers and tags, (passive low-frequency (LF), high-frequency (HF), and ultra-high frequency (UHF) RFID tags, as well as active RFID tags), sensors that communicate using Zigbee, and location tracking tags based on Wi-Fi. We associate a software module with each AIDC hardware, in order to abstract the hardware-specific details involved in directly interfacing with that device. These device driver modules follow an object-oriented methodology and implement the same abstract interface and this enables us to expose a uniform API even when our platform is extended to accommodate future M2M technologies. The device driver modules send the raw data they collect from the smart devices to their respective DAMs. The DAM module is in charge of importing the AIDC data that it receives from the various readers that interact with a plethora of sensors within a facility. The DAM then preprocesses that data to filter duplicates and forwards the filtered data to the BSM that is operating in a remote data center. The DAM also serves as an interface for configuring various AIDC devices. We currently deploy one DAM at each facility, although this can be extended to multiple DAMs in the case of large facilities.

We now discuss some of the key challenges that we had to address in the design of the DAM. First, the different AIDC technologies supply data in disparate formats. Before forwarding the data to the BSM, the DAM resolves the data heterogeneity by converting different read records into a normalized representation, which includes the facility ID, sensor address within the facility, and the data. Similarly, when the DAM receives control messages from the application server to be forwarded to the AIDC devices, it needs to address them to the respective device driver modules in the facility, which then execute the device-specific commands. Second, the addressing formats of the AIDC devices may vary. For example, the RFID readers have IP addresses, while the tags are identified by a serial number. If these tags are EPC-compliant [5], they have a globally unique ID, in accordance with the Object Naming Standard (ONS). On the other hand, the Zigbee enabled motes have to be assigned a local identifier, since they do not have a standard address. The device driver modules handle these differences in address formats. Third, some events need to be processed with a low latency. For example, in the case of an access control application, the entry needs to be granted within a preset response time. In this case, sending the employee credentials to a remote data center for validation may result in long and variable response times. To process such low-latency events, the DAM uses an event cache module, which is a small set of rules that process data locally within a facility. Fourth, in order to handle transient disruptions in the communication and provide reliable data transmission, the DAM locally buffers the data it transmits to the BSM, until the BSM acknowledges the receipt of the data.

Communication Manager (COM): The COM module extends across the client facilities and business premises and has a dual role to play. The COM is responsible for relaying AIDC data and configuration commands between the DAM in the customer premises and the BSM in the data center. Additionally, the COM module is used by the BSM to send notifications about situations (business events) to end users and external applications. Currently, our COM module is simple and supports wired and wireless (Wi-Fi) communications for long and short ranges across Internet and Intranet networking infrastructures. We have mainly focused on managing end-to-end sessions in a robust manner. In the future, when we extend the COM to support more communication options, we will have to deal with more challenges and we now enumerate some of those challenges.

There are multiple network connectivity choices for a general M2M solution, including wired ethernet, wireless (802.11x, Bluetooth, 802.15 etc), cellular (GSM/SMS, GSM/GPRS, CDMA, etc.). The use of these communication options in an M2M system for facility management has different pros and cons, with wireless communication being more flexible for large-scale deployment across facilities. These communication options also provide different levels of security, with the cellular transport providing higher levels of security compared to Ethernet. So an important challenge that needs to be addressed when multiple communication methods are involved, is to design a secure M2M solution that adapts according to the security level provided by the underlying network. Another challenge is to choose the communication method according to the network traffic. For example, in the case of GSM networks, GPRS is more appropriate for relaying larger volume of data, while SMS is more appropriate for alerts and smaller volume of data. In addition, when the network is shared, the COM module will need to find ways to prioritize the traffic and improve network utilization, by channeling the M2M traffic with lower priority during off-peak hours. Currently, the choice of communication method and the frequency of end-user notification is left to the user. However, in the future, such decisions can be made in the COM module, based on the nature of the M2M traffic.

Business Service Manager (BSM): The BSM receives AIDC raw events collected and transmitted by one or more DAMs. The goal of the BSM is to store, process, and analyze the AIDC low-level events, thereby inferring and generating alerts about new or changing situations (business events), such as equipment malfunctions, transactions, and service requests. The event processing in the BSM is consistent with EPCGlobal's standard on Application Level Events (ALE) [5]. The BSM architecture uses two key components for this purpose, as shown in Figure 1(b): the complex event processor (CEP) and the event post-processor. The CEP is a rule engine that converts low-level events from devices to business-level events. The BSM currently uses IBM's AMiT as the

CEP engine [6]. When a rule is processed, it results in the triggering of an action. In some cases, the action may determine that the event needs to undergo further application-specific analysis, in which case it is sent to the event post-processor. The post-processor includes modules, such as those that perform billing and dependability modeling. We currently use the SHARPE statistical tool [7] to perform dependability modeling, because the AMiT rule engine does not have an advanced in-built statistical library. SHARPE is used for making decisions related to predictive maintenance, such as whether it is better to replace or buy new equipment based on the future likelihood of faults, and when it is time to schedule a maintenance call proactively for a facility. Predictive analysis can also be used in the energy management scenario to predict when it is time to initiate load shedding based on past trends in the energy consumption within a facility. In general, predictive maintenance is useful for facility management.

After a rule is analyzed by the rule engine and inferences are made, the action may trigger a notification or alert. The BSM uses a publish-subscribe model in which subscribers need to register with the BSM to receive notifications. The notification can be sent to human users (such as, facility managers and technicians) or external applications (such as, Enterprise Resource Planning (ERP), Supply Chain Management (SCM), Customer Relationship Management (CRM), etc.). Business events generated by the BSM may be prioritized and are distributed according to the communication method (e.g. email, SMS etc.), as chosen by the subscribers.

Since the BSM needs to handle large volumes of data, one of the challenges is to design it to be scalable. The BSM has two options when it has to process a large number of events simultaneously. It can create multiple instances of only the AMiT CEP engine, so that unrelated events (e.g. those originating at different DAMs) can be processed in parallel by different instances. Alternatively, we can have multiple instances of the entire BSM module running on different servers and assign each BSM to different DAMs. In this case, the DAM needs to specify the address and port of the BSM server assigned to it, so that the COM can forward the event to the appropriate BSM. Although both options can be supported in our middleware, we currently prefer the first option, because it requires fewer resources.

Control Manager (CON): The CON serves as the centralized administrative console for managing the entire M2M system. It provides a hierarchical classification of users into system administrators, programmers, and clients, and presents a web-based front-end with multiple functionalities for different classes of users. For example, system administrators of a facility have full administrative privileges and can shutdown, restart, or reconfigure the BSM, DAM, or COM. Programmers can use the CON to update rules or load new rules for their application in the BSM. General clients can use the CON to query the status of the devices, load their contact information for receiving alerts, and subscribe to events. Figure 2 shows the web-based user interface of the control manager. The high-level menu options appear on the left. From this menu, one can choose the configuration options. The figure shows an example of the configuration options

for the BSM. First, the BSM identifier needs to be chosen for configuration.
The Edit option allows reconfiguration of the settings of the chosen BSM. The
BSM module option is used to load and unload application-specific modules (e.g.
Billing). The URL shows the address of the BSM, which runs as a web service.
The status button shows the current status of the BSM. The DAM list at the
bottom shows the list of DAMs associated with that BSM and their respective
status.

Fig. 2. CON user interface

Since the CON resides in the data center and serves as a basis for hosting the
applications of different customers, one of the challenges that we need to address
in designing the CON when multiple customers are involved, is to ensure that
the customers have access only to the data pertaining to their facility. The CON
can achieve this by virtualization, which results in the separation of the three
managers (COM, BSM, DAM) of each customer facility from those of other
customers, yet allowing them to run on the same physical machine.

4   Middleware Implementation

Having described the design of the M2M middleware modules and some of the
challenges that we addressed, we now provide some implementation details of
our software. We have also developed a testbed and a prototype application that
uses the M2M middleware to realize the facility management scenarios listed in
Section 1.2. We begin this section by first briefly describing the implementation of
the middleware modules, follow that by describing the technologies that we have
selected for our testbed, and conclude this section by providing an operational
view of the middleware for some of the facility management scenarios.

Fig. 3. UHF gate reader for inventory management

4.1   Software Implementation Details

The BSM, COM, and DAM have been implemented using Java and J2SE, because Java is portable. On the other hand, the CON, which provides a web-based front-end with several user interface components has been written in C#, which is part of the .NET 2.0 framework. The .NET framework provides rich and configurable GUI components that were used for the graphical user interface. The different components communicate with each other using web services. The COM component is based on the Apache Tomcat HTTP server and Axis [8], which is an implementation of the SOAP (part of web services). Currently, we have chosen to use SOAP over HTTP for the COM implementation to improve performance and scalability. However, if reliability is more important than scalability, then a better alternative would be to use SOAP over JMS (Java Messaging Service), which serves as a more reliable transport compared to HTTP. The BSM is associated with a database for storing data and this database has been implemented using PostgreSQL. The CON is also associated with a database that has been implemented using Microsoft SQL Server 2005 to provide optimal performance when working with .NET applications. In order to provide persistence, the DAM and BSM store their configuration information in XML files, and use that to recover their state when they restart.

4.2   M2M Technology Selection

Table 1 shows the different M2M technologies that we have included in our testbed to realize the scenarios listed in Section 1.2. We use high-frequency

Table 1. M2M technologies supported in testbed

Scenario	Technology
Access Control	UHF gate (Deister); HF reader (FixedScan)
Inventory mgmt.	UHF Gate (Intermec, Deister); Handheld reader (Psion)
Data logging	Handheld reader with CF (HP iPAQ)
Predictive maintenance	Zigbee motes (Crossbow [9])
Real Time Location	Active RFID tag (ActiveWave); Wi-Fi Tag (Ekahau [10])

(HF) RFID proximity readers and HF tags for the access control scenario; HF RFID tags and handheld readers with compact flash (CF) for the data logging scenario; Zigbee motes that monitor physical parameters (e.g. temperature) for the remote monitoring and predictive maintenance scenario; UHF RFID gate readers (setup shown in Figure 3) and UHF tags for inventory management when metallic elements are involved and HF tags when non-metallic elements are involved. For the real-time location-tracking scenario, we used Wi-Fi location tracking tags from Ekahau and active RFID technology. Table 1 also shows the vendors we used for different technologies in parentheses and in some cases, we purchased the same technology from different vendors (e.g. UHF gate reader from both Intermec and Deister), as this helped us to evaluate the compatibility and compare the performance.

4.3   Operational View of the Middleware

We now show the operational view of the middleware in the context of two prototype facility management scenarios, which were actually executed in the server room of a technical facility. This server room had several racks with servers, computers, and other technical equipment. In the first scenario, which involves predictive maintenance, we used Zigbee-enabled sensor nodes [9] to monitor the temperature of server racks and report the readings to a Stargate gateway node [11], which in turn forwarded the information to the DAM. When the temperature in the racks raised above a certain threshold, it resulted in a high priority event at the BSM. The event processing in the rule engine used the SHARPE tool for predictive modeling of the likelihood of a risk and then triggered an email notification to the technician in the facility, as shown in Figure 4(a). The notification shows the risk probability and informs the technician that the airflow needs to be turned on. Our middleware is also linked to the actuators that control the cooling unit and the rule

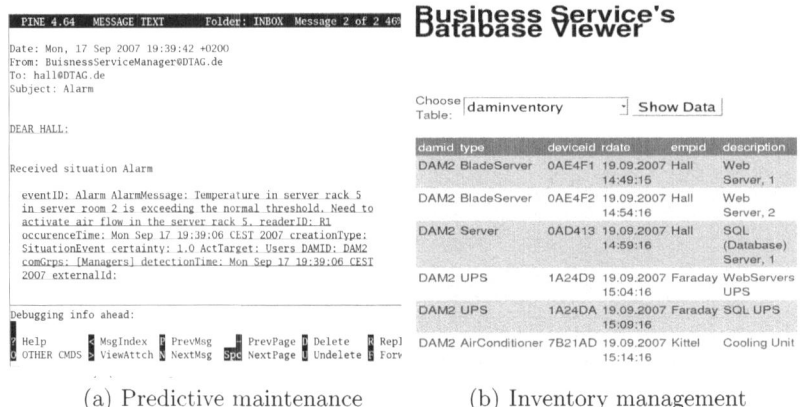

(a) Predictive maintenance   (b) Inventory management

Fig. 4. Outcome after executing facility management scenarios

engine in the BSM triggers an action to turn on the airflow automatically based on the risk probability, without involving the technician. In the second scenario, we show the use of the middleware for inventory management. The blade servers were brought in as new inventory into the server room by a worker. The UHF RFID tags on the packages were recorded by the UHF gate reader at the entrance (shown in Figure 3). The event was relayed to the BSM by the DAM, which in turn updated its database to record the inventory, as shown in Figure 4(b). The remaining entries in the database show an update of existing inventory. The inventory data includes, in order from left to right, the identifier of the DAM to which the inventory device is registered, the type of the inventory, the inventory ID (which is obtained from the RFID tags), the time-stamp when the last inventory update was done, the ID of the worker who took the inventory, and some additional description of the inventory.

5   Performance Evaluation

In this section, we present some of the results of the experiments that we conducted to evaluate the performance of our middleware using our testbed equipment that was listed in Table 1.

5.1   Measurement of One-Way Latency

The goal of our first experiment was to measure the end-to-end latency as well as the hop-by-hop latency involved in transmitting data from the device to the BSM, as this provides a measure of the overhead at each hop. The DAM and BSM ran on different servers. We used a Wi-Fi location-tracking tag that was attached to a moving asset as the source device to trigger events. The real-time location (RTL) engine reported the location to the DAM, which in turn relayed the information to the BSM. We instrumented our code to measure the latency at each hop as well as the end-to-end latency to transmit the data from the RTL at one end to the BSM at the other end. We repeated the experiment by transmitting about 40 different events from the RTL. Table 2 shows the average latency (in milliseconds) for the transmission of a single event.

The DAM processing latency is the time it took the DAM to convert the incoming data to a normalized message format and send the event to the BSM. The DAM $\rightarrow$ BSM includes the time for the DAM authentication and the time for the COM to transmit the event from the the DAM to the BSM by using a

Table 2. Measurement of latency

	DAM Processing	DAM→ BSM	BSM Processing	End to End Latency
Average latency (millisec)	23.11	177.16	4.11	204.37

web service (basically a SOAP call implemented in AXIS). The BSM processing
latency is the amount of time it took the BSM to process the event, pass it
through the complex event processor (rule engine), and generate a business event.
From the results, we see that most of the latency is incurred at the DAM→BSM
hop and one of the reasons for this is that this hop incurs the overhead of a
socket call when the COM transmits the data to the BSM using the web service.
The DAM→BSM latency was also observed to have high variance, which may be
attributed to the network load on the wireless network that we used between the
DAM and BSM, which was a shared network. Although we have tried to improve
the performance and scalability of our middleware by choosing the option of
using SOAP over HTTP for the COM implementation, the latency between the
DAM and BSM may be further reduced by using more efficient implementations
of the web service in the future.

5.2   Measurement of Throughput

We conducted experiments to evaluate how the DAM throughput and system
throughput vary with the rate at which events are generated at the device driver.
As in the previous case, we setup the DAM and the BSM on separate servers. In
this experiment, we used the Zigbee-based sensor motes for generating events.
Each mote generated events at the rate of 10 events/second and to increase the
load, we incrementally increased the number of motes from 1 to 8. All of the
motes reported to a Stargate gateway node, which then forwarded the events
to the reader module in the DAM. Figure 5(a) shows how the throughput of
the DAM and that of the whole system (measured at the BSM) varies with the
rate at which events are read and input to the DAM by the reader module.
The graph shows that both the DAM throughput and the system throughput
follow the event generation rate closely and increase almost linearly with the
event generation rate, indicating that there is minimal bottleneck within the
middleware for the range of load we were able to generate. However, when the
event generation rate exceeds the maximum rate at which the system can handle

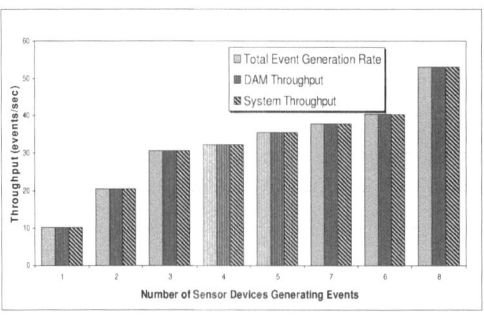

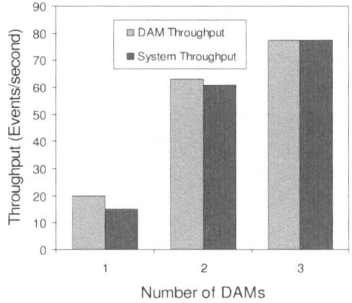

(a) Throughput using single DAM   (b) Effect of distributing the load
                                    across multiple DAMs

Fig. 5. Measurement of throughput

events from the devices, we expect the throughput to reach a steady state. In such a case, we need to ensure that the sizes of the event buffers used to store events in the DAM and BSM are sufficient to avoid events being dropped.

5.3   Effect of Load Distribution

In the previous experiment, all of the events originated at a single DAM, which in our case effectively translates to a single facility, because we currently use a one-one mapping between a DAM and a facility. We now evaluate how the middleware handles input from multiple DAMs (facilities). As in the previous case, we set up the BSM and DAMs on different servers. We varied the DAM from 1 to 3. We used the same set of devices (4 Wi-Fi location tracking tags and 4 Zigbee sensor motes) in all of these experiments. When using multiple DAMs, we distributed the devices evenly across the DAMs, thereby generating events at approximately the same rate from each DAM. The effective load at the BSM was approximately the same across all of the experiments. Figure 5(b) shows the DAM throughput and the system throughput (i.e. the rate at which BSM generates business events after processing the raw events from the DAMs), as the load is distributed across facilities. From the results, we see that the throughput of the system improves, as the load is distributed across multiple DAMs. So the middleware is able to efficiently handle events originating at different physical facilities, even though we use a single BSM and CEP to handle all of the events. As in the case of the experiments in Figure 5(a), we see that the system throughput is comparable to the DAM throughput, indicating that for the range of load we considered, any overheads incurred between the DAM and BSM did not pose a significant bottleneck and for this range, the system throughput is limited more by the rate at which we are able to generate events at the DAM.

5.4   Effect of Heterogeneity

To study the impact of multiple types of events and sensors on the middleware performance, we conducted experiments in three parts. In Part A, we generated events using 5 Wi-Fi location tracking tags (RTL) and measured the system throughput and one-way latency (from the time the event was received at the DAM until the time it triggered an action at the BSM). We then gradually increased the load by adding 5 Zigbee sensor motes to generate additional events in Part B. In Part C, we added 5 UHF tags to the pool and effectively created events of three different types. We used a single DAM and BSM in all of these experiments and measured the throughput and latency. Figure 6(a) shows the total throughput as well as the throughput based on the event type. The total rate of incoming events (without distinguishing between the event types) as measured by the DAM was around 103 events/second in part A, 153 events/second in part B, and 160 events/second in part C. From the plot we

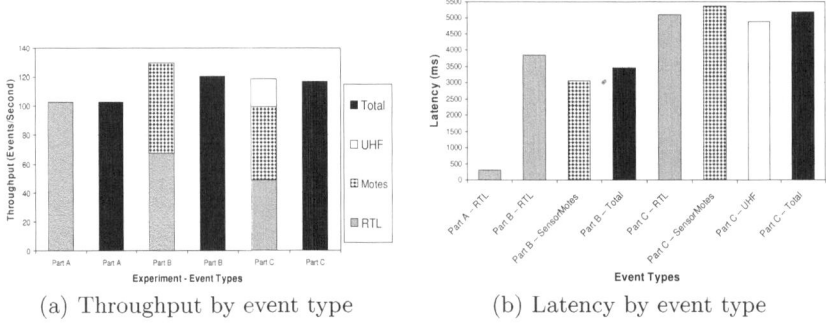

(a) Throughput by event type   (b) Latency by event type

Fig. 6. Effect of heterogeneous events

see that the system throughput measured was around 102 events/sec in part A, which is close to the event generation rate. However, the measured system throughput was 120 events/sec in part B, and 117 events/sec in part C, which are lower than the source event generation rate. We no longer see the linear increase in the system throughput that we saw in the previous experiments in Section 5.2, because our system with a single DAM and BSM reaches its maximum throughput around 120 events/sec. Further increase in the throughput may be achieved either by distributing the load across multiple DAMs (as discussed in the previous subsection) or by increasing the number of BSMs (deferred for future work) after we perform more analysis to determine whether the bottleneck is at the DAM or at the BSM.

Figure 6(b) shows the latency distribution for each event type, as well as the average latency without distinguishing between the event types (labeled as "Total" in the figure). In this set of experiments, all types of events had the same priority. From Figure 6(b), we see that the latency increases as the number of events increases. However, within each part, different types of events experience latencies of the same order of magnitude, which shows that the middleware is capable of handling different types of M2M devices and events simultaneously and in a similar manner, without causing an adverse impact on each other.

5.5   Summary of Experimental Results

Based on our experiments, we conclude that most of the overhead in our system is caused by the communication overhead resulting from the web service calls between the DAM and the BSM. Our middleware is capable of scalably and reliably handling multiple events of the same type as well as of different types. Currently, with a single DAM and BSM, our system has a maximum throughput of around 120 events/second. However, the scalability of our system can be further improved by distributing the load across multiple data acquisition managers and business service managers.

6   Conclusions

Current M2M developments represent opportunities for small and medium-sized enterprises (SMEs). Nevertheless, limited budgets, lack of in-house expertise, and lack of access to the newest technologies are but a few of the significant barriers faced by SMEs. Many entrants that are newly thrust into RFID technology assessments and selections view RFID as primarily consisting of only tags and readers. However, the full scope of technologies often needed to complement RFID in a domain is wider and more complex. In addition to tags and readers, SMEs need to consider appropriate sensors, computers, middleware, database systems, enterprise applications, networking, and business process management tools. The key contribution of this work is that it describes the software architecture and practical experience gained from developing an M2M middleware from ground up, in order to assist SMEs in overcoming the hurdles and complexities involved in enabling seamless interaction between diverse M2M technologies that are relevant to the facility management domain.

We conclude by summarizing some of the issues that we learned from building an M2M middleware from ground up, including the evaluation and acquisition of M2M technologies from different vendors. First, the selection of right devices to fit the different scenarios in a facility management application is particularly challenging because some of the environments are harsh and involve metallic elements. Issues that we had to consider for selecting the technologies are range, read rate, positioning of tags, metallic reflections and electromagnetic interferences with the server equipment. We chose special UHF tags for dealing with metallic elements. In some cases when an item is large, attaching multiple tags to an item may enhance readability. Second, we were able to use multiple M2M devices with disparate technologies, such as RFID, motes, and Wi-Fi location-tracking tags simultaneously. Since the interfaces and protocols used by these different technologies and vendors were often incompatible, we had to develop a normalized format, so that the higher-level middleware modules can remain agnostic to those differences and treat the events generated by the different underlying technologies uniformly. However, this normalization did not adversely impact the performance of our system, as shown by our experimental results. From a software perspective, since most of the software was developed in-house, we were able to customize it to our needs. However, the rule engine in the BSM was third-party software and the RFID reader software from the vendors was proprietary. Using third-party software (e.g. AMiT, Intermec, SHARPE, Active-Wave samples) greatly reduced our development time and enabled the use of the third party's expertise, knowledge, resources, and experience in a specific field. However, we were sometimes limited by their lack of easy extensibility, because they were not open-source.

References

1. Stouffer, K., Falco, J., Kent, K.: Guide to Supervisory Control and Data Acquisition (SCADA) and Industrial Control Systems Security. National Institute of Standards and Technology, Tech. Rep (September 2006)
2. Legner, C., Thiesse, F.: RFID-based Facility Maintenance at Frankfurt Airport. IEEE Pervasive Computing 5(1), 34–39 (2006)
3. Pakanen, J.E., Hakkarainen, K., Karhukorpi, K., Jokela, P., Peltola, T., Sundstrom, J.: A Low-Cost Internet Connection for Intelligent Appliances of Buildings. Electronic Journal of Information Technology in Construction 7 (2002)
4. Watson, D.S., Piette, M.A., Sezgen, O., Motegi, N.: Machine to Machine (M2M) Technology in Demand Responsive Commercial Buildings. In: Proceedings from the ACEEE 2004 Summer Study on Energy Efficiency in Buildings: Breaking out of the Box, pp. 22–27 (August 2004)
5. EPC Global, http://www.epcglobalinc.org/home
6. IBM Active Middleware Technology, http://domino.watson.ibm.com/comm/research.nsf/pages/r.datamgmt.innovation.cep.html
7. The SHARPE Tool and Interface, http://www.ee.duke.edu/~chirel/IRISA/sharpeGui.html
8. Web Services - Axis, http://ws.apache.org/axis/
9. MicaZ, Crossbow Inc., www.xbow.com/Products/Product_pdf_files/Wireless_pdf/6020-0060-01_A_MICAz.pdf
10. Ekahau RTLS, http://www.ekahau.com
11. STARGATE: X-Scale Processor Platform, Crossbow Technologies, http://www.xbow.com/Products/Product_pdf_files/Wireless_pdf/6020-0049-01_A_STARGATE.pdf

The Software Fabric for the Internet of Things

Jan S. Rellermeyer[1], Michael Duller[1], Ken Gilmer[2],
Damianos Maragkos[1], Dimitrios Papageorgiou[1], and Gustavo Alonso[1]

[1] ETH Zurich, Department of Computer Science,
8092 Zurich, Switzerland
{rellermeyer,michael.duller,alonso}@inf.ethz.ch
{dmaragko,dpapageo}@student.ethz.ch
[2] Bug Labs Inc.
New York, NY 10010
ken@buglabs.net

Abstract. One of the most important challenges that need to be solved before the "Internet of Things" becomes a reality is the lack of a scalable model to develop and deploy applications atop such a heterogeneous collection of ubiquitous devices. In practice, families of hardware devices or of software platforms have intrinsic characteristics that make it very cumbersome to write applications where arbitrary devices and platforms interact. In this paper we explore constructing the software fabric for the "Internet of Things" as an extension of the ideas already in use for modular software development. In particular, we suggest to generalize the OSGi model to turn the "Internet of Things" into a collection of loosely coupled software modules interacting through service interfaces. Since OSGi is Java-based, in the paper we describe how to use OSGi concepts in other contexts and how to turn non-Java capable devices and platforms into OSGi-like services. In doing this, the resulting software fabric looks and feels like well known development environments and hides the problems related to distribution and heterogeneity behind the better understood concept of modular software design.

1   Introduction

The notion of an "Internet of Things" refers to the possibility of endowing every day objects with the ability to identify themselves, communicate with other objects, and possibly compute. This immediately raises the possibility to exchange information with such objects and to federate them to build complex composite systems where the objects directly interact with each other. These composite systems exhibit very interesting properties and can have a wide range of applications. Yet, they are complex to develop, build, and maintain in spite of the purported spontaneity of the interactions between the objects. A crucial reason for these difficulties is the lack of a common software fabric underlying the "Internet of Things", i.e., the lack of a model of how the software in these different objects can be combined to build larger, composite systems.

The problem in itself is neither new nor restricted to pervasive computing. For instance, a great deal of effort has been invested in the last decades to

C. Floerkemeier et al. (Eds.): IOT 2008, LNCS 4952, pp. 87–104, 2008.

develop concepts for modular software design. There, the problem is how to build a coherent application out of a possibly large collection of unrelated software modules. Conceptually, it is intriguing to consider the problem of building the software fabric for the "Internet of Things" as another version of the problem of modular software design. Both problems are closely related, in spite of addressing different contexts, and being able to establish a connection between them will mean that we can bring all the experiences and tools available for one to bear on the other.

This is the premise that we pursue in this paper. We start from OSGi [1], a standard specification for the design and management of Java software modules. Despite its recent success for server-side application, OSGi was originally developed for managing software modules in embedded systems. Taking advantage of recent developments that have extended OSGi to distributed environments (R-OSGi [2]), we show how to turn small and ubiquitous devices into OSGi-like services.

Many aspects of the OSGi and R-OSGi model are perfectly matching the requirements of applications on the "Internet of Things". First, the devices involved in these networks form a heterogeneous set. Different hardware platforms and operating systems are available on the market and in use. The abstractions provided by the Java VM dramatically simplify the development of software for these devices. Furthermore, ubiquitous networks often involve a large quantity of devices that have to be managed by the user. The module lifecycle management of OSGi allows to consistently update software modules among all devices.

However, these devices are highly resource-constrained and thus often unable to run a Java virtual machine. They also often lack a TCP/IP interface and use other, less expensive communication protocols (e.g., Bluetooth [3], ZigBee [4]). To accommodate these characteristics, in the paper we describe extensions to R-OSGi that:

- make communications to and from services independent of the transport protocol (in the paper we show this with Bluetooth and 802.15.4 [5]),
- implement an OSGi-like interface that does not require standard Java or no Java at all (in the paper we demonstrate this for CLDC [6], embedded Linux, and TinyOS [7]), and
- support arbitrary data streams to and from services.

Through these contributions, the resulting fabric for the "Internet of Things" can be treated and manipulated as an OSGi framework that is also applicable to small devices, alternative transport protocols, and non-Java applications. This means that we can treat devices as composable services with a standard interface, services can be discovered at runtime and changed dynamically, data exchanges can use the Java types even for non-Java applications, and failures can be masked behind service withdrawal events. Application developers see a collection of software modules with a homogeneous interface rather than a collection of devices. In the paper we give an example of an extensible hardware platform – commercially available – where each hardware module is treated as an OSGi module, thereby proving the feasibility of the premise that we are pursuing in this paper.

The paper is organized as follows. In Section 2, we review the main ideas behind OSGi and motivate our approach. In Section 3, we discuss how to make OSGi frameworks independent of the transport protocol. In Section 4, we describe how to provide OSGi-like services on platforms either running reduced versions of the JVM, programmed in languages others than Java, or lacking a full-fledged operating system. In Section 5, we present an extension to R-OSGi to support data streams. In Section 6, we describe how all these extensions make the OSGi model general enough to become the software fabric for the "Internet of Things".

2   Background on OSGi and Motivation

2.1   The OSGi Model

OSGi [1] is an open standard for developing and managing modular Java applications. Efforts around OSGi are driven by the OSGi Alliance, with contributions from software manufacturers like IBM, BEA, or ORACLE, as well as device vendors like NOKIA, SIEMENS, or BOSCH. OSGi is already in use in a wide range of systems (including the development environment ECLIPSE, but also in embedded systems such as those in the BMW 3 series car).

In OSGi, software modules are called *bundles*. Bundles can export packages and import shared packages of other bundles. The runtime of OSGi (called *framework*) handles the dependencies between the bundles and allows the active control of every bundle's lifecycle. Each bundle is loaded through a separate Java classloader and shared code is handled by a delegation model. It is possible to extend an application by installing new bundles at runtime as well as updating existing bundles or unloading them. To avoid link-level dependencies between code of bundles, which would limit the flexibility of a modular architecture, OSGi provides a *service* abstraction. A service is an ordinary Java object which is published to the framework's *service registry* under the names of the interfaces it implements. Additionally, a service can be described in more detail by attaching properties to the service's registration. These properties can also be dynamic, i.e., change at runtime. Other bundles can make use of services by requesting the implementation from the framework. Since service clients only have to know the interface but not the actual implementation, this leads to a loose coupling of the bundles. Use of services relies on common, published interfaces but the implementation of each service remains a black box. It is, thus, possible to exchange the implementation of a service at runtime. It is also possible to have several alternative implementations of a service which can then be matched at a finer level through the use of properties.

Existing OSGi frameworks include commercial implementations such as IBM's SMF [8] as well as open source projects such as ECLIPSE EQUINOX [9], APACHE FELIX [10], CONCIERGE [11], or KNOPFLERFISH [12].

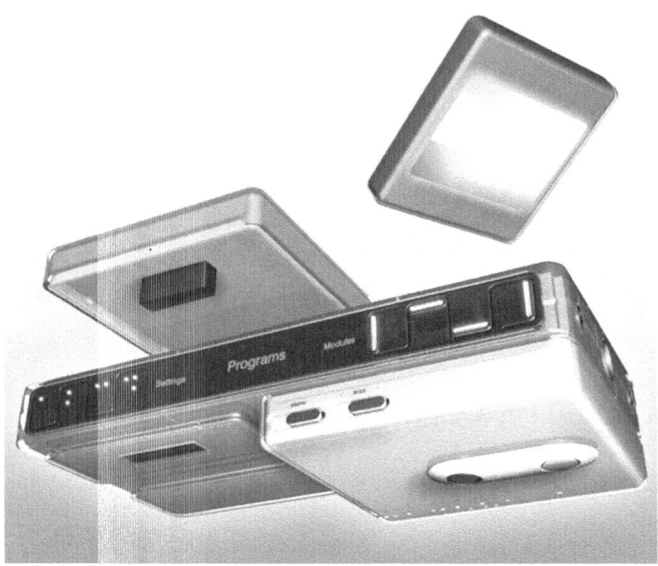

Fig. 1. The BUG with several modules

2.2   Example: The BUG Platform

One can question the wisdom of trying to generalize OSGi beyond module management in a centralized setting. However, this has been done already and quite successfully. An example of a commercial device which internally uses OSGi to utilize modular hardware is the BUG (Figure 1), developed by the New York based company Bug Labs Inc. The device consists of a base unit (computer) and hardware modules that allow the system to be extended. The base unit is a handheld ARM-based GNU/Linux computer that runs the Concierge OSGi framework on a CDC VM.

A primary feature of BUG is that it's expandable. A dedicated 4-port bus known as the *Bug Module Interface* (BMI), allows external devices (*hardware modules*) to interface with the base unit. The BMI interface is hot pluggable both in hardware and software, meaning that modules can be added to and removed from the base unit without restarting the computer or manually managing applications that depend on specific modules. The BMI interface is comprised of eight standard interfaces such as USB, SPI, and GPIO which are all available through each connector. Each hardware module exposes via OSGi a set of services that local or remote applications can utilize.

At the software level, BMI behaves like any other hot-pluggable bus, such as PCI and SB. Hardware modules can be added to and removed from the BUG base unit without restarting the device. Based on what modules are connected, OSGi services are available that can be utilized by local and remote OSGi-based applications. Some examples of hardware modules are LCD screen, barcode scanner, weather station, and motion detector.

In a way, the BUG device can be considered as a physical actualization of the OSGi service model: each hardware module contains devices that, when connected to a base unit, register themselves within the OSGi runtime as a service. When a user attaches a hardware module to the base unit, one or more hardware service bundles are started in the OSGi runtime, in addition to any application bundles that depend on those services. Similarly, when modules are removed, the same bundles are removed from the system. Since OSGi was designed to support dynamic services, applications written for BUG work seamlessly with the highly variable nature of the hardware.

The BUG implements the idea of a service-oriented platform for a collection of hardware devices plugged to each other. The vision we want to articulate in this paper is one in which OSGi plays a similar role but in the context of the "Internet of Things" rather than for hardware modules. In order to do this, however, OSGi needs to be extended in several ways. First, it has to be able to support distributed applications which the current specification does not [2]. Second, a framework is needed capable of running on small devices, something that only a few of the existing frameworks can do but at a considerable performance penalty [11]. These two points have been covered by recent developments (see the remainder of this section). From the next section on, we tackle the rest of the necessary extensions to make the OSGi model suitable for the "Internet of Things": making the distribution protocol independent, removing the dependency on a regular Java Virtual Machine, removing the dependency on Java altogether, and extending it to support continuous media streams.

2.3   R-OSGi

OSGi can only be effectively used in centralized setups where all bundles and services reside within the same Java Virtual Machine. Not only the dependency handling requires all modules to reside on the same machine, the service registry is a centralized construct that is not designed for distribution.

R-OSGi [2] is an open source project designed to address these limitation and provide a distributed version of the OSGi model. In R-OSGi, service information is registered with a network service discovery protocol so that a node can locate services residing on other nodes. The original implementation of R-OSGi was targeting TCP/IP networks and used RFC 2608, the SERVICE LOCATION PROTOCOL (SLP) [13,14]. The notion of services in OSGi and SLP are very close to each other and SLP can run in both managed networks where a *Directory Agent* acts as a central service broker, and in unmanaged networks by using multicast discovery. The connection between two applications through an R-OSGi service is called a *network channel*. To access remote services, R-OSGi generates a dynamic proxy for the service on the caller's side, which makes the service look like a local service. The proxy itself is implemented as an ordinary OSGi bundle. To deal with dependencies, R-OSGi determines the minimal set of classes which are required to make the service interface resolvable. Those classes from this set that do not belong to the VM classpath are additionally transmitted to the client and injected into the proxy bundle. Method calls from the proxy to the remote service are executed by exchanging appropriate messages between the R-OSGi instances running in the proxy's framework and running in the remote service's framework. Arguments to the remote method are serialized at the caller's site and sent with the call request. At the callee's site the arguments are deserialized again and the method is invoked with the deserialized arguments. If the method returns a return value, the value is serialized and sent to the caller's site together with the acknowledgment of the successful method invocation. Finally, the return value is deserialized at the caller's site and returned to the proxy.

In addition to method calls to the remote services, R-OSGi also transparently supports event-based interaction between services as specified in OSGi's Event Admin service specification [1].

R-OSGi can be effectively used to build applications from a federation of distributed services. However, it works only with TCP/IP and requires a (at least J2ME CDC compliant) Java VM with an OSGi framework implementation. Finally, R-OSGi has no support for data streams such as those produced by, for instance, RFID readers. Thus, it is not suitable for the applications we have in mind as part of the "Internet of Things". In what follows we show how this general model we have just described (starting from R-OSGi) can be extended so as to build the underlying software fabric for the "Internet of Things".

3   Extending R-OSGi to Other Transport Protocols

The current Internet is based on TCP/IP both for wired and wireless communication (802.11 wireless LAN). In contrast, the "Internet of Things" often involves communication over low-power, short-range networks such as Bluetooth [3] or ZigBee [4]. R-OSGi, however, is limited to TCP/IP and not suited well for the context of the "Internet of Things". Thus we have modified R-OSGi, which is available as open source, to make it independent of the underlying transport protocol. We do this by describing the necessary extensions to support Bluetooth and 802.15.4 protocols.

There are two design properties that prevent R-OSGi to be effectively used with transports others than TCP/IP. First, the SLP service discovery is an integral part of R-OSGi and pervades all the layers. For instance, remote services are identified by their SLP service URL. The SLP protocol, however, is tightly bound to IP networks. Second, the network channel model [2] of R-OSGi allows to plug in alternative transports for outgoing communication but the acceptor for incoming connections only supports TCP/IP sockets. In the model of R-OSGi, it is assumed that alternative transports would bridge to TCP over local loop, like the HTTP transport provided by R-OSGi. This approach works for desktop computers with powerful and versatile operating systems but is unfeasible for ubiquitous devices communicating over Bluetooth or ZigBee.

3.1   Bluetooth Transport

To support R-OSGi services over Bluetooth radios, we need to separate the service identifier from the underlying transport. Hence, instead of SLP service URLs, we use opaque URIs. The URI of a service can use any schema and thereby identify which transport implementation it requires. By doing this, the address of a service now also contains information on the transport protocol it supports. For instance, the URI of a service accessible through TCP/IP can be: *r-osgi://some.host#21*. The network channel through which the service can be accessed is identified by the URI's schema, host, and port components (if differing from the default port). The fragment part of the URI describes the local identifier that the service has on the other node. The same service through Bluetooth transport would be, e.g., *btspp://0010DCE96CB8:1#21*.

Bluetooth [3] is a network solution covering the lower five layers of the OSI model [15]. In our implementation we use RFCOMM, which provides a reliable end-to-end connection, as transport protocol. We modified the channel model of R-OSGi so that every implementation of a transport protocol that can accept incoming connections can natively create and register a channel endpoint for them. Thus, messages received via Bluetooth do not have to be bridged over TCP but can be processed directly by the corresponding channel endpoint.

To evaluate the performance of the Bluetooth implementation, we tested the system with the JavaParty benchmark. The JavaParty benchmark was originally developed by the University of Karlsruhe as part of the JavaParty [16] project to test the performance of an alternative RMI implementation. In the OSGi version of the benchmark, a test client calls different methods of a remote service with arguments of increasing size. Figure 2 shows the results of the benchmark on Bluetooth, compared to the baseline (first column), which is an ICMP ping over Bluetooth PAN. The invocation of the void method with no arguments takes only slightly more time than the ICMP ping. This shows that R-OSGi adds little overhead. Furthermore, when the size of the arguments increases, the invocation times scales relative to the size of the arguments.

3.2   Bluetooth Service Discovery

R-OSGi supports service discovery through SLP but only over an IP-based network transport. We have changed service discovery to make it an orthogonal concern. SLP discovery is still supported in the form of a separate and optional service that gives hints to the application when certain services are found. Alternative discovery protocols can now be used as well since the service discovery operates behind a common discovery handler interface that is not restricted to SLP. To match the Bluetooth transport we implemented a corresponding discovery handler, which offers Bluetooth service discovery via Bluetooth's own *Service Discovery Protocol* (SDP) [3].

A major challenge of the OSGi service discovery with SDP is the granularity mismatch. In the Bluetooth service model, R-OSGi itself is the Bluetooth service and the OSGi services it offers are conceptually subentities of the R-OSGi service.

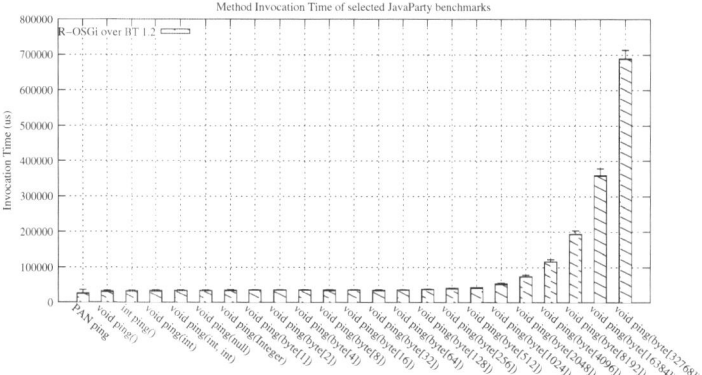

Fig. 2. Javaparty Benchmark over Bluetooth 1.2 Transport

However, SDP has no notion of service hierarchy. Furthermore, SDP is very limited in the way how services can be described, which leaves a gap between Bluetooth services and OSGi services. SDP supports queries only on UUIDs (Universally Unique Identifiers) [17] whereas OSGi allows semantically rich and complex LDAP filter strings [18]. In addition, SDP data types do not suffice to express the properties of OSGi services. This can lead to problems when selecting services as the limited expressibility in SDP will result in many more services answering a request than is necessary.

Standard Attributes	0x0000	Standard Attribute (ServiceRecordHandle)
	0x0001	Standard Attribute (ServiceClassIDList)
	0x0002	Standard Attribute (ServiceRecordState)
		...
Record Header	0x0200	DATALT{UUID1, UUID2, ..., UUIDn}
	0x0201	DATALT{ID1, ID2, ..., IDn}
	0x0202	DATALT{DATALT{iface1A. ..., iface1Z},DATALT{iface2A, ...} ...}
	0x0203	INT_4{hash(blocks)}
Block 1	0x0204	INT_4{mod_timestamp(block_1)}
	0x0205	DATSEQ{len, INT_16,INT_16, ..., INT_16}
		...
Block n	0x0204+2*n	INT_4{mod_timestamp(block_n)}
	0x0205+2*n	DATSEQ{len, INT_16,INT_16, ..., INT_16}

Fig. 3. SDP Service Record for R-OSGi Services

To provide OSGi-like service discovery over Bluetooth, we keep track of services that are marked to be remotely accessible through SDP. The most common form of service filter involves the service interfaces. In order to support searches similar to those in OSGi, the UUID of every service interface is adapted so that filters over service interfaces can be expressed as SDP service searches over the corresponding UUID. In our implementation, the UUID is a hash over the interface name. For each service, the serialized properties are stored as sequences of bytes in the service record. These properties are retrieved lazily and only if a filter expression poses constraints over these particular properties. A modification time-stamp indicates the version of the properties so that clients can cache them for future filters and easily check if the cached version is still valid. A hash value over all service blocks is used to detect if new services were added or some were removed since the last search. Through this structure, OSGi services can be fully described directly in the Bluetooth service discovery mechanisms.

3.3   802.15.4 Transport

The IEEE 802.15.4 standard [5] defines PHY and MAC layers for short range, low power radios. 802.15.4 is the basis for the lower layers of ZigBee [4], which represents a full network solution including upper layers not defined by the 802.15.4 standard.

In contrast to TCP/IP and Bluetooth, 802.15.4 only defines the lowest two layers of the OSI model and thus does not provide a reliable transport on its own. While ZigBee would provide the missing layers and reliable transport, we wanted a more lightweight solution that can even run on, e.g., TinyOS powered devices. Hence, we have implemented a simple transport layer similar to TCP that uses acknowledgments, timeouts, and retransmissions to provide reliable end-to-end communication on top of 802.15.4. Our implementation of the transport layer represents a proof of concept only as we did not optimize it in any way. Reliable transport over packet oriented lower layers is a research topic on its own and we will eventually exchange our implementation with a more sophisticated algorithm tailored to the characteristics of 802.15.4 radios.

Atop of this transport we implemented an 802.15.4 network channel for R-OSGi, similar to the implementation for Bluetooth. Since the message format used is compatible to TinyOS' active messages, the URI schema for this transport is *tos*. An example of a URI for this transport is *tos://100#5* with *tos* being the schema, *100* an example for a node address, and *5* being the service identifier on the peer. The transport behaves like Bluetooth or TCP transports. Once a connection of the underlying layer is detected, a logical R-OSGi channel to this peer is established over the connection and message exchange can start.

4   OSGi-Like Services on Non-standard Java

One obvious limitation of the R-OSGi implementation is that it requires at least a Java CDC VM and an OSGi framework to run on the participating nodes. For small devices, we had to address this issue and come up with solutions for devices with limited computation power and resources. This also includes embedded systems on microcontroller architectures with no Java VM and limited operating system support.

When two R-OSGi peers establish a connection they exchange a list of the services they provide. Each entry in this list comprises the names of the interfaces that the service implements and its properties. When a caller invokes a remote service, the instance of R-OSGi on the callee side sends the full interfaces including all methods to the caller. On the caller, a proxy that implements the interfaces of the remote service is built on the fly and registered with the local OSGi framework like other local services. We exploit the fact that the service is created and registered as a local proxy service on the caller side to hide non-OSGi services behind an OSGi-like interface. The bridging between the caller – which must be a full OSGi framework – and the callee is done at the proxy.

The first platform we adapt to running OSGi is the Java CLDC [6]. It lacks certain properties that are required for running OSGi, for instance, user-defined classloading. The second platform we adapt are embedded Linux devices running services written in C or C++, thereby removing the dependency on Java. Finally, the third implementation works on top of TinyOS running on Tmote Sky [19] nodes featuring 10kB of RAM and 48kB of code storage, thus removing the dependency on a full fledged operating system.

4.1   CLDC

CLDC is the smallest version of the Java Virtual Machine. It ships with almost every mobile phone and is also used on embedded devices with very little resources (minimum requirements are $160kB$ non-volatile memory as storage and $32kB$ volatile RAM). The VM for CLDC is hence called the *Kilobyte Virtual Machine* (KVM) [6]. On top of the VM, different *profiles* can run. The profile defines which subset of the standard Java classes is available, which CLDC-specific classes are provided, and which packages can be optionally included. For instance, the MIDP [20] profile is the most widespread on mobile phones and allows to run applications packages as MIDlets.

The reason why OSGi does not run on CLDC devices is mainly the sandbox model in the underlying KVM. For instance, in the MIDP profile, MIDlets cannot access code of other MIDlets and no new classes can be loaded that are not part of the MIDlet itself. For such devices, we locally run an adapter that understands the R-OSGi protocol and communicates with the local services.

The adapter is implemented as a single MIDlet that contains a stripped-down implementation of the R-OSGi protocol (since it only needs to deal with incoming calls) and the services that the device offers. Since the KVM does not support reflection, the dispatching of remote service invocation calls to the service methods has to be done explicitly. A further limitation of the KVM is the missing Java serialization. This is a severe limitation because not only the arguments of service methods might involve complex Java objects, also the R-OSGi protocol itself has some messages that ship reference types, for instance, the properties of the service. The solution to the problem is to generate custom serialization and deserialization methods for every used complex type ahead of time when the MIDlet is assembled. The generation procedure uses bytecode analysis to inspect the classes and generates a serialization for every complex type used in a service interface. Since the arguments of service methods are known at the time where the MIDlet is generated, arbitrary types can be serialized. Also the customized method calls are generated at build-time so that the service does not have to be changed in order to run on CLDC. Clearly, the statically assembled MIDlet does not support all features that an OSGi service running on a framework supports. Although it is possible to enable and disable services through a management service, the update of a single service requires an update of the whole MIDlet.

4.2 Embedded Linux

Supporting services written in languages other than Java is mainly a problem of adapting the data types. OSGi, that is Java, allows to use very expressive classes which cannot be easily transformed to other languages. The usual way to invoke native code from Java code is through the Java Native Interface (JNI) [21], an interface that exposes an API that can be used to create and access objects and to invoke methods.

In the R-OSGi C/C++ implementation, we developed a stand alone library (JNI Runtime) that implements the JNI interface but without actually being a Java VM (Figure 4). The JNI runtime library does not have the overhead of interpreting any Java bytecode and operating on a virtual stack machine. Instead, it implements only the *JNIEnv* which is normally the handle to the JNI interface of a Java VM. The runtime library maintains internal data structures for classes, fields, methods, and object instances. Methods are implemented by C functions operating on the data structures. Common classes like *java.lang.String* or *java.lang.Integer* are already implemented and part of the library. Service implementations can register new models of Java classes through an extension mechanism, if they require other classes.

With the help of the runtime, JNI code can run without a Java VM. The R-OSGi protocol is implemented in the *R-OSGi daemon*. The daemon accepts incoming TCP/IP connections and handles the requests. It is possible to implement R-OSGi daemons for different transports as well. JNI service implementations are registered with the R-OSGi daemon through a configuration file. In this file, the developer also has to specify the location of a Java interface implementation of the service. This interface is not used for the execution of the JNI service but only as a static resource exchanged in the R-OSGi protocol. Furthermore,

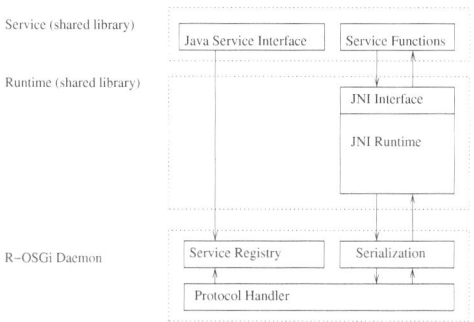

Fig. 4. R-OSGi for Native Code Services

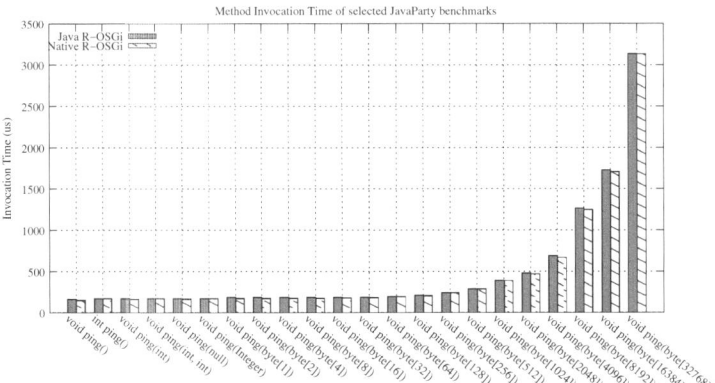

Fig. 5. Javaparty Benchmark on Linux

the daemon also implements the Java serialization mechanism. If the class of an incoming object is known to the JNI runtime (either by default or provided by the service), the R-OSGi daemon can create a corresponding instance in the JNI runtime which can then be used by the service implementation in the same way a corresponding Java class inside a VM would be used. Hence, arguments for service calls can be deserialized (and return values serialized again) to preserve the full expressiveness of Java/OSGi but without the overhead of a Java VM.

To quantify the performance of our solution, we again use the JavaParty benchmarks. The first series of measurements were taken on two Linux notebooks using Java 5. Figure 5 shows that the native implementation of R-OSGi is slightly faster in some cases but no significant performance gains can be observed. This can be expected since the limiting factor on devices with huge resources is the network bandwidth and not the processing. Furthermore, the HotSpot VM benefits from just-in-time compilation. The picture changes when the same experiment is repeated on a Linksys NSLU2, an embedded Linux device with a 133 MHz Intel XScale IPX420 CPU (Figure 6). This time, a JamVM [22] is used as a reference. The native R-OSGi implementation performs significantly better than the interpreting VM in all cases. Even the call with $32kB$ of arguments completes in less than $20ms$.

4.3   TinyOS

Our implementation of R-OSGi services for TinyOS [7] is based on the 802.15.4 transport presented in the preceding section and, like all TinyOS applications, implemented in NesC [23]. NesC adds the concept of components, interfaces, and wiring of components to the C programming language. A TinyOS application consists of several components that are wired together. TinyOS provides a collection of ready to use components, mainly hardware drivers. The Tmote sky nodes implement the TelosB [24] platform of TinyOS and all hardware components that this platform is composed of are supported in the current TinyOS release

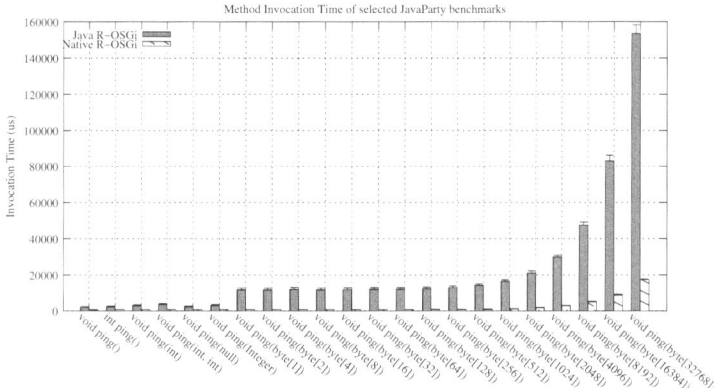

Fig. 6. Javaparty Benchmark on the Slug

2.0.2. Therefore and because of previous experience with Tmote sky nodes and TinyOS, we chose to implement R-OSGi services for sensor nodes using TinyOS.

Conceptually, this implementation is more similar to the implementation for CLDC than the implementation for embedded Linux, despite the fact that NesC is just an extension of C and the programming languages are thus very similar. Dynamic loading of shared libraries is not available on TinyOS and resources are very limited. Additionally, the capabilities and therefore the services provided by a sensor node are usually static. Therefore, the approach of emulating a full JNI environment is not feasible for sensor nodes and TinyOS. We rather compose the services provided by the sensor node at build time, like the MIDlet for CLDC, and also include the appropriate interfaces and data types in advance.

The R-OSGi application for TinyOS uses the 802.15.4 transport to listen for incoming connections. When a connection is established, the application composes the lease message containing the interfaces and properties of all services available by collecting the data from the installed services. Services implement our (NesC) *Service* interface which provides commands for getting the serialized (Java) service interfaces, getting the serialized, built-in properties of the service, and executing a method on the service. Furthermore, it defines an event that will be signaled upon completion of the method invocation. This way it is easily possible to have multiple, different services installed in one node as they can be managed uniformly by the R-OSGi application. A configuration file that is used at compile time defines the services that are available on the node.

When a remote R-OSGi peer invokes a method, the R-OSGi application dispatches the call to the correct service by invoking its *invoke* command with the signature of the method and the serialized method arguments as arguments. The service then dispatches the invocation internally according to the signature. For example, a temperature service might offer *getRaw()* as well as *getCelsius()* and *getFahrenheit()*. The invocation is implemented as split-phase operation. Once the invocation has completed, *invokeDone* is signaled and the return value passed as argument. The serialized return value is then sent to the calling peer as response to the invocation of the method. To deserialize the arguments and serialize the return value, services use our *Serializer* component.

5   Supporting Streaming Data Services

R-OSGi was designed for remote method invocation which corresponds to a conventional request-response interaction. However, many small and ubiquitous devices only deliver or consume data streams, e.g., a sensor node that periodically reports the temperature. Also services offered by more powerful devices can require the exchange of streaming data, e.g., a CCTV camera.

Typically, exchange of streaming data in Java is implemented using instances of *InputStream* and *OutputStream* of the *java.io* package and its subclasses. In the case of interaction through services, the service returns an instance of a stream to the client and data is then transmitted between the service and the client through the stream. However, plain remote method invocation is not sufficient to transparently remote these services. A stream object is backed by a file, a buffer, or a network channel. When it is returned to a remote client, it has to be serialized and the backing is lost. Therefore, plain service remoting with method invocation is not sufficient to deliver streaming data produced or consumed by many ubiquitous devices. We present an extension to R-OGSi that transparently supports streams. We also discuss quality of service issues related to streaming data.

5.1   Transparent Streams on Java

To allow for transparent remote access to streams we make R-OSGi stream aware by intercepting the exchange of stream objects and introducing a separate mechanism for exchanging data remotely over streams.

When R-OSGi executes a remote method call that returns an instance of *InputStream* or *OutputStream*, this instance is replaced by a handle in the result message that is sent to the calling peer. There the handle is used to create a proxy that extends *InputStream* or *OutputStream* and this proxy is returned as result to the method invocation. Likewise, instances of *InputStream* or *OutputStream* that are passed as arguments to remote method calls are replaced with handles and represented by stream proxies in the same way.

Calls to the stream proxy will be converted to R-OSGi stream request messages which will be parsed and result in respective calls to the actual stream. The data returned (in case of input streams) is sent back as R-OSGi stream result messages and returned to the read call to the stream proxy.

This mechanism allows to transparently exchange and use streams that are *per se* not serializable. Stream proxies look and feel like any input or output stream and thus can be used like any input or output stream, e.g., attaching a buffered stream or an object stream. Furthermore, the overhead of stream request and stream result messages is even less than the already low overhead of method invocation messages. This provides the best possible performance which is crucial as streams are often used to continuously transmit significant amounts of data compared to method invocation which occurs sporadically and usually with a small argument size.

5.2 Streams on Non-OSGi Devices

Similarly to the implementation of OSGi-like services in devices without OSGi, we provide support for data streaming in such devices through the adapters.

CLDC already supports *InputStream* and *OutputStream* types. The R-OSGi adapter for CLDC simply replaces instances of streams with handles in the same way R-OSGi does. Incoming stream request messages result in calls to the actual stream and the data returned is sent back as stream result messages. The same applies for embedded Linux platforms with the only difference that the stream, like every other object, is implemented in JNI instead of Java.

For TinyOS, the *Service* interface is extended with two additional commands and two additional events for reading from and writing to streams. When a TinyOS service returns a serialized stream handle as result, it also passes a local handle to the R-OSGi adapter. This handle will be passed back to the service's commands for reading and writing when a stream request message arrives. The service then dispatches the request internally. Once data has become available (in case of input streams) or been processed (in case of output streams), the corresponding events are signaled and the result is passed as argument. The result is then sent to the service caller as response.

5.3 Quality of Service

The possibility to stream large amounts of data requires careful considerations of the possible impact on the whole system. R-OSGi maintains one communications channel to every remote host it is connected to. Through this channel all communication takes place in the form of messages. The data exchanged by streams is transmitted in stream request and stream result messages which are exchanged over the same communication channel over which all other messages like method invocation or lease renewal are exchanged.

To ensure correct and smooth operation of R-OSGi, messages of other types are prioritized over stream request and stream result messages to avoid clogging of the communication channel by a burst of stream messages. On the other hand, R-OSGi-aware applications can choose to prioritize traffic of specific streams over other streams or even over the other message types. This is useful for real-time streams like, e.g., audio or video streams, which are in general more time-critical than method invocations.

6 Discussion

OSGi simplifies constructing applications out of independent software modules. Moreover, it provides facilities for dynamically managing the insertion and withdrawal of software modules, keeping track of the dependencies between modules, enforcing dependencies, and interaction through loosely-coupled service interfaces. With the extensions to R-OSGi that we have just presented, we have a software fabric for the "Internet of Things" that gives us exactly the same features but for distributed applications residing in heterogeneous devices.

Having this support implies, for instance, that spontaneous interactions between two appliances become service invocations within a distributed OSGi framework. Switching one of the appliances off appears as a module withdrawal and the framework – the software fabric – handles the cleanup of all dependencies and informing users of this service. It is also possible to dynamically replace the implementation of a service, or even the device itself, without the caller being aware of the swap.

Service discovery can be made more precise and efficient by using the filters provided by OSGi. That way, the amount of information exchanged during a spontaneous interaction can be minimized, an important factor when communication is through a low bandwidth radio link. The same mechanism is helpful to prevent that, when several devices are present, all of them have to reply to indiscriminate broadcast looking for services.

In the software fabric we just described, all devices independently of language, platform, or size present a uniform frontend which makes developing applications far easier than when programmers have to cope with a wide range of technologies and paradigms. It is also important to note that this common frontend does not reduce the exchanges to the minimum common functionality of all devices involved. Finally, note that this common frontend and standardized interfaces apply not only to software but, as the example with the BUG illustrates, also to hardware.

We are not aware of any other solution of the many that have been proposed that is capable of bridging heterogeneity as well as the software fabric we propose. In particular, with the very low overhead that it introduces.

7   Related Work

The idea of constructing distributed systems out of cooperating services has been pioneered by the JINI [25,26] project. In JINI, every Java object can be a service. The information about the services around is maintained by a central *Lookup Server* which has to be initially discovered by each peer in order to communicate. The later communication between services is point-to-point and does not require the lookup server any more. JINI does not pose restrictions on the protocol that this device to device communication follows. However, since Sun's standard implementation of JINI uses RMI for the communication between devices and the lookup server, most applications use RMI also for the device interactions. Although spontaneity was one of the main goals of JINI, the requirement of a lookup server restricts the applications of JINI to networks with some management. In contrast, the approach presented in this paper does not require a central component and also works in opportunistic networks with no management, such as ad-hoc networks. This is crucial for applications around the "Internet of Things", where interaction is often driven by the momentum of opportunity. Furthermore, JINI is tailored to TCP/IP networks and cannot be easily run over other transports.

Other projects that are discussing how to contruct applications from cooperating services and provide specifications for design and interaction of these are ODP [27], CORBA [28], and DCE [29].

For consumer devices, UPnP [30] is a standard for discovering and interacting with equipment. UPnP is able to discover new devices through the multicast protocol SSDP [31] and even includes an addressing schema derived from Zeroconf [32,33] for networks where no infrastructure is available which assigns IP addresses. Devices and services are described by XML descriptors which can be used to select devices and services based on fine-granular properties. The communication between devices uses SOAP-based XML-RPC. Since SOAP can basically run over any transport, UPnP is technically not restricted to TCP/IP networks. However, the high verbosity of the XML involved requires networks with sufficient bandwidth and limits the feasibility for low-power communication. Furthermore, UPnP limits the arguments of calls to a small set of specified types. General type mappers for complex objects as known from web services are not supported.

Recently, the idea of *Service Oriented Device Architecture* (SODA [34]) has gained momentum. SODA tries to incorporate principles of SOA into the world of devices to facilitate their integration into enterprise systems. The work presented in this paper can be seen as an implementation of this idea. Instead of an enterprise bus based on web services, OSGi and R-OSGi are used for the communication between services on devices. Since the R-OSGi protocol is far more lightweight than web services, it appears to be more amenable to implement the "Internet of Things" in resource constrained devices.

8   Conclusions

In this paper we have outlined a software fabric for the "Internet of Things" that is based on the ideas borrowed from the field of modular software design. The next steps we intend to pursue include developing applications on this fabric to better test its properties, continue extending the fabric to encompass a wider range of devices and protocols, developing tools that will simplify the development of applications that federate collections of services provided by ubiquitous devices, and explore bridging of heterogeneous network architectures as well as multi-hop scenarios in general.

References

1. OSGi Alliance: OSGi Service Platform - Release 4 (2005)
2. Rellermeyer, J.S., Alonso, G., Roscoe, T.: R-OSGi: Distributed Applications Through Software Modularization. In: Cerqueira, R., Campbell, R.H. (eds.) Middleware 2007. LNCS, vol. 4834, pp. 1–20. Springer, Heidelberg (2007)
3. Bluetooth Special Interest Group: Bluetooth (1998), http://www.bluetooth.com
4. ZigBee Alliance: Zigbee (2002), http://www.zigbee.org
5. IEEE: 802.15.4-2006 Part 15.4: Wireless MAC and PHY Specifications for Low Rate Wireless Personal Area Networks (WPANs) (2006)

6. Sun Microsystems, Inc.: J2ME Building Blocks for Mobile Devices: White Paper on KVM and the Connected, Limited Device Configuration (CLDC) (2000)
7. Hill, J., Szewczyk, R., Woo, A., Hollar, S., Culler, D.E., Pister, K.S.J.: System Architecture Directions for Networked Sensors. In: Architectural Support for Programming Languages and Operating Systems (2000)
8. I.B.M.: IBM Service Management Framework (2004), http://www.ibm.com/software/wireless/smf/
9. Eclipse Foundation: Eclipse Equinox (2003), http://www.eclipse.org/equinox/
10. Apache Foundation: Apache Felix (2007), http://felix.apache.org
11. Rellermeyer, J.S., Alonso, G.: Concierge: A Service Platform for Resource-Constrained Devices. In: Proceedings of the 2007 ACM EuroSys Conference (2007)
12. Makewave AB: Knopflerfish OSGi (2004), http://www.knopflerfish.org
13. Guttman, E., Perkins, C., Veizades, J.: RFC 2608: Service Location Protocol v2. In: IETF (1999)
14. Guttman, E.: Service Location Protocol: Automatic Discovery of IP Network Services. IEEE Internet Computing 3 (1999)
15. Zimmermann, H.: OSI Reference Model-The ISO Model of Architecture for Open Systems Interconnection. IEEE Transactions on Communications 28 (1980)
16. Haumacher, B., Moschny, T., Philippsen, M.: The Javaparty Project (2005), http://www.ipd.uka.de/JavaParty
17. Leach, P., Mealling, M., Salz, R.: A Universally Unique IDentifier (UUID) URN Namespace. RFC 4122 (2005)
18. Howes, T.: A String Representation of LDAP Search Filters. RFC 1960 (1996)
19. Moteiv: Tmote sky (2005), http://www.moteiv.com/
20. Sun Microsystems: Java Mobile Information Device Profile (1994)
21. Sun Microsystems: Java Native Interface (2004), http://java.sun.com/j2se/1.5.0/docs/guide/jni/index.html
22. Lougher, R.: JamVM (2003), http://jamvm.sourceforge.net
23. Gay, D., Levis, P., von Behren, R., Welsh, M., Brewer, E., Culler, D.: The nesC language: A holistic approach to networked embedded systems (2003)
24. Crossbow Technology: The TelosB Platform (2004), http://www.xbow.com/Products/productdetails.aspx?sid=252
25. Sun Microsystems: Jini Specifications v2.1 (2005)
26. Waldo, J.: The Jini architecture for network-centric computing. Communications of the ACM 42 (1999)
27. Linington, P.F.: Introduction to the open distributed processing basic reference model. Open Distributed Processing (1991)
28. Object Management Group, Inc.: Common Object Request Broker Architecture: Core Specification (2004)
29. CORPORATE Open Software Foundation. In: Introduction to OSF DCE (rev. 1.0), Prentice-Hall, Inc., Upper Saddle River (1992)
30. UPnP Forum: Universal Plug and Play Device Architecture (2000)
31. Goland, Y.Y., Cai, T., Leach, P., Gu, Y., Albright, S.: Simple Service Discovery Protocol (Expired Internet Draft). In: IETF (1999)
32. Cheshire, S., Aboba, B., Guttman, E.: RFC 3927: Dynamic Configuration of IPv4 Link-Local Addresses. In: IETF (2005)
33. Guttman, E.: Autoconfiguration for IP Networking: Enabling Local Communication. Internet Computing, IEEE 5 (2001)
34. de Deugd, S., Carroll, R., Kelly, K., Millett, B., Ricker, J.: Soda: Service oriented device architecture. IEEE Pervasive Computing 5 (2006)

The Benefits of Embedded Intelligence – Tasks and Applications for Ubiquitous Computing in Logistics

Reiner Jedermann and Walter Lang

Institute for Microsensors, actors and systems, University of Bremen, FB1, Otto-Hahn-Allee
NW1, D-28359 Bremen, Germany
{rjedermann,wlang}@imsas.uni-bremen.de

Abstract. The concept of autonomous cooperating processes is an approach to solve complex logistical planning tasks by representing each object in the transport chain by a separate independent software unit. In general, these software units or agents are applied in a server network. Technologies from the field of the Internet of Things like wireless communication and RFID enable that software execution can be shifted to deeper system layers, even at the level of single freight items. This article examines the ancillary conditions and consequences of this shift. It focuses on whether the introduction of the intelligent parcel or vehicle is advantageous compared to server based planning. The second half of this article describes transport logistic examples for networks of autonomous objects with embedded intelligence.

1 Introduction

The idea of the intelligent parcel searching the way through the supply chain by itself has been discussed for several years. It looks for the best transport route and reacts to changed traffic situations or orders. In an extended scenario a parcel equipped with sensors supervises its content and environment. If it detects a danger threatening the quality of the content, the parcel looks for a solution.

The general technical questions are solved. Passive UHF RFID systems can identify items at transshipment points. Semi-passive RFID data loggers allow temperature recording at reasonable costs. Active tags or wireless sensor networks can communicate between freight, containers, warehouses and vehicles. Real time location systems can keep track of the location of a container.

Despite the technical feasibility, the possibilities for the practical implementation are restricted to mere identification or tracing tasks. Pallets are tagged with passive RFID labels by the biggest food suppliers, sea containers are identified by active tags. However, sensor applications on palette or item level are still in the prototype state.

1.1 Non-processing Sensors or Intelligent Objects?

These RFID and sensor systems are essential to monitor the transport chain. So far, they have mostly been regarded as information providers, instead of participants in planning processes or decision making. But should we constrain the implementation

C. Floerkemeier et al. (Eds.): IOT 2008, LNCS 4952, pp. 105–122, 2008.

of planning processes to remote server platforms or is it more beneficial to equip the embedded devices with more localized computation power and intelligence?

This idea of shifting the intelligence into networked logistic objects has been a tempting vision in the past years, but the real benefits have not been figured out so far. How can the performance of the supply chain be developed without having to make more effort than in case of using a centralized planning system?

Passive RFID tags, active sensors and processing units on vehicle or container level will be considered as alternate platforms to realize the system intelligence. Aim of this contribution is to analyze the ancillary conditions and consequences of the transition from server based solutions to a network of distributed embedded processing units.

1.2 Outline

The first part of this paper presents a framework to classify and compare implementations intelligent objects on different system layers. **Section 2** examines the idea of intelligent objects in the broader context of autonomous cooperating processes. **Section 3** enumerates an extended set of criteria to evaluate the performance of a logistical planning system. The degree of decision freedom will be introduced in **section 4** as a means to categorize the amount of intelligence of an embedded object. The available computation power and additional costs to implement intelligence on different hardware layers are also discussed in this section. The current costs of the complete system are mainly determined by the extent of necessary communication. The concept of the 'length of the information path' will be presented in **section 5** in order to compare different levels of implementation in regard to the effectiveness of communication.

The second part comprises several case studies and demonstration examples for local intelligence on vehicle, sensor or tag level. **Section 6** presents the shelf life model as a basic approach to evaluate the effects of temperature deviation on the quality of the product. These models could be used by an intelligent container (**section 7**) to send automated notifications on predictable quality losses. In addition to the quality evaluation, parts of the route planning can be shifted onto the level of the means of transport (**section 8**).

Finally, a case study is described in the focus of **section 9** illustrating that it is not only economically worthwhile to invest into item level sensor supervision, but it is also necessary to process the sensor data in-place by intelligent objects.

2   Autonomous Cooperation of Logistic Processes

The autonomous cooperation of logistic processes is based on the paradigm to divide the solution of a complex logistical planning task into distributed processes [1]. Ideally, each object that participates in the supply or production chain is represented by its own software program, which autonomously searches a partial solution that is beneficial for the local object. In a transport scenario, the packages collect information, make decisions and negotiate with other entities to enhance the fulfillment of their common goal to perform all deliveries with minimal transport costs.

Autonomous Control has been defined by a workgroup of the Collaborative Research Centre SFB637 as *processes of decentralized decision-making in heterarchical structures. … The objective of Autonomous Control is the achievement of increased robustness and positive emergence of the total system due to distributed and flexible coping with dynamics and complexity.* [2].

Böse and Windt compiled a catalogue of thirteen criteria to characterize logistic systems based on the level of autonomous control [3]. The location the decision making is considered as the most important criterion. Individual decision processes for single objects do not necessarily require that a microcontroller integrated into an object should be in charge of implementing the decision algorithm. The software entities could run on a single or on a divided multi-server platform.

2.1   Representation of Logistical Objects by Software Agents

A common approach to represent autonomous objects is the implementation by software agents, which are a concept to run independent program units in a distributed network. These units autonomously perform a task on behalf of their owner. A more detailed description of software agents can be found in [4] and [5]. The most common environment to test and implement agents is the JavaAgentDEvelopment [6] framework (JADE). Each participant of the supply chain, like freight items, vehicles, sensors, traffic and logistic broker services can be represented by an agent [7].

3   Criteria for Performance Evaluation

In order to be able to benefit from embedded autonomous objects, it is necessary to provide dynamic planning. If everything is already determined before transportation is begun, there is no need for intelligence. But most complex transport optimization tasks need dynamic re-planning of route and freight allocation because of changes in traffic, unexpected delays or new incoming orders. A developed transport planning would also handle freight damage that occurs during transportation or predictable quality losses.

The second precondition is rather psychological: The freight owner has to yield control. Decisions are not made in a controlling room under the direct supervision of a human operator, but in the real world, on the road itself.

Furthermore, the criteria to evaluate the performance of a planning process have to be extended. Good planning cannot only be assessed in terms of finding the shortest route, the least transportation costs or the highest number of punctually delivered orders.

- The system should be flexible enough to react immediately to sudden situation changes.
- The network and the organization of planning entities should be robust to continue to work even in case of a communication failure.
- The privacy of internal planning strategies has to be kept. It should be kept confidential which customers are served first, and which are the ones whose deliveries are postponed in case of a shortage.

- Most information has to be transmitted over mobile networks. The communication costs should be kept low.
- A thorough search for the optimal route to deliver a great deal of items to several costumers could take hours or days. The computation time should also be taken into account. An optimal solution might quickly become obsolete due to changes in the traffic situation; therefore a recalculation should be possible with minimal computation by heuristic algorithms. In many cases it is possible to split up heuristic algorithm into independent distributed processes without the need for central collection of all the information.

The increase of robustness and ability to react to sudden changes overlaps with the goals of autonomous control [3], the criteria privacy, communication costs and computation time result from the implementation of embedded systems.

4   The Implementation Level of Decision Instances

Several factors should be taken into consideration in order to be able to decide on which hardware level the autonomy of the system should be implemented. The first one is the extent of intelligence that should be assigned to the embedded objects. As an alternative to the fuzzy term 'intelligence' possible solutions are classified on the basis of how much freedom an object has to make decisions by itself. Secondly, restrictions have to be taken into account in regard to computation power of the decision platforms and related costs. Effects on communication are examined in a separate section.

4.1   Degree of Decision Freedom

The objects in the supply chain can be equipped with a certain degree of decision freedom (**Table 1**). In a server based or central approach objects are only seen as sources of information, they only execute decisions of the central planning instance. On the next level of autonomy, freight items handle incoming data by local pre-processing. They observe and evaluate the conditions of their surroundings and decide when it is necessary to send an alarm notification to the planning system.

Table 1. Degree of decision freedom for mobile logistical objects

Class	Decision scope	Example
No freedom	None	Objects only executes centrally made decisions
Drawing conclu-sions from data	Evaluation of local sensor information	Object observes its environment and decides whether measured deviations form a risk for the freight quality
Adjust solution	Adaptive route planning	Freight might change transport route or swap ve-hicle by own decision
Adjust goal	Order management	Freight might changes its destination, according to new orders or changed quality state

In case of an adaptive route planning system a freight item or its software representation could additionally decide to alter the transport route or vehicle if necessary. If the freight item might also change its destination, triggered by new orders or an unexpected quality loss, the maximum degree of decision freedom for an autonomous freight item is reached.

4.2 System Layers

Shifting the autonomy into the network does not necessarily mean to add new hardware layers to the system, but rather equipping an existing layer with enhanced computation facilities so that it will be able to implement decision algorithms. Therefore, cost calculations should be based on the extra costs of additional processing power rather than on the basic costs of the related layer. The possible hardware layers range from item-level tags to powerful servers. The available computation power on the lower levels could be less than 0.1% compared to a PC. **Table 2** gives a summary of these factors as well as the current and future applications on different hardware levels.

In current solutions all information is collected on **server** level. An Object Name Server (ONS) links the Electronic Product Code (EPC) to a related data base. This data base contains an XLM description of the object or ideally even a web-page depicting the object properties and user instructions. Routing problems are commonly also handled at server level as well. A lot of research is being carried out in order to solve vehicle routing problems by multiple agent systems [8], [9].

Several telemetric systems can be currently found on the level of the **means of transport**. Equipped with mobile communication and GPS they measure the temperature, tire pressure and trigger an alarm if the doors are opened outside a specified area where it is allowed.

Supposing that the truck or container is already equipped with sensors and communication, the extra hardware costs for intelligence are about 100 Euro for an embedded processing module. The ARM processor used in our demonstrator for an intelligent container with a clock rate of 400 MHz reaches about 2% of the performance of a PC. An autonomous system on this level could pre-process sensor data or participate in the transport planning process.

Systems below the level of the means of transport have to operate without a fixed energy source; therefore, they are mostly battery powered. One main design goal is to reduce energy consumption in order to extend their lifetime, while communication can be active or passive.

Active devices usually use the ZigBee or the directly the basic IEEE 802.15.4 protocol in the 2.4 GHz range. Alternate proprietary protocols operate at 866 MHz or 433 MHz. This group of devices is comprised by the Auto-ID Labs as EPC class 4 active tags, which can communicate peer-to-peer with other active tags [10]. Tags for container identification, which are partly equipped with an additional temperature sensor, are a common example for the application of active devices.

Other research groups examine small battery powered devices with active communication under the term of **wireless sensor networks**. The focus lays on automated (Ad-hoc) networking and forwarding messages in order to extend the range of communication. The sensor nodes are equipped with an ultra-low-power microcontroller, typically the MSP430 from Texas Instruments. In combination with additional memory resources this processor might also be used as decision platform. Future wireless sensors could provide a spatial supervision of the cargo space so that local deviations of environmental parameters can be detected in time. Because of the components required for active communication and batteries, prices below 10 Euro will be hard to achieve.

Devices with **passive RFID** communication are not restricted to mere identification tasks. Semi-passive tags to record the temperature during transportation are already available on the market [11]. To operate in the absence of the reader field a small battery is required to power the device. The currently available RFID data loggers use the 13.56 MHz HF frequency range with limited reading distance. But UHF devices are expected to be available on the market in 2008. Future intelligent RFID will pre-process sensor data in the chip. The computation power will be even more restricted than in active devices due to the smaller energy budget. Because of the necessary extra components like sensors, processor and battery in addition to passive standard RFID tags, the price cannot go below 1 Euro.

Table 2. System layers as platform for decision processes

Location	Current application	Future applications	Computation power	Basic costs	Extra costs
Server networks	Objects representation by global database	Multi agent vehicle routing	100%	> 1000 €	low
Means of transport	Telemetric supervision, GPS	Intelligent Container	~2 %	< 1000 €	~ 100 €
Active communication devices	Active tags attached to containers	Spatial supervision by wireless sensors networks	~0.1 %	> 10 €	~ 1 €
(semi-) passive RFID tags	Identification / temperature logging	Intelligent RFID	<< 0.1 %	> 1 €	~ 1 €

5 Communication as a Limiting Factor

Wireless communication is one major pre-condition for the 'internet of things'. Whereas stationary servers can use internet connection almost free of charge, the wireless communication of battery powered devices is restricted in terms of data rate, volume and accessibility. **Table 3** summarizes different types and their limiting factors. Shifting the decision making system from a server level solution into the network of autonomous objects could largely reduce the amount of communication, as well as related risks and costs. Implementation on different hardware layers can be compared by the resulting 'length of the communication path'.

Passive RFID is the type of communication that is the most sensitive to interferences. Studies on the reading rate of tags mounted on single boxes inside a palette show a high dependency on the product type [12]. Because of the damping of UHF electromagnetically fields by liquids and metals, passive tags could be hardly read inside a palette or container that is packed with foodstuff. Access is only possible offline, at the end of transport. The items are typically scanned by a gate reader during the transshipment of the goods. The RFID tags are only visible for the second that a forklift needs to pass the reader gate during the unloading of the container.

Active devices allow higher reading range and data sizes. But the total volume of communication is limited by the energy budget. Commercial global mobile networks allow almost unlimited transmission of data at the price of higher energy and volume costs.

Table 3. Types and limits of wireless communication for battery powered devices

Type	Range	Limiting factor	Access
Passive RFID	~3 m	Access only during passage through reader gate	Offline
Active Wireless	~ 100 m	Volume limited by energy budget	Permanent
Commercial Networks	Global	Mobile connections costs (GPRS, UMTS, Satellite)	Permanent

The energy consumption of networked objects comprises the following three components: thinking, sensing and communication. Communication is normally the costliest factor; an example calculation for the TmoteSky wireless sensor node [13] shows that sending a message usually uses 16.5 mJ of energy, whereas reading temperature and humidity sensors requires 0.1 mJ and 200 ms of thinking costs 1 mJ [14]. The first aim in system design should be the reduction of communication by data pre-processing. If the costs for batteries and service are taken into account, embedded intelligence could even reduce the hardware costs.

5.1 Communication Sources and Sinks

To be able to compare the advantages of different implementation levels of intelligence in a logistical setting, it is necessary to asses how the shift of data processing to embedded devices affects the communication costs. The whole information path has to be taken into consideration: Which system component provides the data (**source**), who processes them and who needs the results (**data sink**) to execute an appropriate action.

Information sources can be a GPS module, sensors attached to the freight, RFID readers for object identification and traffic or order taking services. The data size ranges from 8 Bytes for a GPS position to several Kbyte (**Table 4**). To evaluate the costs, the data size has to be multiplied by the update rate.

Table 4. Information sources

Type of information	Location of data provision	Data size	Updates per hour
Vehicle location	GPS at truck or container	8 Byte	> 1
Influences to freight quality	Wireless sensors attached to freight or cargo space	~ 10 Bytes per object	>> 1
Object identification	Reader at vehicle or loading platform	16 Byte per object	<< 1
Traffic information	Global or regional information service	~ 1000 Byte	~ 1
New orders	Central or regional order taking service	~ 1000 Byte per order	~ 1

In case of RFID identification, the data is provided on the reader, not on the tag side. The warehouse knows what goods are delivered, but the tag does not know where it is. But it would be possible by some protocol extension to provide the tag with an additional reader identification number or location code.

After assessing dynamic changes in the supply network it could be advisable to change the route of a vehicle or to re-load a freight item to another truck or container. Because an intelligent parcel has no wheels it cannot carry out the recommended action itself. The results of the decision process have to be communicated to the driver of the vehicle, a forklift or the warehouse staff as data destination or sink.

5.2 Length of the Information Path

Most of the criteria to evaluate the performance of the planning process depend directly on the length of the information path from data source over processing to the execution unit:

- Some communication channels could be temporarily unavailable. A short information path increases the robustness of the system. The shortest possible path is found by integrating the decision system into the platform of the information source, its destination or inside an existing network hub between data source and sink. The autonomy of the system is increased if the decision task is divided into sub-tasks that can continue to operate independently if one of the information sources cannot be contacted.

- The outcome of a decision process can be communicated with few Bytes, whereas the required information could comprise the tenth of a Kbyte, e.g. the temperature history of a product combined with traffic information. By shifting the data processing close to the beginning of the information path the total volume and communication cost can be reduced. Data can be pre-processed directly at their point of origin if the related task is separated from the remainder of the decision system.

- The above mentioned points are also important in terms of the flexibility of the system. New information should be quickly communicated through short channels.

- A short information path also increases the level of privacy. The transmission of data over third party networks should be avoided.

The shift of the decision making system onto embedded hardware layers could either shorten or extend the length of the communication path, as the following example shows.

Adaptive planning according to new orders and traffic factors
As first example a transport planning system is considered that is responsible for processing information about new orders and traffic disruptions. Other source of information is the GPS position, available in the vehicle and RFID identification at loading or unloading, available at a door reader mounted inside the vehicle or at the warehouse gate. As data sink the vehicle driver has to be informed about necessary detours. Splitting the decision system into single entities per freight item leads to the idea of the intelligent parcel. It searches at each transshipment point for the next 'lift' by a vehicle that brings it nearer to its final destination. Following principles of existing routing protocols from internet data communication, packages and vehicles decide about the best route only on the basis of information provided by the local and neighboring hubs [15].

The implementation layer of these decision systems could be either inside the vehicle or on servers at the distribution centre or transshipment points. A system on vehicle level has to query traffic and order information from a global service provider, whereas a central decision system can get direct access to traffic and order information. However, it has to contact the vehicle to obtain freight identification and to send the results of the decision. The freight item itself is neither data source nor sink; the implementation of autonomy on this layer would add an additional edge to the information path. In this example the shift of the decision system directly into an intelligent parcel would result in total costs that are even higher than the ones generated by applying the central solution.

Supervision of sensitive goods
If the freight item produces information by itself, the situation is different from the previously described example. The supervision of sensitive goods as in the food sector, for instance produces high amount of data for monitoring deviations of the environmental parameters with local sensors. Even small differences of the temperature can have a major effect on the quality of the product.

Sensor-tags on item level or wireless sensor networks for monitoring the cargo space provide an information source in addition to the example above. The decision system consists of two sub-tasks: Sensor data evaluation and route adaptation. The information path for this scenario is depicted in **figure 1**.

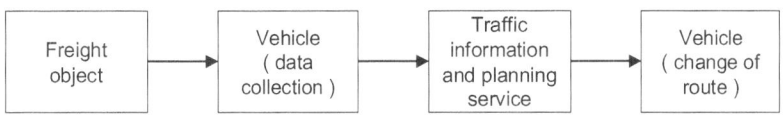

Fig. 1. Information path for the supervision of sensitive goods

The direct pre-processing of the sensor data at their origin by an intelligent parcel could largely reduce communication volume and costs. In another setting the vehicle assesses the data provided by a wireless sensor network. The following second part of this article presents several case studies for the implementation of transport supervision and planning systems for this sensor supervision scenario.

6 The Supervision of Quality Changes

The assessment of quality changes is the most important task for an intelligent freight object in the transportation of sensitive goods. Unfortunately, quality parameters like firmness or taste of fruits cannot directly be measured during transportation. Quality losses can be predicted only by taking the changes in the environmental conditions into account.

6.1 Ubiquitous Sensors

An intelligent object has to measure these conditions with appropriate sensors. Only few sensor types are suitable for integration into ubiquitous devices. The limiting factors are not only costs, but mainly the energy consumption. Temperature and humidity sensors consume the least energy; measurements have only to be taken in intervals of several minutes. Proper ventilation of cooling air can be checked with miniaturized thermoelectric flow sensors. Shock or acceleration sensors are more demanding, they have to operate with a duty cycle of 100%; otherwise they might miss a shock event. Typical metal oxide gas sensors consume the most energy, because heating for several minutes above 300 °C is required until all chemical reactions are balanced.

6.2 Shelf Life Modeling

The quality that is lost by a deviation from the recommended transport conditions depends on its duration and magnitude as well as the individual sensitivity of the product type. Among other parameters like humidity, light and atmosphere the temperature has the largest effect on the quality of food products. These effects are well examined especially in agricultural science.

The 'keeping quality' or 'remaining shelf life' model [16] is based on the Arrhenius law of reaction kinetics. The sensitiveness of some fruits to chilling injuries can be modeled by adding a second Arrhenius function with negative activation energy. The original static model was transferred into a form to calculate the curse of a quality index for fluctuating temperature conditions.

Figure 2 provides the acceleration factor for decay processes as a function of temperature related to a standard temperature ($T_{St} = 15$ °C). This value indicates how many days of shelf life have to be subtracted from the initial value Q_0 per day of transport at a certain temperature. The description of the quality curve can be compressed into a set of five parameters and stored on an item level tag as additional information. The curves in figure 2 are based on the parameters estimated by Tijskens [16] and Schouten [17].

In cases where the temperature sensitivity cannot be approximated by an Arrhenius type model, the quality changes can be estimated by interpolation of reverence curves [18]. The curse of certain quality conditions has to be recorded under laboratory conditions for several temperatures. To upload the model, the complete curves have to be transmitted to the evaluation device.

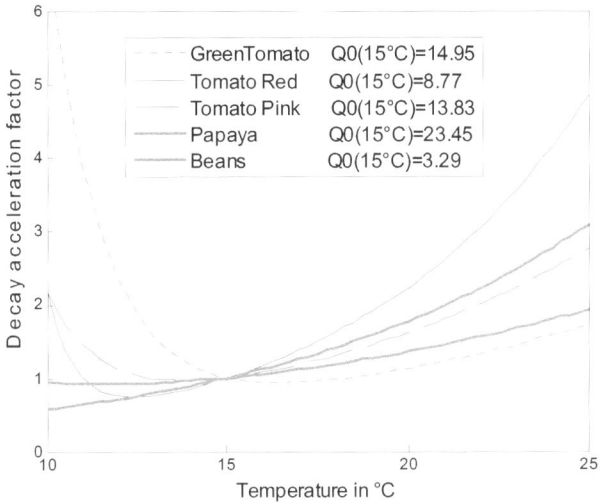

Fig. 2. Relative loss in shelf life in relation to a standard temperature

6.3   The Advantages of Local Sensor Data Evaluation

The major advantage of the localized evaluation of the environmental parameters by shelf life modeling is the minimal data update rate. The processing instance only has to submit a warning message if the quality falls below an acceptance limit or if it foresees a pending quality loss [14]. Instead of the sensor raw data, which needs several updates per hour, only one message has to be sent in case a problem arises. Furthermore, the robustness of the system is increased. The permanent quality supervision can continue during periods when external communication is not available.

7   Processing on Vehicle Level

The first implementation study shows how the pre-processing of sensor data can be realized on the level of the means of transport. Our demonstrator of this 'intelligent container' has been described in detail in earlier papers [19], [20]; this section gives a brief summary and relates it to the previous defined criteria to assess the benefits of embedded intelligence.

One key feature is the automatic adaptation of the supervision system to new freight items and their related quality models for specific kinds of goods. In contrast to common solutions the description of freight objects is not stored on a fixed server, but travels in parallel to the physical object through the supply chain. A mobile software agent carries out the individual supervision program on the computation platform of the current means of transport or warehouse.

The hardware of the reduced scale prototype consists of an RFID reader used to scan for new loaded items, a wireless sensor network for spatial supervision of the cargo space, a module for external communication and an embedded processor module as agent platform. If a new freight item is detected, a request is sent through the communication network to receive the corresponding freight agent.

7.1 Communication Path and Costs

The wireless sensor nodes provide as data source the temperature and humidity measurements by active communication. The limiting factor is the battery reserves of the sensor nodes. The processing unit on vehicle or container level compresses this data to single 'change of quality' notifications and sends it thorough an external network to the route and transport planning service as data sink.

The system costs mainly comprise single investments into the infrastructure like the processing unit and the wireless sensors belonging to the truck or container. This investment reduces the amount of constant expenses for communication and increases robustness. The additional hardware costs per transport unit consist only of one disposable standard RFID tag per freight item.

7.2 Future Research

One bottleneck is the transmission of the complete sensor raw data from the sensor nodes to the processing unit. Shifting the pre-processing into the sensor network could extend the battery lifetime. The related quality model has to be transmitted to all sensors in the surrounding of the freight item under supervision. Furthermore, the measurement period of the sensors could be adapted to the current situation: if a sensor node detects that it provides exactly the same information as its neighbors, it could set itself into sleep mode.

Another limiting factor is the performance of the agent platform on embedded systems. Although additional optimizations have been carried out on the JADE LEAP (Lightweight Extensible Agent Platform) [21], the transmission of one mobile software agent still takes several seconds. The performance of the supervision software might be increased by switching to another framework like OSGi, formerly known as the 'Open Services Gateway initiative' [22], which was developed for industrial applications. Special services for new kinds of products can be remotely installed, started, stopped or updated on the embedded Java runtime of the processing unit of the vehicle or container in form of software bundles.

8 Dividing the Route Planning Process

This section serves as an example for a setting with a higher degree of decision freedom by presenting an approach to how the route planning can be gradually shifted to vehicle level. The vehicle does not only draw conclusions about quality changes by the sensor data but also adjusts its solution for the goal to deliver as many items in proper quality as possible.

This approach depicts a further important aspect of distributed planning processes. Local entities often do not have full access to all information. The following software simulation shows how a processing unit at vehicle level could make best use of limited traveling information provided by a remote server. The route planning process is divided into two instances:

- A remote server holds information about **distances and traveling speed** based on the current traffic situation. This information is public, because it does not contain any confidential data.
- The vehicle assesses the **current quality** of the loaded items. In case of unexpected quality loss the originally planned route has to be modified. This freight state data has to be handled as a confidential piece of information.

Considered for a single vehicle the common task can be categorized as Traveling Salesman Problem with Time Window (TSPTW). The time window for each product starts with zero for immediate delivery and ends with its expiration of shelf life.

For a test case to deliver N_0 perishable items to N_0 different destination towns the route is modified on the basis of the following principle: During the stopovers at the customers the vehicle evaluates for the N undelivered items their remaining shelf life. The vehicle sends a request for a short round trip to cover the N missing towns. The remote traffic information server generates a round trip without accessing product quality data. Based on this proposal the vehicle generates 2*N route suggestions, which enter the round trip at different points and follow it clockwise or counter-clockwise. The route suggestions are assessed based on the degree to which they have fulfilled their goal to minimize the total driving time and strongly avoid zero shelf life products. A first performance test has shown that it is feasible to organize to route planning by this divided planning approach [23].

8.1 Reaction to Dynamic Quality and Traffic Changes

To test the ability to react to sudden quality and traffic changes the test setting has been expanded. Before leaving the depot the freight items are assigned to the vehicles. This first multi-vehicle optimization step can be done by a server at the depot. After leaving the depot the vehicles cannot swap freight items any longer. The traffic information server can only be contacted over mobile communication. After the first two deliveries the shelf life of the remaining items is reduced by an unexpected event. This could be either a change in temperature or a delay by traffic disruptions. The ability of different algorithms to find alternative routes which avoid late deliveries has been tested in software experiments (**Table 5**) for a setting with $N_0 = 20$ customers.

Table 5. Performance of different planning strategies (500 software experiments)

Method	Delivered Packages	Driving time	Improvement
Full re-planning	16.41	76.81 hours	100%
Local vehicle planning	15.66	76.82 hours	64.5%
Repeated vehicle planning	15.75	75.80 hours	68.6%
Unchanged route	14.30	74.68 hours	0%

Without changing the route 14.3 packages in average could be delivered in proper quality. The divided planning approach slightly increases the driving time because of the necessary detours. The number of average delivered items has increased to 15.66

packages. An extensive server based search for the best route gives a result of 16.41 packages. According to these figures the divided planning approach achieved 64.5 % of the possible improvement of the number of properly delivered packages. By allowing the vehicle planner to repeat the route request up to three times in the first step after a disruption, the performance has improved to 68.6 %.

8.2 Evaluation of the Communication Path

Freight quality information is kept completely confidential inside the vehicle. For this kind of data the vehicle is information source, sink and processing unit at the same time. The only piece of external communication is a route request per stop and the server's answer. The size of the answer is about 100 Bytes to transmit the sorted list of towns and the driving distances along a round trip. In case of a communication failure the vehicle can continue its planning based on already known round trips.

Although the performance of the above presented simple algorithm is not fully satisfying, this example depicts typical obstacles and questions of autonomous objects. In cases where the local object does not find an acceptable solution, the object would re-delegate the problem together with all relevant quality information to a server, which has direct access to all traffic data, at the price of higher communication costs and reduced privacy. Limits in communication bandwidth, availability and allowed response time make it in most supply chain application impossible to achieve the same performance as the ideal solution. However, it needs to be questioned whether the object makes use of the available information to the greatest extent. Failures to deliver orders in time should be assessed in the light of certain aspects like flexibility towards dynamic situation changes, robustness against communication failures, hardware and communications costs.

9 Data Processing on Item Level

The idea of data processing on item level is introduced by a study on potential cost saving achieved by extended temperature and quality monitoring. A new approach to warehouse management implements the following principle: Items with low remaining shelf life are sent to nearby stores for immediate sale, whereas items with longer usage expectancy are used for export or distant retail stores. By replacing the standard 'First In First Out' (FIFO) policy with this 'First Expires First Out' (FEFO) approach the amount of waste due to product expiration before delivery is reduced [24].

9.1 Case Study on Strawberries

A case study carried out by Ingersoll-Rand Climate Control and the University of Florida [25] has estimated how much less waste is produced when the above described FEFO approach is used. 24 pallets of strawberries were equipped with temperature loggers. The shelf life was calculated before and after the transport based on the temperature history of each palette. Two palettes were rejected before transport because of their bad temperature history. The remaining 22 palettes had a predicted shelf life between 0 and 3 days at the end of transport (**Table 6**).

Table 6. Waste with and without FEFO approach. Figures according to Emond [26], [18]

Number of palettes	Waste on random retail	Estimated shelf life	FEFO recommendation	Resulting waste
2	100%	-	Reject immediately	(rejected)
2	91.7%	0	Reject at arrival	(rejected)
5	53 %	1	Sell immediately	(25%)
8	36.7%	2	Nearby stores	(13.3%)
7	10%	3	Remote stores	(10%)

After arrival, the truck-load was divided and sent to retail stores with different transport distances. The pallets were opened in the shops; boxes with strawberries of low quality were sorted out. Column 2 of table 6 reports the measured waste grouped by the predicted palette shelf life. The current solution without quality based assignment results in a loss rate between 10% and 91.7% per palette. Column 5 gives the results of a thought experiment: If the palettes were assigned according to a FEFO recommendation to match driving distance and remaining shelf life for each retail shop, the waste could be reduced to 25 % at most. Considered for the whole truck load a loss of 2300 $ could be turned into a profit of 13000 $. If we assume a price of 10 $ per recording unit, the investment for additional hardware is still less than 10% of the profit, even if 4 loggers are attached to each palette.

9.2 Necessity and Feasibility of Data Processing on Tag Level

The above described study has been carried out by manual handling of the data loggers at end of delivery. But this handling has to be converted into an automated process before this approach can be applied in practice. The interface of the data loggers is restricted to passive RFID to enable low cost solutions. But due to the very limited data rate of RFID transmission, it is not possible to transmit the full temperature protocols of several measurement points during unloading. The tags are only visible to the reader for less than a second while the pallets are passed through the gate.

Because of this communication bottleneck, an automated evaluation of the temperature data can be achieved only if the system intelligence is shifted into the network. By on-chip pre-processing the data volume can be compressed into one Byte per item to transmit either the current quality state or just on state bit discerning between 'green' as 'quality state ok' and 'red' as 'item needs further checking'.

The crucial point in the integration of the shelf life modeling into a semi-passive RFID data logger is the required computation power. The model has to be updated after each temperature measurement. The required calculation time for different shelf life modeling approaches was measured for the low power microcontroller MSP430 from Texas Instruments [18]. For all model types the required CPU time per model step was below 1.2 ms equivalent to an energy consumption of 7μJ per step or 20 mJ per month at an update period of 15 minutes. Compared with the capacity of 80 J for a miniature zinc oxide battery as used in the Turbo-Tag data loggers the energy for model calculation can be neglected. As a CPU facility only a 16 bit integer multiplication is required. The shelf life modeling will add only very low processing requirements to an RFID chip that already contains temperature sensor, clock and data storage. For demonstration purposes the modeling has been programmed into the processor of existing wireless sensor nodes, although this platform is too expensive for commercial applications. The sensor node has been programmed to behave as RFID: on request it transmits its identification code and the current shelf life state.

But future research will enable tags with integrated shelf live modeling as semi-passive devices at reasonable costs and surely reach the business case.

10 Summary and Conclusions

The aim of this paper was to analyze the boundary conditions and consequences of the shift of intelligence to ubiquitous devices. There are already many applications where RFID are used in a beneficial way, but the business opportunities for ubiquitous sensors in logistical applications have not been clearly figured out so far. There are only very few studies available on this topic, but the one quoted in the previous sections indicates that sensing on box or palette level can make a high return on investment.

The central question of this article was to determine in which of the above mentioned cases with a need for ubiquitous sensors and RFID it is worth, to make the additional investment for local intelligence. The implementation of local data analyzes and decision making needs extra computation power. However, the increase in hardware costs for additional CPU resources is still moderate in comparison to the primary costs of active wireless sensors and telemetric units. But communications restrictions related to costs and bandwidth are the major obstacle of the implementation of embedded intelligence. The concept of the 'length of the communication path' between information source and sink was introduced as a means to compare the effectiveness of different implementation levels.

The application field of the 'intelligent parcel' is mainly limited to settings where a high amount of local information has to be processed, like in the supervision of perishable products. In other applications it needs to be questioned whether it is worth extending the communication path by exchanging information and the results of the decision process with the parcel. This might be only the case when company information should be kept confidential inside the processing unit of the parcel. Other tasks, for instance parts of the route planning, can be better performed on the level of the means of transport.

The implementation of embedded intelligence is most beneficial in cases, where the object not only provides identification but also additional sensor information. Furthermore, there are also cases in which an automated supervision of the transport chain is only feasible if the sensor data is processed by an embedded system as in the example of the intelligent RFID. A share of the processing power, which is required to implement a logistical planning system, was shifted onto objects that are parts of the network. The equipment of logistical objects with embedded intelligence brings advantages in terms of reduced communication costs, higher robustness against communication failures and flexibility to react to unforeseen events.

Acknowledgments. This research was supported by the German Research Foundation (DFG) as part of the Collaborative Research Centre 637 "Autonomous Cooperating Logistic Processes".

References

1. Freitag, M., Herzog, O., Scholz-Reiter, B.: Selbststeuerung logistischer Prozesse - Ein Paradigmenwechsel und seine Grenzen. Industrie Management 20(1), 23–27 (2004)
2. Windt, K., Hülsmann, M.: Changing Paradigms in Logistics - Understanding the Shift from Conventional Control to Autonomous Cooperation and Control. In: Hülsmann, M., Windt, K. (eds.) Understanding Autonomous Cooperation & Control - The Impact of Autonomy on Management, Information, Communication, and Material Flow, pp. 4–16. Springer, Berlin (2007)
3. Böse, F., Windt, K.: Catalogue of Criteria for Autonomous Control. In: Hülsmann, M., Windt, K. (eds.) Understanding Autonomous Cooperation and Control in Logistics – The Impact on Management, Information and Communication and Material Flow, pp. 57–72. Springer, Berlin (2007)
4. Langer, H., Timm, I.J., Schönberger, J., Kopfer, H.: Integration von Software-Agenten und Soft-Computing-Methoden für die Transportplanung. In: Nissen, V., Petsch, M. (eds.) (Hrsg.): Softwareagenten und Soft Computing im Geschäftsprozessmanagement. Innovative Methoden und Werkzeuge zur Gestaltung, Steuerung und Kontrolle von Geschäftsprozessen in Dienstleistung, Verwaltung und Industrie, pp. 39–51. Cuvillier Verlag, Göttingen (2006)
5. Wooldridge, M., Jennings, N.R.: Intelligent Agents: Theory and Practice. The Knowledge Engineering Review 10(2), 115–152 (1995)
6. Bellifemine, F., Caire, G., Poggi, A., Rimassa, G.: Jade – a white paper. In: TILAB EXP in search of innovation, Italy, vol. 3 (2003)
7. Lorenz, M., Ober-Blöbaum, C., Herzog, O.: Planning for Autonomous Decision-Making in a Logistic Scenario. In: Proceedings of the 21st European Conference on Modelling and Simulation, Prague, CZ, pp. 140–145 (2007)
8. Thangiah, S.R., Shmygelska, O., Mennell, W.: An agent architecture for vehicle routing problems. In: Proceedings of the 2001 ACM symposium on Applied computing, pp. 517–521. ACM Press, New York (2001), http://doi.acm.org/10.1145/372202.372445
9. Leong, H.W., Liu, M.: A multi-agent algorithm for vehicle routing problem with time window. In: Proceedings of the 2006 ACM symposium on Applied computing, Dijon, France, pp. 106–111 (2006), http://doi.acm.org/10.1145/1141277.1141301
10. Sarma, S., Engels, D.W.: On the Future of RFID Tags and Protocols, technical report, Auto ID Center MIT-AUTOID-TR-018 (2003)
11. Turbo-Tag Data Logger by Sealed Air Coperation, USA, http://www.turbo-tag.com
12. Clarke, R.H., Twede, D., Tazelaar, J.R., Boyer, K.K.: Radio frequency identification (RFID) performance: the effect of tag orientation and package contents. Packaging Technology and Science 19(1), 45–54 (2005), http://dx.doi.org/10.1002/ pts.714
13. TmoteSky sensor node from Moteiv, http://www.moteiv.com/products/tmotesky.php
14. Jedermann, R., Behrens, C., Westphal, D., Lang, W.: Applying autonomous sensor systems in logistics; Combining Sensor Networks, RFIDs and Software Agents. Sensors and Actuators A (Physical) 132(8), 370–375 (2006), http://dx.doi.org/10. 1016/j.sna.2006.02.008

15. Scholz-Reiter, B., Rekersbrink, H., Freitag, M.: Internet routing protocols as an autonomous control approach for transport networks. In: Teti, R. (ed.) Proc. 5th CIRP international seminar on intelligent computation in manufacturing engineering, pp. 341–345 (2006)
16. Tijskens, L.M.M., Polderdijk, J.J.: A generic model for keeping quality of vegetable produce during storage and distribution. Agricultural Systems 51(4), 431–452 (2006)
17. Schouten, R.E., Huijben, T.P.M., Tijskens, L.M.M., van Kooten, O.: Managing biological variance. Acta Hort (ISHS) 712, 131–138 (2006), http://www.actahort.org/ books/712/712_12.htm
18. Jedermann, R., Edmond, J.P., Lang, W.: Shelf life prediction by intelligent RFID. In: Dynamics in Logistics - First International Conference, LDIC 2007, Bremen, Germany, August 2007, Proceedings, pp. 229–237. Springer, Berlin (to appear, 2008)
19. Jedermann, R., Behrens, C., Laur, R., Lang, W.: Intelligent containers and sensor networks, Approaches to apply autonomous cooperation on systems with limited resources. In: Hülsmann, M., Windt, K. (eds.) Understanding Autonomous Cooperation & Control in Logistics - The Impact on Management, Information and Communication and Material Flow, pp. 365–392. Springer, Berlin (2007)
20. Homepage of the project 'intelligent Container', http:// www.intelligentcontainer.com
21. Moreno, A., Valls, A., Viejo, A.: Using JADE-LEAP to implement agents in mobile devices. In: TILAB EXP in search of innovation, Italy (2003), http://jade.tilab.com/papers-exp.htm
22. OSGI Alliance Website, http://www.osgi.org
23. Jedermann, R., Antunez, L.J., Lang, W., Lorenz, M., Gehrke, J.D., Herzog, O.: Dynamic Decision making on Embedded Platforms in Transport Logistics. In: Dynamics in Logistics - First International Conference, LDIC 2007 Bremen, Germany, August 2007. Proceedings, pp. 189–197. Springer, Berlin (to appear, 2008)
24. Scheer, P.P.: Optimising supply chains using traceability systems. In: Smith, I., Furness, A. (eds.) Improving traceability in food processing and distribution. Woodhead publishing limited, Cambridge, England, pp. 52–64 (2006)
25. Pelletier, W., Emond, J.P., Chau, K.V.: Effects of post harvest temperature regimes on quality of strawberries. Report 1. University of Florida, p. 48 (2006)
26. Emond, J.P.: Quantifying RFID's Cold Chain Benefits. In: Fifth RFID Academic Convocation, Orlando, Florida (2007)

User Acceptance of the Intelligent Fridge: Empirical Results from a Simulation

Matthias Rothensee

Humboldt University Berlin
rothensee@wiwi.hu-berlin.de

.

Abstract. The smart fridge has often been considered a prototypical example of applications of the Internet of Things for the home. However, very little research has been conducted on functions desired by prospective users, and how users will eventually use the fridge. A simulation of a smart fridge was developed and tested within a controlled laboratory between-subjects experiment with 105 participants. Four different assistance functions were tested. It was found that generally a smart fridge is evaluated as moderately useful, easy to use and people would tend to buy it, if it was already available. Emotional responses differed between the assistance functions. Displaying information on durability of products, as well as giving feedback on nutrition health and economics are the most appreciated applications. Structurally, overall usefulness ratings of the device are the strongest predictors for the intention to use a smart fridge, but the emotional response to the product was also an important explanatory variable. Results are not influenced by technical competence, gender, or sense of presence in the simulation. Regression models confirmed that the simulation-based results explained 20% more variance in product acceptance than written scenarios. An outlook is given on future questions to be answered using the simulation.

1   Introduction

When speaking of Ubiquitous Computing (UbiComp), or, the Internet of Things, it is often regarded a central point that information and communication technology leaves the workplace and enters the home [1]. In fact, nowadays, every household in industrial countries is packed with a number of devices that contain all kinds of information technology [2].

Ethnographic studies have found that a large part of the social life of people takes place in the kitchen [3]. This place is normally not only used for food preparation but also serves important communication and social bonding purposes [4]. Research on UbiComp applications in the kitchen in the past years has focussed on nutrition and dietary support [5], cooking, recipe planning [6] and communications [7]. The bulk of research, however, are prototype applications that are rarely tested empirically with users. Within the kitchen the fridge is considered especially important because of its ubiquity and format. Every

C. Floerkemeier et al. (Eds.): IOT 2008, LNCS 4952, pp. 123–139, 2008.

household in the western sphere owns a fridge, and almost everyone has contact with it daily. Furthermore, the fridge offers large, flat surfaces that can be used for user interfaces. So it is only natural to give the fridge a prime position when it comes to digitally augmenting kitchen activities. In fact, the intelligent fridge has become a prototypical example for UbiComp applications in the home of the future. Multimedia fridges are already on the market today, but offer no product-based services [8]. Equipped with RFID and sensor technology it will soon be capable to correctly identify all products that are stored in a household [9]. Fridges that are aware of their contents are still a future scenario, albeit a highly desirable one from the retailer's point of view [10].

In this paper we assume that in the not so distant future all technical barriers to the introduction of a smart fridge will have been overcome. Taking a functionalist perspective we envision the smart fridge as a bundle of assistance functions. We concentrate on user-supporting assistance functions [11], as opposed to, say increased monitoring capabilities for a retailer. Furthermore we assume that the most basic function of the smart fridge, namely sensing its content, is hardly of any value to a prospective customer. Therefore, the underlying question of this research is, which intelligent functions will be appreciated, whether people differ in their acceptance of product-based assistance functions, and how its perceived attributes will influence the acceptance of the device.

1.1 Affective Factors in Household Technology Acceptance

Throughout the past years there has been an upsurge of empirical and theoretical works that emphasize the importance of affective appeal of a product for its acceptance [12]. It has also been proposed as a user interface [13,14] or product design aspect [15]. This view, however is in stark contrast to traditional technology acceptance models (TAM) that focus on the workplace [16]. In this tradition a product's usefulness (along with its ease of use) has repeatedly been identified as the core explanatory variable for its acceptance. Some integrations of the disparate traditions have been tried [17,18]. The consensus of the integrating approaches, however, is that the relative weight of affective vs. utilitarian factors in product acceptance is context dependent. It stands to reason that products for private use underlie different acceptance dynamics than products for office use. [19] found that the fun of using a PC for home use positively influenced the participant's intention to use it. [20] classified world wide web users as either work- or entertainment-oriented, and confirmed that usefulness has a greater impact on acceptance in work-oriented web usage, a point that is also mentioned by [21] in the discussion of their results concerning work related world wide web surfing. [22] confirmed the diminished importance of usefulness in favor of ease of use and enjoyment when it comes to hedonic as opposed to utilitarian information systems. [18] recently included work- vs. leisure context as a moderating variable in their unified technology acceptance model. Whether this is true for smart home technology such as the smart fridge as well, remains an open question. Research on the acceptance of smart home technology is rather scarce, owing to the fact that only very few of these products are yet developed. A recent position paper pointed out one of the major questions for the future: "How can we learn about users' experiences in hybrid contexts of everyday life when technology is not yet mature enough to enter users' real everyday life environments?" ([23], p.51). In principle, there are three ways to achieve this: scenarios, simulation, and prototypes. Given that in the case of the smart fridge, a prototype is not yet available, the researcher has to choose between scenario and simulation based research. The following section contrasts these approaches. Conceptually, however, scenario and simulation are not exclusive but hierarchical in that the interactive simulation is based on a scenario of use as described in section 2.2.

1.2   Scenario vs. Simulation

While written scenarios of future technologies are a feasible methodology from an economic standpoint, they have their drawbacks. The main problem with scenarios is their limited external validity. It has been shown that attitudes formed on the basis of direct experience have a stronger influence on the intention to behave toward the attitude object [24]. Therefore one can expect the attitudes formed after direct exposure to a smart fridge to be more valid as predictors of eventual intention to use such a system. Furthermore, several studies showed that people make incorrect predictions when asked about their future feelings in hypothetical situations [25,26]. Therefore, asking people how they would feel interacting with a smart fridge only after a scenario description also bears the risk of limited validity. The simulation approach partly remedies these problems by providing a realistic experience from which people can judge their attitudes. Simulation studies can combine the advantages of a controlled laboratory setting with added realism of a real world experience [27]. Most likely it is the quality of simulations that will determine how valid people's evaluations are in the end.

In the field of workplace technology acceptance, Davis et al. [28] challenged the view that scenario-based predictions can suffer limited predictive validity. They argued that usefulness and intention to use as opposed to ease-of-use predictions are accurate even without hands-on experience with a mainframe computer system and empirically tested this view. Their model, as depicted in a simplified form in Figure 1 was empirically supported allowing for the interpretation that acceptance ratings can validly be obtained already after reading a scenario without the need of further prototype testing. We argue that in the context of radically innovative household products predictions are less valid on the basis of simple scenarios, because it could be difficult for people to imagine the usefulness and emotional factors in interaction with such a system. Therefore the Venkatesh et. al. model will be replicated in this study.

2   The Smart Fridge Simulation

2.1   The Program

The smart fridge simulation is a PHP-based database system with access to a MySQL-database. The database stores characteristics of 350 groceries. These

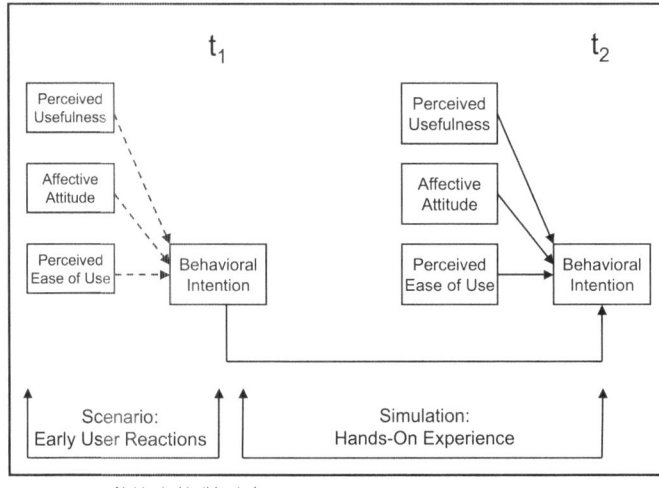

Fig. 1. Research Model, simplified and modified after [28]

groceries can be added to a person's household and subsequently be consumed. The simulation proceeds in rapid motion. A user indicates for each meal the groceries that she wants to eat. Having assembled the respective amounts of food, the consumption is analyzed, feedback is given and the next meal follows. Participants specify their meals for two consecutive weeks in the simulation, which takes about one hour real-time. Every Monday and Thursday participants go shopping in a virtual supermarket, which is designed as a typical online shopping site. After the purchase of an article the respective amount of food is integrated into the household and can be subsequently consumed. The simulation is described in more detail in [44].

2.2   Assistance Functions

The assistance functions offered in the simulation have been derived from a taxonomy in [11]. Wandke deduces these categories from the phases of human action: motivation and activation, perception, information integration, decision making, action execution, and feedback of results. These assistance function categories were mapped to the example of the smart fridge in order to derive meaningful potential assistance functions. Furthermore, the mapping of assistance functions to feasible designs considered the existing literature on smart fridge prototype applications [29,30,31] in order to maximize realism of the assistance functions. The results of the mapping are displayed in Table 1.

From this list of potential assistance functions, four were selected and underwent empirical testing: Best-before dates were displayed in a separate column when participants selected groceries for consumption, feedback about health and economy of the actual consumption were displayed on a separate screen

Table 1. Assistance Functions

Action stage	Assistance Function	Used in Simulation
Motivation, Activation	Recipe Planer	Group 2
Perception	Display of best-before dates	Group 3
Information integration	Categorization and Summation of Items	Group 4
Decision making	Selection of Auto-Replenishment Items	No
Action Execution	Auto-Replenishment	all Groups
Feedback of Results	Nutrition Health and Economy Information	Group 5

after each consumption. The recipe planer was available at shopping trips and on weekends in order to select meals to prepare. For further information on the implementation of the assistance functions, see [44].

3 Study

3.1 Study Design / Procedure

In order to test the perception and evaluation of the various assistance functions, a five-group between-subjects design was chosen. Participants were randomly assigned to one of these groups. The groups differed in terms of the assistance function offered. Every group was introduced to only one assistance function, the mapping of the groups to assistance functions is displayed in Table 1. Group 1 interacted with a simulation that had no assistance function included in order to serve as a baseline model against which to evaluate all other models. For the study, participants arrived at the Institute of Psychology at Humboldt University Berlin. Before interacting with the simulation they filled in questionnaires about their technical proficiency and general attitudes about technology. After that, participants read a scenario about the smart fridge and answered all evaluative questions (t_1). The scenarios and questionnaire items are provided in the appendix. Afterwards they were introduced to the simulation. An introduction explained in detail how to use the simulation. Then, participants interacted with the simulation until the simulated two-week period was over. Subsequently all attitude measures as presented in 3.3 were answered again (t_2). The participants returned for a second interaction session, those results are not included in this report.

3.2 Participants

Participants were recruited via courses and mailing lists at Humboldt University Berlin, as well as via smalladvertisements on the Internet. 105 subjects participated in the study. The age of the participants was restricted to be between 20 and 30 years, a gender distribution of 50 % female and 50 % male was established in every group. Participants were highly educated (90 per cent having obtained A-levels and higher), had low to medium incomes (88 % had less than 1000 €/month net income), lived mostly in shared flats (50 %) or alone (28 %), and were single or in unmarried partnership (95 %).

3.3   Measures

Participants were surveyed in a number of measures concerning their evaluation of the smart fridge at two times during the course of the experiment. Most of the measures were taken from the literature on technology acceptance. All items are given in the appendix.

Usefulness and Ease of Use. The perceived usefulness and ease of use of the technologies were measured using adapted items from the original TAM research studies [16]. Both constructs were measured by three 5-point Likert-scale items respectively.

Intention to use. Intention to use the system was measured using a set of three self-developed items, each on a 5-point Likert-scale. The items were based on the results of a qualitative investigation of people's coping with new technologies [32]. The items were formulated directly encompassing the perceived intention to actively use vs. reject the respective technology. They have been shown to be reliable in former studies [45].

Affective Attitude. Affective attitude toward the technology was measured using three 9-point semantic differential scales adapted from [33], representing the pleasure scale in their model of reactions to environments.

Covariates General technical competence (tech) was tested using the KUT questionnaire [34]. The scale consists of 8 items on five point Likert scales. Participants' sense of presence in interaction with the simulation was tested using the IPQ [35], a questionnaire measuring the presence components spatial presence(SP), involvement(INV), and realness(REAL) on separate scales. A slightly shortened version was used comprising 10 items.

4   Results

All measures of a given construct were tested on their internal consistency using the Cronbach alpha coefficient [36]. As shown in Table 2, all constructs exhibit satisfying levels of internal consistency. The alpha values of the Intention to Use scales are diminished but according to [37] satisfying for exploratory research.

4.1   Absolute Evaluations of Assistance Functions

The evaluations of the smart fridge simulation as well as results of Analyses of Variance with the factor "experimental group" are given in Tables 3 and 4. Generally all groups are neutral to positive about the smart fridge. They regard the system as useful, easy to use, and would slightly tend to use it, if already on the market. As can be seen in Table 4 the evaluations of the different smart fridge models do not differ in terms of usefulness, ease of use, and intention to use but they differ significantly between groups in terms of affective attitude. The group "best-before dates" and "nutrition feedback" felt positive interacting with

Table 2. Cronbach alpha coefficients of constructs

Measure	t1 (after scenario)	t2 (after simulation)
Usefulness	.7979	.8465
Ease of Use	.8841	.9067
Affective Attitude	.7618	.8601
Intention to Use	.6136	.6125
KUT	.8995	

the fridge. Though statistically not significant, group 2 (recipe planer) tended to evaluate the simulation worse than group 1 (baseline) in the intention to use, which means that integrating the recipe planer has a negative effect on overall product appreciation.

Table 3. Evaluations of the Smart Fridge after the Scenario (t1)

Group	1		2		3		4		5		ANOVA	
	M	SD	M	SD	M	SD	M	SD	M	SD	F	Sig.
Usefulness	4.02	0.84	3.88	0.84	4.14	0.68	4.10	0.88	4.17	0.54	.486	.746
Ease of Use	4.03	0.82	3.92	0.64	3.89	0.65	4.17	0.84	4.10	0.63	.568	.687
Affective Attitude	6.18	1.35	6.27	1.13	6.67	0.71	6.73	1.47	6.70	1.09	1.02	.401
Intention to Use	3.67	0.75	3.21	0.80	3.93	0.61	3.33	0.76	3.61	0.71	3.24	.015

Note: df for all analyses of variance = 4

Table 4. Evaluations of the Smart Fridge after the Interaction (t2)

Group	1		2		3		4		5		ANOVA	
	M	SD	M	SD	M	SD	M	SD	M	SD	F	Sig.
Usefulness	3.67	0.86	3.58	0.95	3.42	1.09	3.70	1.20	3.92	0.85	.676	.610
Ease of Use	4.10	0.81	3.95	0.88	4.17	0.52	4.29	0.75	4.30	0.80	.748	.561
Affective Attitude	5.14	1.08	5.29	1.42	6.26	1.52	5.81	1.53	6.13	1.55	2.56	.043
Intention to Use	3.62	0.66	3.20	0.83	3.74	0.71	3.38	0.79	3.49	0.65	1.76	.142

Note: df for all analyses of variance = 4

4.2   Scenario vs. Simulation

Comparing Tables 3 and 4 shows that evaluations of the smart fridge dropped after the interaction. People judge the fridge less useful ($t = 5.38, df = 105, p = .00$) and especially their affective reactions turn out to be worse than expected beforehand ($t = 5.97, df = 103, p = .00$). On the other hand, the evaluations concerning ease of use increase, which hints to the fact that most people imagine the smart fridge more difficult to use than it actually is. This difference, however, is very small and consequently not significant. The intention to use the smart fridge is not significantly affected by interacting with the simulation.

4.3 Structural Relationships

To test whether the evaluations of the smart fridge after the scenario are related to their counterparts after interaction with the simulation a multiple regression analysis was computed including the factors from Figure 1. The results are displayed in the leftmost section of Table 5 (Model 1). Predictors included were the intention to use the system as specified after having read the scenario and the usefulness, ease of use, and affective attitude ratings that were made after interaction with the simulation. The most important factor explaining the intention to use the smart fridge is the intention to use it at t_1. Unlike [28], however, the evaluative statements obtained after the simulation are also predictive of the intention to use. The most important of these variables is perceived usefulness. At the same time, the pleasure a person feels while interacting with the simulation is also a significant predictor of the intention to use. Ease of use of the system, however does not influence the intention to use. Without these three variables the regression model accounted for 40.7% variance in the intention to use, including usefulness, pleasure, and ease of use results in 58.8% variance explained, an increase of 19.2%.

Table 5. Multiple Regression Analysis

	B	SE B	β		B	SE B	β
Model 1				Model 2			
Constant	.178	.344		Constant	-.123	.387	
IntUse (t1)	.469	.070	.469**	IntUse (t1)	.441	.072	.442**
Usefulness	.209	.064	.288**	Usefulness	.221	.066	.304**
Pleasure	.092	.043	.187*	Pleasure	.072	.046	.148
Ease of Use	.082	.072	.081	Ease of Use	.070	.086	.069
				Gender	.071	.104	.049
				Tech	.001	.078	.001
				SP	.045	.059	.063
				INV	.064	.053	.087
				REAL	.052	.073	.056

Note: $R^2 = .588$; $\Delta R^2 = .025$; $*p < .05$; $**p < .01$

4.4 Covariates

Gender, general technical competence, and sense of presence in using the simulation were tested for their moderating effects on the evaluations of the smart fridge after the simulation. T-Tests for independent samples tested whether there are significant gender differences in any of the variables. None of variables showed significant differences between female and male users of the smart fridge. Furthermore we tested for general technical competence in order to find out, whether this has an effect on the appreciation of the smart fridge. Technical competence was higher for men ($M = 4.14, SD = .63$) than for women ($M = 3.49, SD = .80; t = -4.59, df = 104, p < .01$). Finally, sense of presence was tested for its effect on the evaluation of the smart fridge after interaction with the simulation. Factor analysis of the items resulted in three factors, see the appendix for factor loadings. The items pertaining to each factor were tested for their internal consistency to form a scale ($\alpha_1 = .8367, \alpha_2 = .8617, \alpha_3 = .7528$) and subsequently averaged to preserve original scale metrics. The three resulting factors were named "spatial presence"(SP), "involvement"(INV) and "realness"(REAL), following [35]. They were moderately correlated ($r_{1/2} = .40, r_{1/3} = .49, r_{2/3} = .24$). An analysis of variance with the five-level factor "experimental group" (see Table 6) confirmed that there were significant differences between the groups concerning the factor "involvement"(INV). The factor "realness"(REAL) showed almost significant differences between groups and the factor "spatial presence"(SP) approached significance.

Table 6. Sense of Presence by Experimental Group

Group	1		2		3		4		5		ANOVA	
	M	SD	M	SD	M	SD	M	SD	M	SD	F	Sig.
Factor												
SP	2.79	1.14	2.29	0.82	3.02	0.98	2.35	1.02	2.65	1.01	2.03	.09
INV	2.80	1.10	2.32	0.88	3.20	0.99	2.59	0.83	2.72	0.93	2.46	.05
REAL	2.07	0.76	1.95	0.72	2.52	0.95	1.86	0.73	2.12	0.54	2.40	.06

Note: df for all analyses of variance = 4

In order to find out whether the covariates affect the intention to use, the multiple regression reported in section 4.3 was repeated with gender, technological competence, and sense of presence as additional predictors as shown in the rightmost part of Table 5. The R^2-change was non-significant, confirming that the inclusion of the covariates indeed did not increase the explanatory power of the regression model. Consequently, none of the adjusted beta-coefficients in the regression equation is statistically significant.

5   Discussion

This study investigated people's evaluations of a smart fridge offering different assistance functions to them. Generally, participants were neutral to positive about the smart fridge. They regarded the system as useful, easy to use, and would slightly tend to use it, if already on the market. Participants estimated their likely reactions to a smart fridge, both, before and after interacting with a simulation of it. Results have shown that despite the fact that the intention to use such a system remains stable after interacting with the simulation, usefulness and

Table 7. Presence Items with Factor Loadings

| | | Factor Loadings | | |
	Item	INV	SP	REAL
SP1	Somehow I felt that the virtual world surrounded me.		.797	
SP2*	I did not feel present in the virtual space.		.851	
SP3	I felt present in the virtual space.		.775	
INV1*	How aware were you of the real world surrounding while navigating in the virtual world? (i.e. sounds, room temperature, other people, etc.)?	.823		
INV2	I was not aware of my real environment.	.855		
INV3*	I still paid attention to the real environment.	.859		
INV4	I was completely captivated by the virtual world.	.734		
REAL1*	How much did your experience in the virtual environment seem consistent with your real world experience ?			.817
REAL2*	How real did the virtual world seem to you?			.838
REAL3	The virtual world seemed more realistic than the real world.			.701

Note: varimax-rotated PCA, factor loadings < .4 are not displayed, from [35]

affective reactions are negatively affected by interacting with it. This reaction can be interpreted as the participants' disappointment about the apparent dullness of the smart fridge. Because they were confronted with only one assistance function, their expectations might not have been fulfilled. It can be hoped that with a model including *all* the functions under focus the appreciation would increase. Of course, interaction effects could come into play then, resulting in a diminished overall appreciation, because the product is overwhelmingly complex.

The question has been investigated, whether the information contained in a scenario suffices to explain intention to use after interaction with the simulation. This is not the case. Participants' experiences in the simulation contribute nearly 20 % to the explanation of the behavioral intention. This stands in contrast to [28]. We suspect that this difference is due to the fact that the smart fridge, as most other smart home technologies is a "really new product" that can only be insufficiently judged before direct contact. Furthermore, it is a product for voluntary use outside the workplace that also has to be purchased prior to usage. All these differences render it unlikely that users' acceptance can be validly forecasted by help of scenario methodology. Rijsdijk and Hultink [38] see the same limitation in their scenario based evaluation of smart home devices.

Turning to structural relationships, the present study showed on the one hand that usefulness remains the most important predictive variable for the acceptance of the smart fridge, as in traditional workplace technology acceptance literature [16]. On the other hand, however, we learned that pleasure felt during interaction with the simulation is also a valuable predictor, underlining the importance of emotion in the acceptance of household technology. Furthermore it was found that ease of use's impact vanishes completely, even after interaction with the simulation. This probably is due to the characteristics of the so called "calm computing" [39], the acting of technology in the background.

Interestingly, people's evaluations differed between the groups, confirming the hypothesis that smart fridge functions are differently appreciated. Nutrition and healthy lifestyle feedback are evaluated most positively, whereas the recipe planer flops. An anecdotic finding can be added here: in preparation of the smart fridge simulation extensive interviews were lead with households of various family and age compositions. Especially older participants were impressed by a feature offering them easy monitoring of their medically prescribed diets (e.g. diabetic). This feature was valued so highly that it completely outweighed older interviewee's reservations concerning another "high-tech" device in their home.

Integrating the recipe planer resulted in a more negatively evaluated smart fridge model than the baseline model. In the baseline model, the amounts of groceries appeared unordered, without best-before dates and nutrition information on the screen. However, this model was still rated better than the one offering recipe planing functions. This surprising result might have occurred out of two reasons: firstly because of the limited flexibility of the recipe planer. The system included relatively few, namely 274 recipes. Furthermore, recipes were only provided if the exact ingredients required for preparation were available in the household, but individual ingredients, e.g. a certain type of vegetable could not be substituted by, say, similar vegetables. This could be a reason for the comparatively negative evaluation of this feature. Secondly it could be that participants saw this model as one that offers a superfluous extra function, whilst leaving out the basic functions that would be really necessary. Clearly, this is a point for further investigation. Furthermore it has to be kept in mind that this result only approaches statistical significance ($t = 1.84, df = 41, p = .07$) and could therefore be due to random variation between the groups.

Presence was included as a covariate in the regression analyses, because it was expected that the feeling of being immersed into the simulation could play a role in evaluating the fridge. This proved not to be the case. Even though there was some variation between the experimental groups in the different aspects of sense of presence, this variation did not impact the evaluation of the simulation. This is a promising result in evaluating the external validity of experimentally simulating the smart fridge. The presence ratings ranged from low to medium. From our point of view, this result is satisfying given the fact that the system simulated only the logistic aspects of daily food consumption and shopping, but not the processes of food preparation and eating. [40] measured sense of presence as a moderator of the impression of self-motion in a photorealistic environment. Their presence scores, ranging from 2.5 to 4 are somewhat higher than in our study. On the other hand, however, their virtual reality environment was not interactive and had no function rather than displaying a photographic scene. It is acknowledged that the present approach remains an artificial procedure compared to field tests of prototypes. The strongest argument for field testing such a device may be long-term effects of adaptation between user and device - in both directions - that cannot be captured by the present methodology. From the author's point of view, however, the gain in reliability of the findings by using a controlled, laboratory procedure and testing more than 100 prospective users outweighs this drawback.

A limitation of the present research is its exclusive reliance on a highly educated, homogeneous sample. Thereby the results of this study may not be generalized to the entire population. However, it is very likely that with the present course of development, this group of persons will be the ones that will have the opportunity to buy a smart fridge in their 30s or 40s, when they also have sufficient economic backgrounds. These aspects render the group of 20-30 year old people an attractive target group for smart home acceptance research. It would be very desirable, however, for future studies to intent to replicate the findings of the present study with a sample that is representative of the population.

5.1 Marketing Implications

From a marketing perspective, however, it should be kept in mind that the smart fridge to many people may be a so-called "really new product", a radical innovation from traditional practices. For this class of products [41] proposed to use "information acceleration" strategies, among them product simulation, in order to receive valid product appreciation data from customers. Evaluation of such products has been shown to be facilitated by giving clues to multiple analog categories [42]. In the case of the smart fridge, it would therefore be helpful to present the device as a mixture of fridge, storage management system (like in stock keeping in logistics) and health and lifestyle companion. Generally, results of the present study suggest that the fridge is valued for several different reasons by its prospective users, and this even in a rather homogeneous group of people. This implies that *one prototypical* smart fridge is not a feasible option for development. It would be more promising to develop a variety of models for distinct target groups. It is estimated that in the case of the smart fridge the overlap of functions will be much lower compared to mobile phones for example, which offer similar functions to all user groups.

6   Outlook

A major part of the resources of the present research project was spent on programming the smart fridge simulation. By help of this application it is possible to investigate a much broader spectrum of questions than have been the focus of this study. A few next steps are:

- Investigating the reactions of older people.
- Implementing more than one participant and thereby simulating shared usage of the smart fridge, e.g. in families.
- Including consumption and shopping data into the analysis of acceptance: It could for instance be that people who habitually eat very healthy do appreciate a reinforcing feedback mechanism more than people who live rather unhealthy and are constantly parented by the technology.
- Simulating and testing the acceptance of automatic replenishment.

We hope to acquire valid forecasts of smart home technology and to provide guidelines as to how this class of technologies has to be designed to provide the greatest benefits to its prospective users.

References

1. Edwards, W., Keith; Grinter, R.E.: At home with ubiquitous computing: Seven challenges. In: Abowd, G.D., Brumitt, B., Shafer, S. (eds.) UbiComp 2001. LNCS, vol. 2201, Springer, Heidelberg (2001)
2. Venkatesh, A.: New technologies for the home−development of a theoretical model of household adoption and use. Advances in Consumer Research 24, 522–528 (1997)
3. Mateas, M., Salvador, T., Scholtz, J., Sorensen, D.: Engineering ethnography in the home. In: Human Factors in Computing Systems, Vancouver, BC, pp. 283–284 (1996)
4. Swan, L., Taylor, A.: Notes on fridge surfaces (2005). In: Proceedings of: Conference on Human Factors in Computing Systems, pp. 631–639 (2005)
5. Hanson-Smith, V., Wimalasuriya, D., Fortier, A.: NutriStat: Tracking young child nutrition (2006). In: Proceedings of: CHI 2006 - Conference on Human Factors in Computing Systems, pp. 1831–1836 (2006)
6. Ju, W., Hurwitz, R., Judd, T., Lee, B.: Counteractive: an interactive cookbook for the kitchen counter (2001). In: Proceedings of: CHI 2001 - Conference on Human Factors in Computing Systems, pp. 269–270 (2001)
7. Bauer, J., Streefkerk, K., Varick, R.R.: Fridgets: digital refrigerator magnets (2005). In: Proceedings of: CHI 2005 - Conference on Human Factors in Computing Systems, pp. 2060–2064 (2005)
8. LG electronics: Grd-267dtu - digital multimedia side-by-side fridge freezer with LCD display (2007), http://www.lginternetfamily.co.uk/fridge.asp (retrieved: 9/15/2007)
9. Roemer, K., Schoch, T., Mattern, F., Duebendorfer, T.: Smart identification frameworks for ubiquitous computing applications. Wirel. Netw. 10(6), 689–700 (2004)
10. Roussos, G., Tuominen, J., Koukara, L., Seppala, O., Kourouthanasis, P., Giaglis, G., Frissaer, J.: A case study in pervasive retail. In: Proceedings of the 2nd international workshop on Mobile commerce, Atlanta, Georgia, pp. 90–94 (2002)
11. Wandke, H.: Assistance for human-machine interaction: A conceptual framework and a proposal for a taxonomy. Theoretical Issues in Ergonomics Science 6(2), 129–155 (2005)
12. Zhang, P., Li, N.: The importance of affective quality. Communications of the ACM 48(9), 105–108 (2005)
13. Hassenzahl, M., Platz, A., Burmester, M., Lehner, K.: Hedonic and ergonomic quality aspects determine a software's appeal. In: CHI 2000 - Conference on Human Factors in Computing Systems, The Hague, Amsterdam, ACM, New York (2000)
14. Kim, J., Lee, J., Choi, D.: Designing emotionally evocative homepages: An empirical study of the quantitative relations between design factors and emotional dimensions. International Journal of Human-Computer Studies 59(6), 899–940 (2003)
15. Norman, D.A.: Emotional Design - Why we love (or hate) everyday things. Basic Books, New York (2004)
16. Davis, F.: Perceived usefulness, perceived ease of use, and user acceptance of information technology. MIS Quarterly 13(3), 319–340 (1989)

17. Venkatesh, V., Morris, M., Davis, G., Davis, F.: User acceptance of information technology: toward a unified view. MIS Quarterly 27(3), 425–478 (2003)
18. Sun, H., Zhang, P.: The role of moderating factors in user technology acceptance. International Journal of Human-Computer Studies 64, 53–78 (2006)
19. Brown, S., Venkatesh, V.: Model of adoption of technology in households: A baseline model test and extension incorporating household life cycle. MIS Quarterly 29(3), 399–426 (2005)
20. Moon, J.W., Kim, Y.G.: Extending the TAM for a world-wide-web context. Information and Management 38, 217–230 (2001)
21. Lederer, A., Maupin, D., Sena, M., Zhuang, Y.: The technology acceptance model and the world wide web. Decision Support Systems 29(3), 269–282 (2000)
22. van der Heijden, H.: User acceptance of hedonic information systems. MIS Quarterly 28(4), 695–704 (2004)
23. Terrenghi, L.: Sticky, smelly, smoky context: experience design in the kitchen. In: Proceedings of the international workshop in conjunction with AVI 2006 on Context in advanced interfaces, Venice, Italy, ACM Press, New York (2006)
24. Regan, D.T., Fazio, R.: Consistency between attitudes and behavior - look to method of attitude formation. Journal of Experimental Social Psychology 13(1), 28–45 (1977)
25. Loewenstein, G., Schkade, D.: Wouldn't it be nice?: Predicting future feelings. In: Kahneman, D., Diener, E., Schwarz, N. (eds.) Well-being: The foundations of hedonic psychology, Sage, New York, pp. 85–105 (1999)
26. Robinson, M., Clore, G.: Belief and feeling: evidence for an accessibility model of emotional self-report. Psychological Bulletin 128(6), 934–960 (2002)
27. Axelrod, R.: Advancing the Art of Simulation in the Social Sciences. In: Conte, R., Hegselmann, R., Terna, P. (eds.) Simulating Social Phenomena, Berlin, pp. 21–40. Springer, Heidelberg (1997)
28. Davis, F.D., Venkatesh, V.: Toward preprototype user acceptance testing of new information systems: Implications for software project management. IEEE Transactions on Engineering Management 51(1), 31–46 (2004)
29. Bonanni, L., Lee, C.-H., et al.: Attention-based design of augmented reality interfaces. In: CHI 2005 extended abstracts on Human factors in computing systems, Portland, OR, USA, ACM Press, New York (2005)
30. Hamada, R., Okabe, J., et al.: Cooking navi: assistant for daily cooking in kitchen. In: Proceedings of the 13th annual ACM international conference on Multimedia, Hilton, Singapore, ACM Press, New York (2005)
31. Silva, J.M., Zamarripa, S., et al.: Promoting a healthy lifestyle through a virtual specialist solution. In: CHI 2006 extended abstracts on Human factors in computing systems, Montreal, Quebec, Canada, ACM Press, New York (2006)
32. Mick, D.G., Fournier, S.: Paradoxes of technology: Consumer cognizance, emotions, and coping strategies. Journal of Consumer Research 25(2), 123–143 (1998)
33. Mehrabian, A., Russell, J.A.: An approach to environmental psychology. M.I.T. Press, Cambridge (1974)
34. Beier, G.: Kontrollueberzeugungen im Umgang mit Technik (Control beliefs in exposure to technology). Report Psychologie 24(9), 684–693 (1999)
35. Schubert, T.: The sense of presence in virtual environments: A three-component scale measuring spatial presence, involvement, and realness. Zeitschrift fuer Medienpsychologie (Journal of Media Psychology) 15, 69–71 (2003)
36. Cronbach, L.: Coefficient alpha and the internal structure of tests. Psychometrika 16, 297–333 (1951)

37. Nunnally, J.C., Bernstein, I.H.: Psychometric Theory. McGraw-Hill, New York (1994)
38. Rijsdijk, S.A., Hultink, E.J.: Honey, Have you seen our Hamster? Consumer Evaluations of Autonomous Domestic Products. Journal of Product Innovation Management 20, 204–216 (2003)
39. Weiser, M., Brown, J.: The coming age of calm technology (1996), http://nano.xerox.com/hypertext/weiser/acmfuture2endnote.htm (retrieved on: 9/15/2007)
40. Riecke, B.E., Schulte-Pelkum, J., Avraamides, M.N., von der Heyde, M., Buelthoff, H.H.: Scene consistency and spatial presence increase the sensation of self-motion in virtual reality. In: APGV 2005: Proceedings of the 2nd symposium on Applied perception in graphics and visualization, pp. 111–118. ACM Press, New York (2005)
41. Urban, G.L., Weinberg, B.D., Hauser, J.R.: Premarket forecasting of really-new products. Journal of Marketing 60, 47–60 (1996)
42. Moreau, C., Markman, A., Lehmann, D.: what is it?: Categorization flexibility and consumers responses to really new products. Journal of Consumer Research 27, 489–498 (2001)
43. Arning, K., Ziefle, M.: Understanding age differences in PDA acceptance and performance. Computers in Human Behavior 23, 2904–2927 (2007)
44. Rothensee, M.: A high-fidelity simulation of the smart fridge enabling product-based services. In: Proceedings of Intelligent Environments. IET, pp. 529–532 (2007)
45. Rothensee, M., Spiekermann, S.: Between extreme rejection and cautious acceptance: Consumer's reactions to RFID-based information services in retail. Social Science Computer Review 26(1) (2008)

A   Appendix

A.1   Scenario Descriptions

"It is the year 2015... It has become normal that many groceries for daily consumption are not bought in stores anymore, but are delivered to the households. This saves time and effort. It is possible to order all items one would usually buy in the supermarket. For this purpose every fridge contains a smart organizer. The organizer is a small monitor attached by default to every fridge..."

Group 1 "...It recognizes by help of sensors, which groceries are still in stock in my household. The list of groceries in my household is displayed."

Group 2 "...It recognizes by help of sensors, which groceries are still in stock in my household and orders them according to categories. The ordered list of groceries is displayed."

Group 3 "...It recognizes by help of sensors, which groceries are still in stock in my household. Furthermore it recognizes the best-before dates of all groceries. Stock and best-before dates of every item are displayed."

Group 4 "...It recognizes by help of sensors, which groceries are still in stock in my household. Furthermore a recipe planer is included, which can propose recipes to me and include the necessary ingredients in the shopping list. The list of groceries in my household is displayed."

Group 5 "...It recognizes by help of sensors, which groceries are still in stock in my household. Furthermore it analyses with every consumption how healthy and economic my nutrition is and displays the respective informations. The list of groceries in my household is displayed."

A.2   Questionnaire Items

Perceived Usefulness ([16])

1. Using the smart fridge would enable me to accomplish eating and shopping more quickly.
2. Using the smart fridge would make it easier to do manage eating and shopping groceries.
3. I would find the smart fridge useful.

Ease of use ([16])

1. Learning to operate the smart fridge will be easy.
2. It will be easy to interact with the smart fridge.
3. The smart fridge will be easy to use.

Affective Attitude ([33])
"Please indicate how you ... would feel using a smart fridge (t_1) / ... felt interacting with the smart fridge(t_2)."

1. pleased / annoyed
2. happy / unhappy
3. satisfied / unsatisfied

Intention to Use (developed on the basis of [32])

1. I would not want to use such a smart fridge at all.
2. I would naturally adopt the smart fridge.
3. I would thoroughly concern myself with the smart fridge trying to master its operations.

Technical competence ([34], translated into English by [43])

1. Usually, I successfully cope with technical problems.
2. Technical devices are often not transparent and difficult to handle.
3. I really enjoy cracking technical problems.

4. Up to now I managed to solve most of the technical problems, therefore I am not afraid of them in future.
5. I better keep my hands off technical devices because I feel uncomfortable and help- less about them.
6. Even if problems occur, I continue working on technical problems.
7. When I solve a technical problem successfully, it mostly happens by chance.
8. Most technical problems are too complicated to deal with them.

Sensor Applications in the Supply Chain: The Example of Quality-Based Issuing of Perishables

Ali Dada[1,2] and Frédéric Thiesse[1]

[1] Institute of Technology Management (ITEM-HSG), University of St. Gallen,
Dufourstrasse 40a, 9000 St. Gallen, Switzerland
[2] SAP Research CEC St Gallen, Blumenbergplatz 9, 9000 St. Gallen, Switzerland
{ali.dada,frederic.thiesse}@unisg.ch

Abstract. Miniaturization and price decline are increasingly allowing for the use of RFID tags and sensors in inter-organizational supply chain applications. This contribution aims at investigating the potential of sensor-based issuing policies on product quality in the perishables supply chain. We develop a simple simulation model that allows us to study the quality of perishable goods at a retailer under different issuing policies at the distributor. Our results show that policies that rely on automatically collected expiry dates and product quality bear the potential to improve the quality of items in stores with regard to mean quality and standard deviation.

Keywords: RFID, sensors, cool chain, supply chain management, issuing policies, inventory management, retail industry.

1 Introduction

The distribution of perishable goods such as fresh meat and fish, flowers, frozen food of all kinds, etc. poses a major challenge to supply chain management. The complexity of the issue arises from the fact that not only cost efficiency, but also a maximum quality level of products in retail stores is necessary to meet customer expectancies. In reality, however, spoilage because of expired products or interrupted cool chains is a common phenomenon in the industry. On the one hand, fresh products make up about 65% of retail turnover. On the other hand, up to 30% of perishable products are estimated to become subject to spoilage at some point in the supply chain [1]. The resulting financial loss for both retailers and their suppliers is substantial. About 56% of shrinkage in supermarkets is attributed to perishables, which equals several billions of US$ in the United States alone each year [2]. The root cause for many of these problems can be found in the current practices of inventory management, e.g. flawed stock rotation [3,4].

In the context of perishables, the performance of inventory management in the supply chain depends to a large extent on the respective issuing policy that is in place at the echelons between the supplier and the store, e.g. at the retailer's distribution center. The purpose of these policies is to determine which products are picked and sent to a specific store when an order arrives. Rules that can typically be found in

C. Floerkemeier et al. (Eds.): IOT 2008, LNCS 4952, pp. 140–154, 2008.
© Springer-Verlag Berlin Heidelberg 2008

practice are classics such as 'First-In-First-Out (FIFO)', 'Last-In-First-Out (LIFO)', or simply issuing in random order. None of these policies, however, is related to product quality since quality in the sense of perceived optical appearance, microbial safety, etc. cannot be verified effectively for a large number of items or logistical units.

In recent years, however, the ongoing trends of miniaturization and price decline are increasingly allowing for the use of tiny RFID tags and sensors in inter-organizational supply chain applications. These technical artifacts are the foundation for the seamless tracking of containers, pallets, and individual sales units as well as the monitoring of a variety of environmental parameters, e.g. temperature, acceleration, humidity, and so on. These data collection capabilities, again, enable novel issuing policies based on expiry dates and quality-influencing conditions that bear the promise to address the above-mentioned issues [5].

Against this background, this contribution investigates the potential of RFID- and sensor-based issuing policies in the perishables supply chain performance. For this purpose, we develop a simple simulation model that allows us to study the quality of perishable goods at a retailer under different issuing policies at the distributor. The output parameters that we use to measure performance include a) number of unsaleable items and b) the quality of sold units. Furthermore, we consider the impact of the customer's selection criteria when deciding for a specific item.

The remainder of the paper is organized as follows. In the next section, we first provide an overview over sensor technologies in the perishables supply chain. Second, we review the existing body of literature on issuing policies. In section 4, we present our model and numerical results from our simulation experiments including a sensitivity analysis. The paper concludes with a summary and suggestions for further research.

2 Technology Background

The quality of fresh products is affected by a number of factors, including post-harvest treatments (e.g. pre-cooling, heat, ozone), humidity, atmosphere, packaging, etc. The by far most important factor that determines quality, however, is the change in temperature conditions during transport from the manufacturer to the store. From a technical point of view, temperature tracking in supply chains has basically been a well-known issue for many years. A number of different technologies are available on the market that we will shortly present in the following. Furthermore, we discuss the major differences between these classical tools and the possibilities of novel wireless technologies such as RFID and sensor tags.

The traditional means for temperature tracking in logistics is the use of chart recorders as depicted in figure 1. A chart recorder is an electromechanical device that provides a paper printout of the temperature recordings over time. Its main disadvantage – besides cost and size – is in the fact that data is recorded on paper and has to be interpreted manually, which limits its applicability if large amounts of data have to be processed automatically in an information system in real-time.

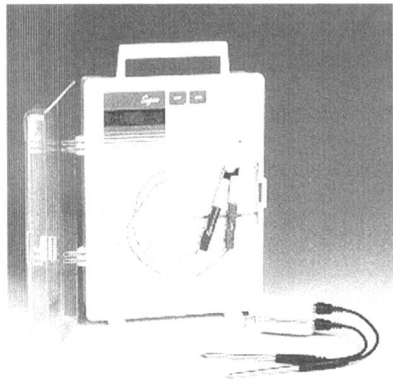

Fig. 1. Chart recorder (source: Linseis)

A second category comprises so-called "data loggers", i.e. digital or analog electronic devices with integrated sensors for measuring and tracking temperature data over time (cf. figure 2). Loggers can easily be started by pressing a key and provide a visual alert upon receiving. In contrast to chart recorders, data is stored digitally in the logger's memory. Unfortunately, data access usually requires a physical connection, e.g. via a serial cable. Accordingly, it is hardly possible to react on unexpected temperature changes in the process without interrupting the workflow. Moreover, data loggers are usually too bulky and expensive to be economically of use in many application settings.

Fig. 2. Temperature logger (source: MadgeTech)

Unlike the above-mentioned technologies, Time-Temperature Indicators (TTI) are based on chemical, physical, or microbiological reactions. TTI are inexpensive labels that show an easily-measurable time- and temperature-dependent change, which cumulatively reflects the time-temperature history of the product (cf. figure 3). The

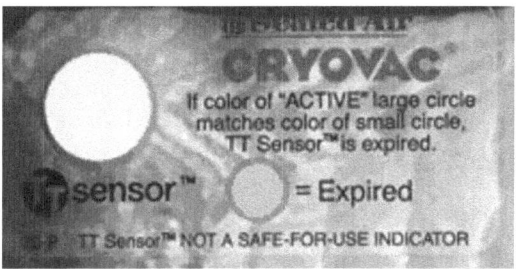

Fig. 3. TTI on a food package (source: SealedAir)

color shift on the TTI label can easily be read and understood and does not require an additional reader device. These main features, however, are also the technology's main disadvantage since the non-digital information reflects only accumulative effects, requires manual examination, and does not allow for remote monitoring.

Driven by the rise of RFID in general, a fourth technology has evolved in recent years, which combines active RFID transponders with temperature sensors. These integrated sensors continuously record temperature readings and store it in the tag's memory (cf. figure 4). As with other RFID data on the tag, the temperature data can be accessed by an RF reader at any point in the process and forwarded to an organization's information systems. In contrast to other tools, RFID-based sensor tags allow for fully automatic data collection in real-time, which in principle enables the retailer to react on environmental changes before products become unsaleable. In today's retail supply chains, deployments of sensor tags are nevertheless still rare. On the one hand, relatively high tag prices are the consequence of the cost of the energy supply that is needed to power the active sensor. On the other hand, the need to reduce power consumption leads to the implementation of low-power sensors, which do not achieve the same level of accuracy as their traditional counterparts. However, both issues are likely to be resolved in the near future such that the use in the context of logistical units (e.g. pallets and cases) becomes economically feasible along the entire supply chain.

Fig. 4. Temperature sensor integrated with an RFID tag (source: KSW-microtec)

3 Perishables in the Supply Chain

The management of perishables in supply chains has been a research issue since the early 1970s in operations management literature and beyond. The background of these contributions is not only in the food supply chain, but also in entirely different application domains such as blood banks. In the following, we give a short overview of related works in this area that are relevant to our research.

Pierskalla and Roach [10] were among the first to discuss issuing policies such as order-based FIFO and LIFO. The authors study issuing policies for a blood bank, with random supply and demand. They show that FIFO is the better policy for most objective functions, maximizing the utility of the system and minimizing stockouts. Jennings [9] sets his focus on inventory management of blood banks as well. He showed the importance of blood rotation among hospitals. In his analysis, he considers stockouts, spoilage, and costs. Cohen and Pekelman [11] analyzed the evolution over time of the age distribution of an inventory under LIFO issuing, periodic review, and stochastic demand. They concentrate on the analysis of stockouts and spoilage.

Wells and Singh [15] introduce an SRSL policy ('shortest remaining shelf life') and compare it to FIFO and LIFO. They take into account that, because of variations in storage temperature, items have different quality deterioration histories, which motivates their use of SRSL. Their results, however, are a little confusing: in the abstract the authors state that SRSL leads to better performance, but the figures in the paper show that SRSL has a better standard deviation but worse average quality.

Goh et al. [12] study two-stage FIFO inventory policies, where a first stage holds fresher items and the second stage holds older items. Liu and Lian [8] focus on replenishment. They consider the inventory level of an (s, S) continuous review perishables inventory, and calculate and optimize the cost functions for s and S. Chande et al. [6] focus on RFID for perishables inventory management, not issuing policies in themselves. They show that dynamic pricing and optimal order in the perishables inventory can be determined with the help of information – such as the production date – stored on RFID tags.

Huq et al. [13] define an issuing model for perishables based on remaining shelf-life and the expected time to sale. They compare it to FIFO and SIRO with regard to revenue and find a better performance in the majority of cases. Donselaar et al. [7] study the difference between perishables and non-perishables and between different categories of perishables using empirical data available from supermarkets. Ferguson and Ketzenberg [14] quantifiy the value of information shared between the supplier and the retailer on the age of the perishable items. The authors propose heuristic policies for the retailer under both conditions: no information sharing and information sharing with the supplier regarding the age of the products.

In contrast to the previous works, our contribution distinguishes between quality-based and expiry-based policies. For this purpose, we compare a total of seven issuing polices using a simple supply chain model that comprises a manufacturer, a distribution center, and a retail store. We measure performance in the sense of mean quality and standard deviation. Furthermore, we account for unsold items in our analysis. We also consider the impact of different patterns of customer behavior while selecting items.

4   Experimental Results

4.1   The Model

The supply chain we model consists of a manufacturer, a distribution center, and the retailer's store as shown in Figure 5. Customer arrival at the retailer is modeled as a Poisson process. The retailer and the distribution center manage their inventories using a (Q, R) replenishment policy, i.e. when the inventory falls below a value R, a replenishment order of Q items is placed. Lead time from the distribution center to the retailer is deterministic. The lead time from the manufacturer to the distribution center is normally distributed. By varying the initial lead times at the distribution center we are simulating different initial ages for products arriving at the distribution center. This is a plausible assumption because of delays in transportation and because of the retail supply chain where items are always manufactured and stored waiting for the orders, as opposed to build-to-order supply chains. By varying initial ages, we can distinguish between issue policies based on time of arrival (FIFO, LIFO) and those based on age (FEFO, LEFO).

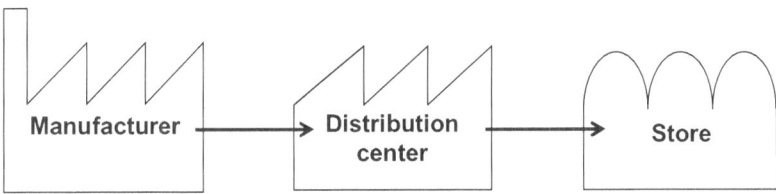

Fig. 5. The supply chain we base our model on

Because of reasons such as deterioration due to cold chain conditions, we also vary the initial product qualities at the manufacturer. We assume that the recording of temperature data by wireless sensors allows for calculating a sufficiently accurate estimate of product quality. Initial qualities are assumed to be normally distributed and then depleted when they arrive to the distribution center based on the initial lead times as discussed above. Production capacity at the manufacturer is unlimited.

We simulate seven different issuing policies at the distribution center. For each issue policy we record the qualities of sold items, calculating at the end their mean and standard deviation. We also record the number of spoiled items. The issue policies we compare are the following:

1. Sequence In Random Order (SIRO). Products in the distribution center are selected randomly and issued to the retailer.
2. First In First Out (FIFO). Products that have been longest in the distribution center are selected first.
3. Last In First Out (LIFO). Products that have been shortest in the distribution center are selected first.
4. First Expiry First Out (FEFO). Products in the distribution center are selected by their age, the items which were manufactured earlier being the first to be issued.

5. Lowest Quality First Out (LQFO). Products are selected by their quality, the items which have the lowest quality being the first to be issued.
6. Latest Expiry First Out (LEFO). Products are selected by their age; the items which were manufactured latest are issued first.
7. Highest Quality First Out (HQFO). Products are selected by their quality; the items which have the highest quality are issued first.

The workflow of the simulation algorithm is shown in figure 6. The simulation comprises a number of runs, each of which simulates all the different issue policies. For each run, we generate a sequence of customer arrivals in advance along with a sequence of initial product qualities and lead times. Thus the different issue policies are compared using the same input. The following main events happen in the supply chain:

- The distribution center and retailer regularly check to see if shipments have arrived in order to replenish their inventories.
- A customer arrives to the retailer and is served an item if the retailer is not out-of-stock
- At the end of the day, the retailer and distribution centers check their inventories and throw any spoiled items.
- When the inventory level at the retailer or distribution center goes below the threshold level, an order is placed.

4.2 Base Case

We implement the model described above in Python and analyze the results in Excel. We consider the following parameters for the base case:

- Demand at the retailer is a Poisson process with $\lambda = 30$ items per day.
- The retailer reorders products with $Q_R =60$ and $R_R =40$ items.
- The distribution center reorders products with $Q_{DC} =120$ and $R_{DC}=80$ items.
- Lead time from manufacturer to distribution center is normally distributed with mean = 1.5 days and standard deviation = 0.8.
- Lead time from the distribution center to the retailer is 1.5 days.
- The initial quality of products upon leaving the manufacturer is normally distributed with mean 90% and standard deviation = 5%.
- Minimum quality below which products are regarded as spoiled is 50%.
- Products deteriorate according to a linear model with a coefficient of 5% quality per day. This implies that the maximum lifetime of the product is 10 days, which is a plausible assumption for products with short shelf-life, such as strawberries and figs stored at 0°C.
- The retailer performs periodic review of the quality of its products at the end of the day.
- We simulate the supply chain for 1000 days per replication.
- We conduct 50 different replications, and for each we generate a new set of customer arrival times, initial product qualities from the manufacturer, and lead times from manufacturer to distribution center.

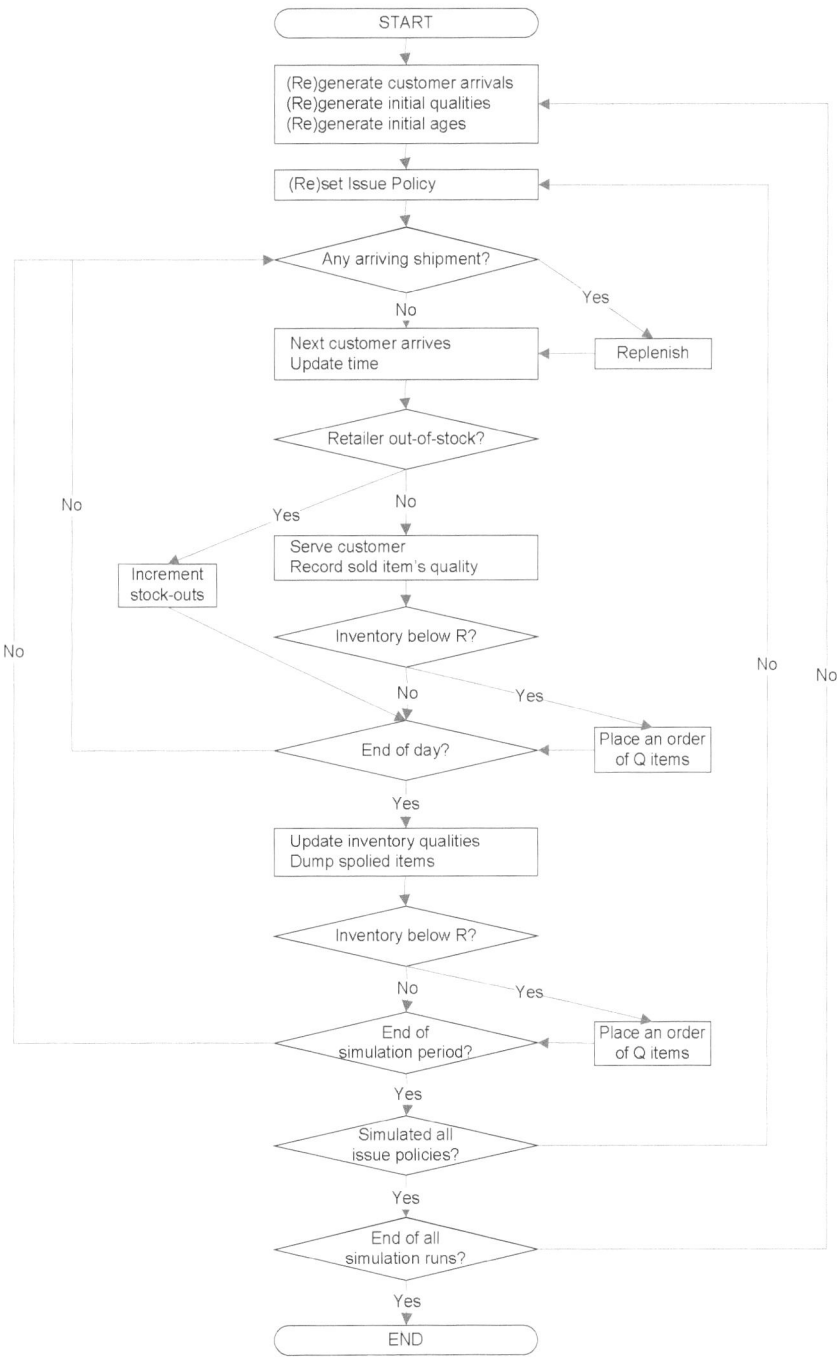

Fig. 6. Flowchart of the simulation algorithm

4.3 Results and Discussion

We first ran the complete simulation assuming that customers select the items randomly from the retailer's store, i.e. the retailer's inventory was modeled as a SIRO inventory. It could be observed that the LEFO and HQFO issue policies had the best average qualities. This was achieved on the expense of having the highest quality deviations and incurring the highest number of spoiled items, which was 25% of all the items. The LIFO issue policy shows slightly better results than LEFO and HQFO. The FIFO, FEFO, and LQFO policies showed the lowest percentage of spoiled items. The quality-based LQFO issue policy showed the least standard deviation of qualities of sold items, which was 4.5%. LQFO also showed the absolute least percentage of spoiled items at 2.6%. Table 1 provides the average qualities of sold items, their standard deviations, the percentage of unsold items due to spoilage, and the percentage of items sold at less than the threshold quality. The quality distribution curves of all issue policies are given in figure 7.

Table 1. Results of the different issue policies for a SIRO retailer inventory

Policy	Average Quality	Std. Dev.	Unsold (%)	Low quality sales (%)
SIRO	67.1	9.6	18.1	3.3
LIFO	70.3	8.6	24.3	**0.6**
FIFO	63.2	8.9	9.5	5.1
FEFO	62.4	7.2	3.4	3.0
LQFO	57.4	**4.5**	**2.6**	4.6
LEFO	71.4	10.0	25.7	1.2
HQFO	**71.9**	10.8	25.6	1.3

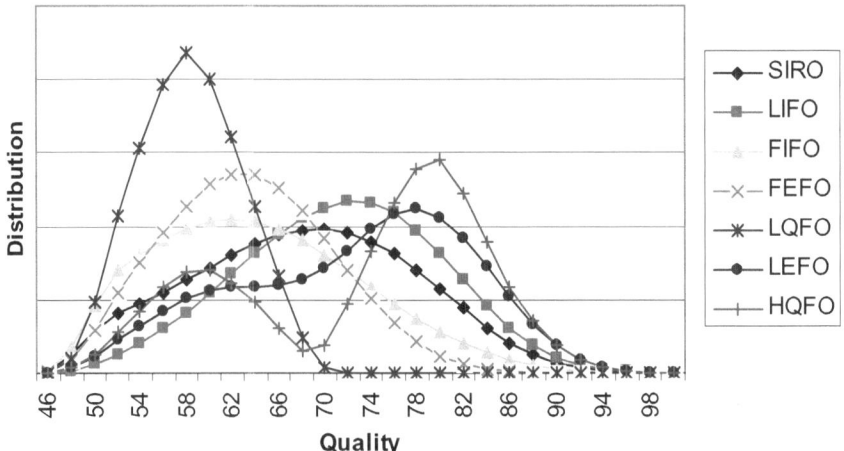

Fig. 7. Quality distribution for the issuing policies for a SIRO retailer inventory

From the results obtained, we conclude that when the primary objective of the retailer is to avoid spoilage and to sell items with qualities that vary as little as possible, the LQFO is the best issue policy. The LIFO, LEFO, and HQFO policies would be considered instead of LQFO if the retailer is willing to incur spoilage given that he can sell more items from the higher quality categories.

We then wanted to see if the results will change if the customers don't select items randomly from the store, but instead pick them based on the basis of age or quality. For this we first ran the simulation assuming that customers select items based on age, basically picking the ones with the latest expiry date first. For simulating this, we model the retailer's inventory as a Latest Expired First Out (LEFO) inventory. The results are shown in table 2 and figure 8 below. The higher standard deviations and percentage of unsold items reveal that all policies perform worse than before despite the higher average qualities of sold items. The LQFO policy is still the one with the least standard deviation of item qualities and with the least percentage of unsold items.

Table 2. Results of the different issue policies for a LEFO retailer inventory

Policy	Average Quality	St. Dev.	Unsold (%)	Low quality sales (%)
SIRO	68.4	10.4	21.7	3.3
LIFO	70.6	9.6	24.9	**1.1**
FIFO	64.2	10.0	13.5	5.7
FEFO	62.7	8.0	5.6	4.1
LQFO	57.6	**5.2**	**4.5**	6.6
LEFO	71.7	10.4	26.3	1.5
HQFO	**72.3**	10.9	26.4	1.8

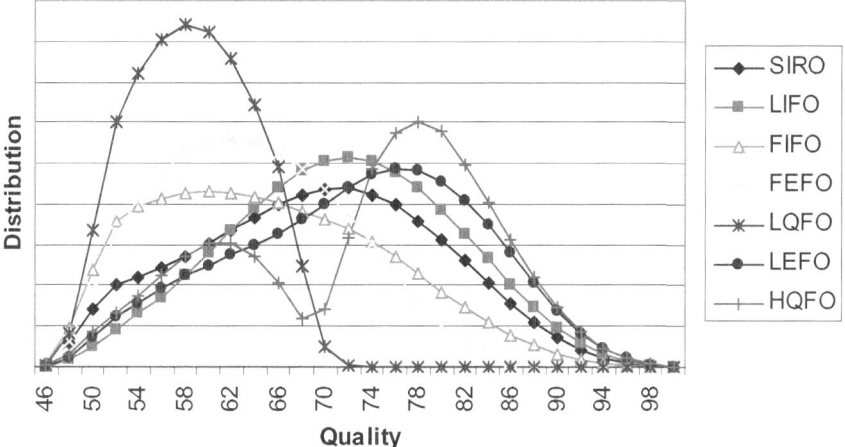

Fig. 8. Quality distribution for the issuing policies for a LEFO retailer inventory

The last customer behavior that we tested was selecting the items based on quality, picking the ones with the highest quality first. For simulating this we model the retailer's inventory as a Highest Quality First Out (HQFO) inventory. The results are shown in table 3 and figure 9. All policies show worse results when compared to the SIRO or LEFO retailer in terms of higher standard deviation, more unsold products, and more products sold that are below the threshold quality. The LQFO issue policy shows again the least standard deviation and unsold items as compared to the other policies.

Table 3. Results of the different issue policies for a HQFO retailer inventory

Policy	Average Quality	St. Dev.	Unsold (%)	Low quality sales (%)
SIRO	68.5	10.7	22.2	3.5
LIFO	70.7	10.1	25.2	**1.3**
FIFO	64.6	10.3	15.3	5.8
FEFO	63.3	8.6	7.8	4.9
LQFO	58	**5.8**	**7.2**	9.4
LEFO	72.0	11.0	27.1	2.0
HQFO	**72.4**	11.3	26.9	2.5

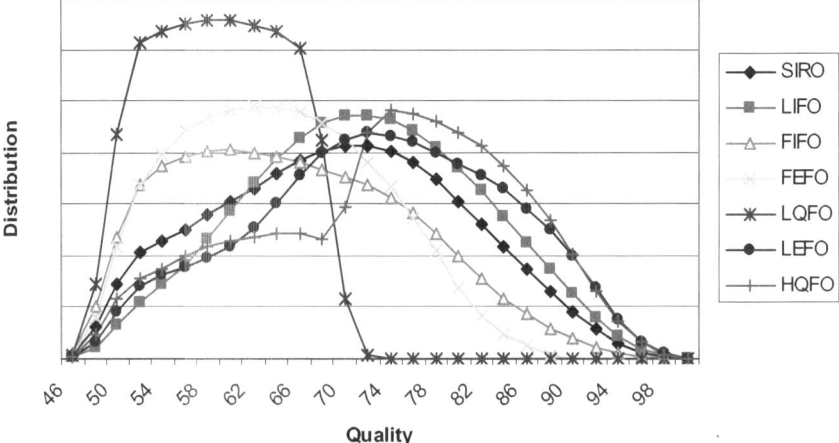

Fig. 9. Quality distribution for the issuing policies for a HQFO retailer inventory

4.4 Sensitivity Analysis

In this section we perform a sensitivity analysis to our simulation study to analyze the effect of varying the initial qualities and the rate of quality deterioration. We conduct all experiments in this section using only a SIRO retailer inventory, not including the HQFO or LEFO customer behaviors.

4.4.1 Varying the Rate of Deterioration

The rate of quality deterioration per day was set to 5% in the base case, which is equivalent to a 10-day lifetime of a product, given that the threshold acceptable quality is 50%. Here we vary the rate of deterioration by ±50%, thus running the simulation with a deterioration rate of 2.5% and 7.5%. We give the results in tables 4 and 5, respectively.

Table 4. Results after decreasing the rate of quality deterioration to 2.5% per day

Policy	Average Quality	St. Dev.	Unsold (%)	Low quality sales (%)
SIRO	76.2	8.4	1.6	0.3
LIFO	79.4	6.2	12.3	0.0
FIFO	73.4	6.2	0.0	0.0
FEFO	73.4	5.5	0.0	0.0
LQFO	73.4	**2.6**	0.0	0.0
LEFO	79.6	6.8	12.4	0.0
HQFO	**80.0**	7.6	13.1	0.0

Table 5. Results after increasing the rate of quality deterioration to 7.5% per day

Policy	Average Quality	St. Dev.	Unsold (%)	Low quality sales (%)
SIRO	62.8	9.9	35.8	9.8
LIFO	62.8	9.3	35.4	7.7
FIFO	59.7	9.3	34.0	14.8
FEFO	56.3	6.7	**28.2**	18.6
LQFO	53.5	**5.0**	30.8	26.2
LEFO	66.3	10	39.1	6.2
HQFO	**68.1**	9.8	41.7	**5.7**

We notice from the results that the LQFO issue policy performs better when the rate of quality deterioration is lower, and as this rate increases, the FEFO policy becomes the better option. When the rate of deterioration is 2.5% (cf. Table 4), the LQFO policy shows a better average quality improvement from the 5% rate (cf. Table 1) compared with the other policies. Having the same average quality as the FIFO and FEFO policies but the least standard deviation makes the LQFO policy the choice for products with low rate of deterioration. But when the rate of deterioration becomes 7.5% (cf. Table 5), the FEFO policy shows a higher average quality and lower percentage of unsold items than LQFO. In addition, the difference between the two policies' standard deviations becomes small, which makes the FEFO policy more suitable for products with high rates of deterioration.

4.4.2 Varying the Standard Deviation of Initial Qualities

The standard deviation of initial product qualities was assumed to be 5% in the base case. We study here the effect of changing this value on the selection of the issuing policy. Thus, we run the simulation twice leaving all the values as in the base case but halving the standard deviation of the initial qualities in one run and doubling it in another. We give the results in tables 6 and 7, respectively.

Table 6. Results after decreasing the initial standard deviation of qualities to 2.5%

Policy	Average Quality	St. Dev.	Unsold (%)	Low quality sales (%)
SIRO	66.9	8.9	17.6	3.2
LIFO	70.3	7.7	24.5	**0.3**
FIFO	62.7	8.5	7.4	5.1
FEFO	58.1	5.1	3.5	4.7
LQFO	57.4	**4.5**	**2.7**	4.6
LEFO	71.2	9.4	25.2	0.6
HQFO	**71.3**	9.6	25.2	0.6

Table 7. Results after increasing the initial standard deviation of qualities to 10%

Policy	Average Quality	St. Dev.	Unsold (%)	Low quality sales (%)
SIRO	67.6	10.7	20.6	4.1
LIFO	70.2	10.3	24.6	**1.9**
FIFO	64.4	9.8	14.5	5.2
FEFO	63.3	8.4	11.1	4.9
LQFO	59.5	**5.5**	**3.1**	3.5
LEFO	71.7	11.3	26.8	2.1
HQFO	**73.0**	12.4	27.2	3.0

Comparing the results shown in tables 6 and 7 with those in table 1 (5% standard deviation) reveals that for low standard deviations of initial product qualities, the FEFO and LQFO issue policies perform very close to each other; however, as the quality deviations increase, the LQFO policy shows significantly better results than FEFO. The advantage of LQFO at higher initial quality variation is due to maintaining its lower quality deviation of sold products and a lower percentage of unsold products as compared to all other policies.

4.4.3 Varying the Mean of Initial Qualities

We specified a mean initial product quality at the manufacturer of 90% for the base case. Here we run two additional simulations, first decreasing this mean to 85% and then increasing it to 95%. We give the results in tables 8 and 9, respectively, and study the changes.

The results show that for higher initial qualities (table 9), LQFO shows the best results both in terms of unsold products and low quality deviation, in addition to a

Table 8. Results after decreasing the initial mean quality to 85%

Policy	Average Quality	St. Dev.	Unsold (%)	Low quality sales (%)
SIRO	63.6	8.7	23.7	5.0
LIFO	65.7	8.2	26.8	**2.2**
FIFO	60.4	8.1	17.9	8.2
FEFO	58.7	6.5	**10.8**	7.5
LQFO	54.3	**3.9**	12.2	13.5
LEFO	67.2	9.4	27.9	3.3
HQFO	**67.9**	10.1	28.3	3.9

Table 9. Results after increasing the initial mean quality to 95%

Policy	Average Quality	St. Dev.	Unsold (%)	Low quality sales (%)
SIRO	70.6	10.3	13.9	2.5
LIFO	74.1	8.4	20.7	0.2
FIFO	66.4	9.5	3.8	2.6
FEFO	63.4	6.5	1.4	1.6
LQFO	61.6	**4.8**	**0.3**	0.4
LEFO	75.6	9.9	23.1	0.2
HQFO	**76.4**	10.5	24.4	**0.1**

reasonable average quality of sold products. For lower initial qualities (table 8), the LQFO policy incurs a higher percentage of unsold products than the FEFO policy but maintains a better standard deviation. In this case, the retailer can adopt either one of these strategies depending on its business priorities.

5 Summary and Outlook

The aim of this paper is to investigate the impact of novel sensor-based issuing policies on product quality. For this purpose, we have conducted a simulation study that compares the performance of policies that rely on quality measurements and expiry date information to classical policies that are in place in today's retail supply chains. The main results from our analysis can be summarized as follows:

- SIRO, LIFO, LEFO, and HQFO policies constantly showed high percentages of spoiled products, so the choice of best-policy was usually among FIFO, FEFO, and LQFO.
- LQFO always showed the smallest standard deviation with regard to the quality of sold items and was usually the policy with the lowest percentage of unsold items, making it the policy of choice in general.
- All policies performed worse when the customers selected items based on quality or age, but LQFO was still better than the other policies.
- The sensitivity analysis showed that FEFO's performance relative to LQFO improves under one or a combination of the following conditions:

- High rates of quality deterioration
- Low variation of initial product quality
- Lower initial product qualities

Regarding further research, our contribution may offer opportunities in various directions. On the one hand, more additional simulations might prove useful that also comprise a comparison of more complex supply chain structures, ordering policies, and product types. On the other hand, we see potential in the development of compound policies that integrate available information on quality, expiry dates, demand, etc. Furthermore, empirical works might be necessary to get a better understanding of current technology requirements in cool chains.

References

1. Scheer, F.: RFID will help keep perishables fresh. Presentation at RFID Live!, Amsterdam, October 25-27 (2006)
2. Edwards, J.: Cold Chain Heats Up RFID Adoption. RFID Journal (April 20, 2007), http://www.rfidjournal.com/article/articleprint/3243/-1/417/
3. Kärkkäinen, M.: Increasing efficiency in the supply chain for short shelf life goods using RFID tagging. International Journal of Retail & Distribution Management 31(10), 529–536 (2003)
4. Lightburn, A.: Unsaleables Benchmark Report. In: Joint Industry Unsaleables Steering Committee, Food Marketing Institute, Grocery Manufacturers of America, Washington, DC (2002)
5. Emond, J.P., Nicometo, M.: Shelf-life Prediction & FEFO Inventory Management with RFID. In: Presentation at Cool Chain Association Workshop, Knivsta, November 13-14 (2006)
6. Chande, A., Dhekane, S., Hemachandra, N., Rangaraj, N.: Perishable inventory management and dynamic pricing using RFID technology. Sadhana 30 (Parts 2 & 3), 445–462 (2005)
7. van Donselaar, K., van Woensel, T., Broekmeulen, R., Fransoo, J.: Inventory Control of Perishables in Supermarkets. International Journal of Production Economincs 104(2), 462–472 (2006)
8. Liu, L., Lian, Z. (s,S) Continuous review models for products with fixed lifetimes. Operations Research 47(1), 150–158 (1999)
9. Jennings, J.: Blood Bank Inventory Control. Management Science 19(6), 637–645 (1973)
10. Pierskalla, W., Roach, C.: Optimal Issuing Policies for Perishable Inventory. Management Science 18(11), 603–614 (1972)
11. Cohen, M., Pekelman, D.: LIFO Inventory Systems. Management Science 24(11), 1150–1163 (1978)
12. Goh, C.-H., Greenberg, B., Matsuo, H.: Two-stage Perishable Inventory Models. Management Science 39(5), 633–649 (1993)
13. Huq, F., Asnani, S., Jones, V., Cutright, K.: Modeling the Influence of Multiple Expiration Dates on Revenue Generation in the Supply Chain. International Journal of Physical Distribution & Logistics Management 35(3/4), 152–160 (2005)
14. Ferguson, M., Ketzenberg, M.: Information Sharing to Improve Retail Product Freshness of Perishables. Production and Operations Management 15(1), 57–73 (2006)
15. Wells, J.H., Singh, R.P.: A Quality-based Inventory Issue Policy for Perishable Foods. Journal of Food Processing and Preservation 12, 271–292 (1989)

Cost-Benefit Model for Smart Items in the Supply Chain

Christian Decker[1], Martin Berchtold[1], Leonardo Weiss F. Chaves[2],
Michael Beigl[3], Daniel Roehr[3], Till Riedel[1], Monty Beuster[3], Thomas Herzog[4],
and Daniel Herzig[4]

[1] Telecooperation Office (TecO), University of Karlsruhe
[2] SAP Research CEC Karlsruhe
[3] DUSLab, Technical University of Braunschweig
[4] University of Karlsruhe
{cdecker,riedel,berch}@teco.edu, leonardo.weiss.f.chaves@sap.com,
{beigl,roehr,beuster}@ibr.cs.tu-bs.de,
{t.herzog,daniel.herzig}@stud.uni-karlsruhe.de

Abstract. The Internet of Things aims to connect networked information systems and real-world business processes. Technologies, such as barcodes, radio transponders (RFID) and wireless sensor networks, which
are directly attached to physical items and assets transform objects into
Smart Items. These Smart Items deliver the data to realize the accurate real-time representation of 'things' within the information systems.
In particular for supply chain applications this allows monitoring and
control throughout the entire process involving suppliers, customers and
shippers. However, the problem remains what Smart Item technology
should be favored in a concrete application in order to implement the
Internet of Things most suitable. This paper analyzes different types of
Smart Item technology within a typical logistics scenario. We develop a
quantification cost model for Smart Items in order to evaluate the different views of the supplier, customer and shipper. Finally, we conclude a
criterion, which supports decision makers to estimate the benefit of the
Smart Items. Our approach is justified using performance numbers from
a supply chain case with perishable goods. Further, we investigate the
model through a selection of model parameters, e.g. the technology price,
fix costs and utility, and illustrate them in a second use case. We also
provide guidelines how to estimate parameters for use in our cost formula
to ensure practical applicability of the model. The overall results reveal
that the model is highly adaptable to various use cases and practical.

1 Introduction

Supply chain scenarios in logistics are an interesting field to apply information
and networking technology to objects or things. Here, embedding technology
into the application results not only in qualitative improvement - e.g. user satisfaction - but also in quantitative improvement, e.g. process optimization. By

C. Floerkemeier et al. (Eds.): IOT 2008, LNCS 4952, pp. 155–172, 2008.
© Springer-Verlag Berlin Heidelberg 2008

implementation of quantitative improvements, the technology of things goes beyond general applicability into the business domain.

This paper is largely inspired by the fact that the use of technology, namely in wireless sensor networks, pervasive computing and ubiquitous computing, allows tighter coupling of information about the overall process and the actual process status itself. This is reflected in Figure 1 showing a status of the information world, and a status of the physical world (figure is adopted from Fleisch, Mattern [1]). More complex technology obviously provides closer matching of both

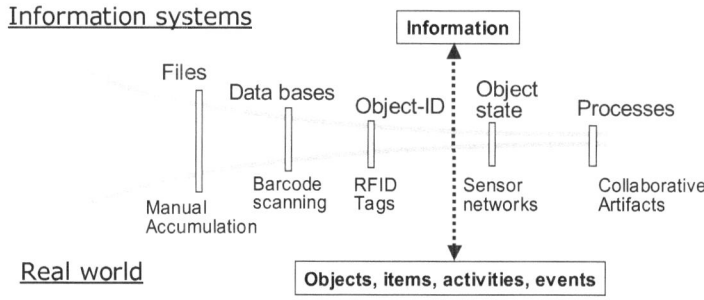

Fig. 1. Bridging the gap between the real world and information systems

worlds, while less complex technology means more fuzzy information. With today's barcode enabled business processes, mostly object-types are collected in databases. Such information offer performance measures for supervision on a process level.

This paper focuses on logistic processes, and the use of information technology in logistic processes. In this business area the use of electronics making objects and processes smart is already a concept used in some settings [2]. The use of RFID-tags for example allows acquiring knowledge about an items location and activities through reading the objects identification. This information is used to accelerate supply chain throughput, thus enabling e.g. lower intermediate inventory stock [3].

A more advanced technology can be attained by the use of sensing technology. A sensing and networking unit is added to each item of a logistic process, e.g. to each container for a chemical good transportation process or to each box of vegetables in a food supply chain process. The electronic device continuously supervises the condition of the item, and reports this information. Reporting can either be carried out continuously or on dedicated synchronization points. The most advanced technology comprises the use of Collaborative Artefacts. Collaborative Artefacts add processing power and smart behaviour to the smart sensor node that is attached to every good or item. They are able to operate independent from an infrastructure and allow spontaneous ad-hoc collaboration with other devices and computer within vicinity. Here, integration of technology allows close collaboration of items and of business processes. One example of such an application is the CoBIs project [4], where items not only deliver information. They also control the business application collaboratively together with business process systems.

1.1   Problem Statement

Such closer control is envisioned to dramatically improve the supply chain process quality. For example, perishable food commodities are prone to post harvest loss in the magnitude of 25% [5], mostly while transport. Sources from the US [6] even report that 40-50% of the products ready for harvest are never consumed - a total sum of several billion dollar per year. Application of smart items into supply chains may therefore be able to save costs in the magnitude of millions or even billions of dollars.

Although such numbers show sheer endless potential for the use of technology in a supply chain process, for any concrete logistic process, benefit has to outweigh cost to be economical feasible. To justify this we require a pre-calculation of cost and benefit. This paper will present a simple, but powerful cost model taking into account overall costs of a logistic process, including the cost for technology, but also the benefit cost when using the technology. The proposed model allows calculating and optimizing the usage of technology in logistic processes for decision makers. The model also enables decision makers to estimate benefits and to justify decisions. E.g., the model can find break-even points at what cost level technology pays-off, and it allows to find the appropriate density of technology usage for a given logistic process.

The paper is driven by applicability, and the model is thus reduced to a set of parameters that are simple to estimate in a technology evaluation process. The paper is focused at supply chain processes and ensures simplicity of use through a black box approach. This allows to only model the most important parameters and views of a supply chain process, and the use of technology therein. The paper will take three different views on the process, which are independently modeled: The supplier, the shipper, and the customer. Each of them may independently optimize it's cost function for the use of smart item technology. The cost formula developed within this paper will enable potential applicants to quantify costs and benefits of use of technology within logistic processes, and especially supply chains. It will also introduce a guideline how to approach the problem of finding parameters for the formula and describe the steps required.

1.2   Paper Overview

The paper first analyses an existing logistic scenario and discusses the use of technology in supply-chains. The scenario is used to develop the parameters used in a cost model for supply chains. In section 3, six cost models are presented, two for each of the major stakeholders in a supply chain: the supplier, the shipper and the customer. The cost model is explained in section 4 using a concrete example. Section 5 provides a short guideline how to estimate and calculate parameters for the cost model in an effective way.

2   Supply-Chain Scenario Analysis

A logistics process in a supply chain consists of planning, implementation and control of a cost efficient transport and storage of goods from a supplier to a customer according to the customer's requirements [7]. The main goal is an increase of the customer's utility while optimizing the performance of the logistics. The basic logistics functions are to transport *the right goods in the right quantity and right quality at the right time to the right place for the right price*. Information systems keep track of the logistics process and implement various techniques to enable the basic functions. Figure 2 associates the functions with the techniques used by information systems. Further it shows different Smart Item technologies and their coverage on the techniques. The information system requires to identify

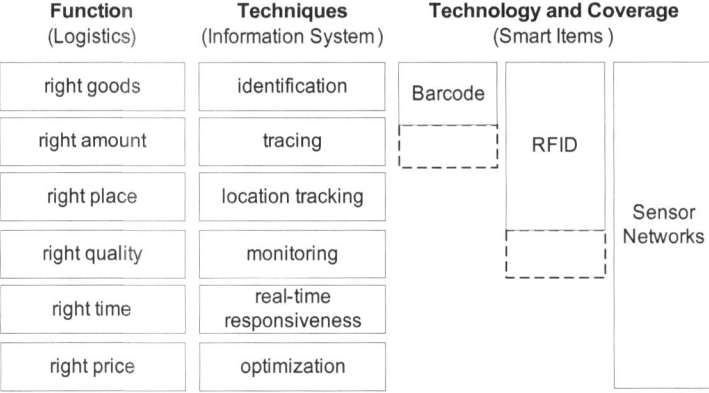

Fig. 2. Techniques of an information system to implement the logistics functions. It also shows how well three Smart Item technologies (barcode, RFID, sensor networks) cover the basic functions. Dashed areas indicate partial coverage.

goods to link electronic processes to the real item. Tracing is necessary to let the system detect when an item gets lost. As a result, it ensures that the right amount of goods is delivered. Location tracking enables the information system to keep track on the transport itself. During the transport the good is not under the control of the supplier. In order to ensure the quality of the delivered goods, an appropriate monitoring of the goods' state is necessary. Having all these data within the information system, the overall logistics process can be observed in very detail. It allows real-time actions to unforeseen events, to determine bottlenecks and it provides the basis for optimization. Finally, this will affect the price accordingly.

Various technologies have been developed for acquiring logistics data electronically directly from the good and process and then delivered to the information system. We refer to this technology as Smart Items. Depending on the technical capabilities (basic to advanced) Smart Items cover different techniques.

Barcodes are a current state-of-the-art technology for electronic identification of goods. A barcode label is attached on the goods and then optically detected by a barcode reader. The reader de-ciphers the printed identification and sends it to the information system, which updates the good's record. Barcodes can support tracing only partly. The line-of-sight requirement makes it impossible to detect single items within a pallet of goods. Solution for in-transit inspections would require a complex infrastructure. As a consequence, barcodes can only be used in loading and unload processes at the ramp at a very coarse-grained scale.

Radio Frequency IDentification (RFID) [8] is a radio-based identification technology. Line-of-sight is not necessary. This allows identification of single items within a box of items. Location tracking and tracing is possible as far as the infrastructure of RFID readers is deployed [9]. A mobile infrastructure, e.g. GSM based readers, allows even a remote identification and tracing while the goods are in transit. Novel RFID transponders acquire sensor information of the goods, e.g. temperature or pressure or shock, during the transport and enable a monitoring of goods' state. However, those sensing capabilities are very limited.

Wireless sensor networks are an upcoming advanced Smart Item technology for logistics processes. Sensor nodes are tiny, embedded sensing and computing systems, which operate collaboratively in a network. In particular, they can be specifically tailored to the requirements of the transported goods. In contrast to previous technology, which delivers data to an information system, sensor networks can executes parts of the processes of an information system in-situ directly on the items. Goods become embedded logistics information systems. For instance, CoBIs [2] presents a sensor network example of storing and in-plant logistics of chemical goods, which covers all identification, tracing, location tracking, monitoring and real-time responsiveness at once.

2.1   A Smart Logistics Example

The following example of the logistics process is derived from previous experiences in [2], [10] and [11]. This example draws a picture of a supply chain process that uses most advanced Smart Item technology. We will first present the example and then analyze the example at the end of the section.

A customer orders chemical substances from a supplier. The supplier subcontracts a shipper for the transport of hazardous chemical substances. The orders and acceptances are recorded in an Enterprise Resource Planning (ERP) system. In this scenario it is important to note that all participants are permanently informed on the state of the transport during the complete process. This is because of legal issues since supplier and shipper are commonly responsible for the safety. This logistics process is very complex because it requires the management of goods in different, potentially unforeseen situations involving different participants. As a consequence, there is a need for smart technology enabling a continuous supervision at any time and place in order to implement this management.

The chemical containers are Smart Items using wireless sensor network technology. The sensor nodes are attached to the containers, identify the containers and constantly monitor their state, e.g. temperature. Further, they establish a network between Smart Items to trace the load of all containers to deliver. The shipper provides a mobile communication infrastructure for the Smart Items with an uplink to a wide area network, e.g. GSM. As a consequence, all participants can query the state and location of their delivery. Figure 3 illustrates the smart logistics process using Smart Items. Following the eSeal approach in [10], the

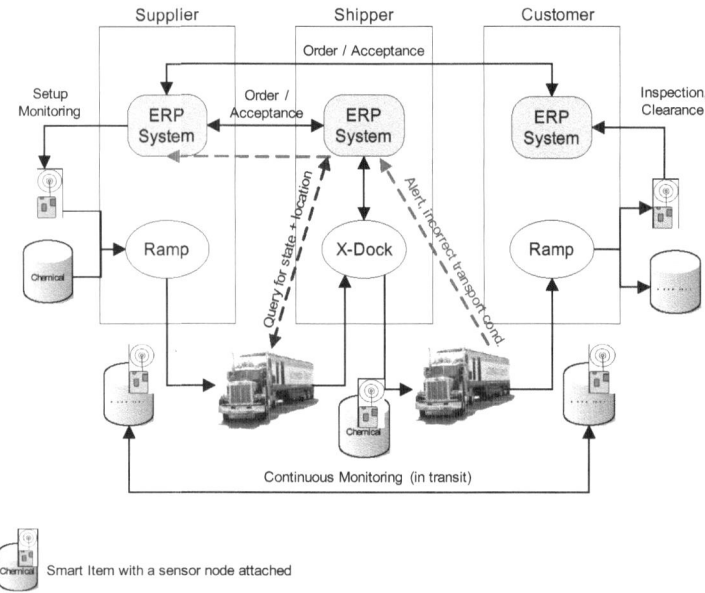

Fig. 3. Smart Items logistics process

supplier first setups and configures all Smart Items with basic transport information. It comprises container identification, destination, transport conditions, size of the delivery, and time to destination.

The shipper plans the routes separately. The different orders are summarized and re-organized in a cross-dock (X-Dock) in order to optimize the utilization of trucks. For instance, loads with the same destination packed into one truck. Other parameters and real-time conditions can also be used for optimising the supply chain. E.g., the supplier and the shipper have to ensure that no hazardous material combination, e.g. flammable and oxidizing substances, is loaded into the truck. Instead of transporting all information to both (supplier and shipper) ERP systems, the Smart Items take care of this by checking the identification of surrounding containers and environment conditions (e.g. temperature). In case of an incompatibility, the items raise an alert.

During transport the Smart Items constantly supervise the transport conditions, e.g. temperature of the chemical containers. Using an uplink to a wide area network the state of Smart Items can directly be queried and checked by the process participants in order to fulfill the safety responsibility. Further, the location can be electronically tracked. Smart Items act proactively and will raise an alert and send out a notification to a participant, if some transport condition does not hold anymore. Appropriate actions can be triggered, e.g. the truck driver is notified immediately to check the load. The Smart Items also trace their packaging. In case that the delivery is accidentally split or a container is left behind, the Smart Items will raise an alert. As a result, Smart Items ensure that the right amount of goods is delivered. All alerts are locally logged for later revision in case that the person in charge is not reached or the alert is ignored.

When the delivery reaches the customer, the Smart Items automatically inform the ERP system on the delivery and the state. Identification and amount are inspected immediately and clearance is given accordingly. If the transport conditions are violated, then a retour process will be initiated and the participant in charge will be determined through the logging data.

Finally, the usage of advanced Smart Items such as wireless sensor networks helped to prevent losses and accidents during the transport. The example shows that Smart Items can accurately supervise the entire smart logistics process.

2.2   Benefits and Deduction of Parameters for a Smart Items Based Supply Chain

We expect major improvements by the usage of Smart Items within the smart logistics process. Based on the benefits we will identify affected parameters for a Smart Items based Supply Chain model:

1. Reduction of critical situations for goods. Smart Items monitor the goods continuously and alert a person in charge when the transport conditions are not appropriate anymore. This leads to a prompt reaction, which prevents further damage or even loss. As a result, we expect a reduced ratio of defective or perished goods and a (positive) change in the
   (a) return costs
   (b) costs for removal of defective goods
   (c) lower transport costs due to lower reshipping rate and higher shipping throughput
2. Clear assignment of responsibilities. If the alert is ignored or the reaction is delayed, the Smart Items will determine and prove the damage or loss of goods. The shipper can accurately be taken into responsibility for the amount of defective goods. This allows for a transparent supply chain process and a clearer separation of costs between supplier and shipper.
3. Since the overall amount of defective and lost goods is known through Smart Items, the supplier is able to accurately conclude on the ratio of defective goods, which is inherent (and previously unknown) in his delivery. This is expected to raise consumer (customer) satisfaction.

As a consequence of the Smart Items usage, each participant in the logistics process can accurately determine its responsibility for the amount of defective goods and transport losses. This enables the potential for process optimization, but involves a complex interplay between the participants and the used technology. In the following section, we break down these relations and we quantify them based on the above analysis of the process.

3 Smart Items Quantification Cost Model

In this section we introduce our model for quantification of Smart Items usage in logistic processes. We describe a cost model of the logistics process and quantify the profit for different technological approaches ranging from simple Smart Items, e.g. barcodes, up to very complex Smart Items, e.g. wireless sensor nodes. In this investigation we adopt three different positions of the three stakeholders: supplier, shipper and customer. For all stakeholders the amount of defective and lost goods determines their profit. Therefore, our approach focuses on how Smart Items relate to lost and defective goods. Important in the model is the complexity of the different technologies from Section 2. We model it by the density ratio ρ. The more complex the technology gets, the larger is the density. Our model describes the shipment scenario in a simplified form and within an ideal environment, e.g. the error free functionality of the Smart Items, the supplier only produces one type of good and the parameters and variables are known or can be determined through a simulation or a field trial. In Table 1 the used variables and parameters are defined and explained.

3.1 Analysis from the Supplier's Point of View

First we define a simplified profit function (Equation 1) for a supplier who uses Smart Items (SI) with limited capabilities, e.g. barcode or passive RFID tags.

$$
\begin{aligned}
\prod_{perShipment}^{SimpleSI,supplier} &= \left((1-\omega)p_{good} - c_{production}\right) \cdot q_{sales} &&\text{turnover} \\
&- \omega \cdot q_{sales} \cdot c_{retour} &&\text{costs for processing} \\
& &&\text{defective good} \\
&+ \psi \cdot s \cdot q_{sales} &&\text{penalty for shipper} \\
& &&\text{for loss} \\
&- C_{fix} &&\text{fixed costs}
\end{aligned}
\tag{1}
$$

The profit $\prod$ (Equation 1) results from margins between the price of the good p_{good} and the costs of production $c_{production}$ per unit multiplied with the amount of sold units q_{sales} less the defective goods ω which were delivered to the customer. The defective goods which were delivered to the customer need to be manually addressed with costs c_{retour}. The shipper has to pay a fee s depending on the price of the good p_{good} for the ratio ψ of goods lost or not delivered in time. The fee is a compensation for costs of damage, customer dissatisfaction and loss of reputation. Additionally the fixed costs C_{fix} get deducted. The profit $\prod$ is interpreted as a profit per shipment. To model the profit under usage of advanced Smart Items (e.g. wireless sensor nodes) Equation 1 is extended to Equation 2.

Table 1. Variables and parameters for the Smart Items cost model

p_{good}	price charged for good
$c_{production}$	variable costs of production per good
q_{sales}	amount of sold/distributed goods
c_{retour}	cost of manual processing of returned goods (defective or perished)
C_{fix}	fixed costs
$C_{fix,SI}$	additional fixed costs using Smart Items (infrastructure)
$c_{operation}$	variable operational costs per Smart Item and shipment (e.g. recharge battery, programming)
c_{SI}	acquisition costs of Smart Item
s	penalty depending on cost of goods (shipper $\Rightarrow$ supplier)
$p_{transport}$	price of shipping per good (to be paid by the customer)
$c_{transport}$	variable transportation costs per good (for shipper)
$p_{special}$	additional shipping charge for usage of Smart Item per good
$c_{capacity}$	costs of capacity loss for reshipping
c_{GSM}	costs of message sent over GSM to ERP-System
F	fleet size of shipper
W	non quantifiable advantage through usage of Smart Items (consumer satisfaction, etc.)
$\rho \in (0, 1]$	factor of density, ratio of Smart Item quantity to quantity of goods
$\nu \in [0, 1]$	factor of maintenance: $\nu = 0$ all Smart Items get shipped back (reusable); $\nu = 1$ no Smart Item is returned
$\omega \in [0, 1]$	ratio of defective goods delivered to customer
$\phi \in [0, 1]$	ratio of triggered Smart Items, $0 \leq \phi \leq \omega \leq 1$
$\psi \in [0, 1]$	ratio of searched (potentially lost) goods during shipping
$\kappa \in [0, 1]$	ratio of recovered goods (previously lost)

$$
\begin{aligned}
\Pi_{perShipment}^{Adv.SI,supplier} = & \; ((1 - \omega)p_{good} - c_{production} - \rho \cdot (c_{SI} \cdot \nu + c_{operation})) \cdot q_{sales} \\
& - (\omega - \phi) \cdot q_{sales} \cdot c_{retour} && \text{costs for processing} \\
& && \text{defective goods} \\
& + \phi \cdot q_{sales} \cdot s && \text{penalty for shipper} \\
& && \text{for damage} \\
& + (1 - \kappa) \cdot \psi \cdot s \cdot q_{sales} && \text{penalty for shipper} \\
& && \text{for loss} \\
& + W && \text{not quantifiable advantage} \\
& - (C_{fix} + C_{fix,SI}) && \text{fixed costs plus SI invest}
\end{aligned}
\tag{2}
$$

An important parameter is the density factor ρ which describes the ratio of goods with Smart Items to the amount of goods without. If every good is equipped with a Smart Item, the density factor will be $\rho = 1$. The density factor is proportionally reduced the higher the number of goods per group which are equipped with a Smart Item. E.g. if there is a pallet with 20 boxes containing each 16 TFTs the resulting density factor would be $\rho = \frac{1}{320}$ if there is only one Smart Item per pallet or $\rho = \frac{1}{20}$ for one Smart Item per box. The assumption is, of course, that the goods are equally grouped and the Smart Items are also equally distributed.

The density factor directly influences the profit, as can be seen in Equation 2. Depending on the density of Smart Items, additional costs for operation, acquisition and maintenance arise. If a Smart Item is not reused, its costs will have to be paid for each shipment, which results in a maintenance factor of $\nu = 1$. In the best case of reuse the maintenance factor is $\nu = 0$, i.e. there is no abrasion or loss. Also new in Equation 2 is the parameter ϕ, which indicates the fraction of Smart Items which trigger at least one alert, i.e. at least one detection of violation of the shipment agreements. So the ratio of defective goods due to improper shipment is expressed through ϕ. The supplier does not adhere for the return costs c_{retour} and gets a penalty s per alerted good paid by the shipper. If only a small amount of goods are equipped with a Smart Item, the penalty for the shipper is high since he needs to cover the whole amount of damaged goods. Through the possibility of locating Smart Items, the ratio of shipment loss is reduced by the parameter κ and accordingly the amount of penalty s. The variable W indicates the not quantifiable advantage resulting of the use of Smart Items, e.g. customer satisfaction, positive reputation as a result of fast deliveries, optimization of the shipment process, etc.

The fixed costs C_{fix} include along the original costs the costs for acquisition of the Smart Items and the equipment for programming and reading them.

3.2   Analysis from the Shipper's Point of View

We model the profit function for the usage of low-performance Smart Items, e.g. barcode and passive RFID-Tags, as follows:

$$
\begin{aligned}
\prod_{perShipment}^{SimpleSI,shipper} = & ((1-\phi)p_{transport} - c_{transport} & \text{penalty paid to} \\
& & \text{the producer} \\
& - \psi \cdot (s + c_{capacity})) \cdot q_{sales} & \text{for loss and loss of} \\
& & \text{capacity for reshipment} \\
& - C_{fix} & \text{fixed costs}
\end{aligned} \tag{3}
$$

The profit $\prod$ per shipment results out of the shipment price $p_{transport}$ the customer has to pay, less the shipment costs $c_{transport}$ and the ratio ψ. Again, the ratio ψ indicates the loss of goods during shipment, which gets multiplied with the penalty s. In addition, the shipper has to do a subsequent delivery of the lost goods, which results in a capacity loss of $c_{capacity}$.

If advanced Smart Items are used, the resulting profit function is modeled as follows:

$$
\begin{aligned}
\prod_{perShipment}^{Adv.SI,shipper} = & ((1-\phi)(p_{special} + p_{transport}) - c_{transport}) \cdot q_{sales} & \text{turnover} \\
& - c_{GSM} \cdot (\phi + \psi \cdot 2 \cdot F) \cdot q_{sales} & \text{penalty for loss} \\
& - \phi \cdot q_{sales} \cdot s & \text{penalty for damage} \\
& - (1-\kappa) \cdot \psi \cdot (s + c_{capacity}) \cdot q_{sales} & \text{comm. costs} \\
& + W & \text{not quantifiable adv.} \\
& - (C_{fix} + C_{fix,SI}) & \text{fixed costs including} \\
& & \text{SI investment}
\end{aligned} \tag{4}
$$

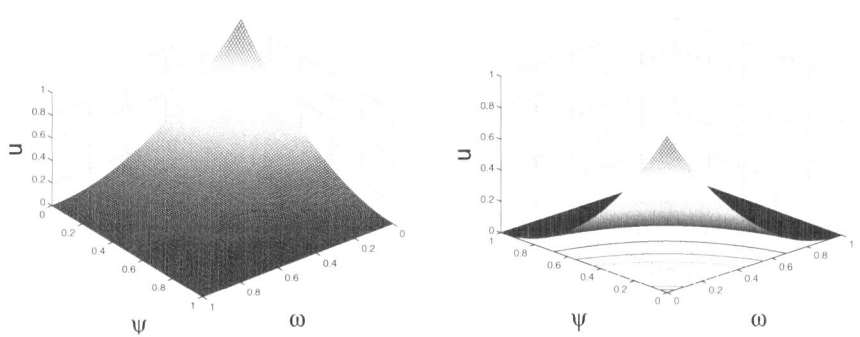

Fig. 4. Gain function $u(\psi, \omega)$ from two perspectives

Because of specialization and higher effort the shipper needs or can demand a higher price $p_{transport} + p_{special}$. The reduction of shipment loss through tracking of lost goods reduces the payments for penalties by a factor of $1 - \kappa$. For detected shipment damage ϕ the penalty s needs to be paid to the supplier. Additionally, the costs c_{GSM} arise for transmitting alerts and tracking ψ of goods. Here the worst case costs are denoted, i.e. all alerts arise outside the reach of an access point at the cross dock stations and for tracking the complete fleet needs to be addressed. The fixed costs C_{fix} comprehend the acquisition of the readers, communication infrastructure and the original fixed costs per shipment. The variable W also indicates the not quantifiable advantage, e.g. customer satisfaction.

3.3   Analysis from the Customer's Point of View

The perspective of the customer is modeled as a profit function with two dimensions, quality and completeness of a shipment aggregating the profit level u. Further values influencing the profit level, e.g. speed of delivery, are omitted for reasons of simplicity. For further extension of the profit function several additional factors can easily be included. The highest profit level is reached at the best case, when the delivery is complete and without defective goods reaching the customer. According to the previous modeling this case occurs when $\psi = 0$ and $\omega = 0$. The result is a normalized Cobb-Douglas function [12] (Equation 5) with its saturation defined in point $u(0,0) = 1$.

$$u(\psi, \omega) = (1 - \psi)^2 \cdot (1 - \omega)^2 \tag{5}$$

The Cobb-Douglas function can be seen from two different perspectives in Figure 4. We assume that the improved amount converges into point $(0,0)$. Presumably, the assignment of the budget (allocation) utilizing basic Smart Items, e.g. barcode, is (ψ', ω') and the customer pays the price $m' = (p_{good} + p_{transport}) \cdot q_{sales}$. The amount of defective goods ω can be reduced by a ratio ϕ through the use of more complex Smart Items. This relationship is shown in Figure 5. Besides, the shipment loss can be reduced through tracking by

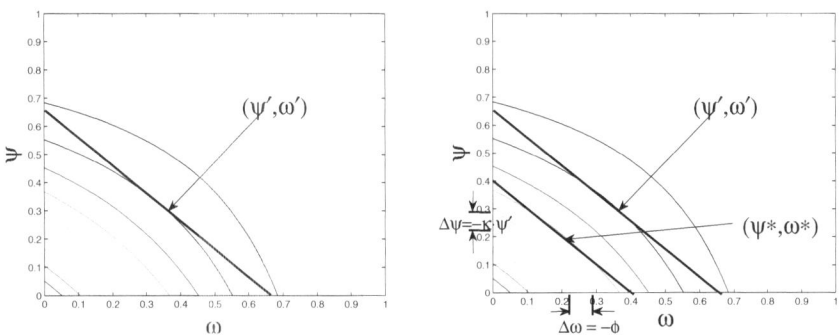

Fig. 5. Indifference curves of $u(\psi, \omega)$

Smart Items in average by $-\kappa \cdot \psi'$, which is also apparent in Figure 5. The allocation is improved from (ψ', ω') to (ψ^*, ω^*). In return, the customer has to pay the increased price $p_{special}$. In sum the costs for the new allocation are $m^* = (p_{good} + p_{transport} + p_{special}) \cdot q_{sales}$.

If the gain of profit through improved allocation is bigger than the loss of usefulness through raised prices, then the customer will have a direct advantage of the use of advanced Smart Items. Let $\widetilde{u}$ be the utility function according to the preferences of the customer that maps monetary units onto a scale comparable with u. If the following inequality evaluates to true, the customer considers the use of advanced Smart Items as beneficial compared to a technology like barcode.

$$\widetilde{u}(m^* - m') < u(\psi^*, \omega^*) - u(\psi', \omega') \tag{6}$$

4 Use Case for Our Model

In this section we will present a simple use case to exemplify the usage of our model. A supplier is selling apples to a customer, which are transported by a shipper. The parameters (e.g. costs) from the model are derived from real world data.

One box of apples holds 18 pounds. In July 2007 a box of apples cost an average of $c_{production} = 5.27\$$ [13] and could be sold for an average of $p_{good} = 17.14\$$ [14] in the United States. We will consider one shipment to be $q_{sales} = 400$ boxes of apples, which is the amount a small truck can transport. From all boxes that arrive at the customer, $\omega' = 20\%$ are rotten. The customer does not pay for boxes with rotten apples and further requires the supplier to pay $c_{retour} = 2\$$ for their disposal. Using Equation (1) we can calculate the supplier's profit when using barcodes for each delivery:

$$\prod_{perShipment}^{barcode, supplier} = 5,484.80\$ - 2,108.00\$ - 160\$ - C_{fix} = 3,216.80\$ - C_{fix} \tag{7}$$

Notice that the supplier looses 20% of his income because of rotten apples. It is also not clear at which point they got rotten (during transport or already at the supplier). To cope with this problem, the supplier decides to track the temperature of the apples during delivery using Smart Items. Every fourth box ($\rho = 25\%$) is equipped with an expensive Smart Item which costs $c_{SI} = 50\$$. The Smart Items are reusable ($\nu = 0$), so the supplier only has to buy them once. Maintenance costs for each Smart Item are $c_{operation} = 0.50\$$ per shipment, e.g. for charging batteries and programming.

Now, the shipper can be held responsible if apples get rotten because of wrong shipping and handling conditions. The tracking with Smart Items further allows the shipper to monitor the apples temperature. Therefore we assume the total amount of rotten apples will fall to $\omega^* = 10\%$. Now, only $\phi = 1\%$ of all apples get rotten because of the shipper, so he has to refund the supplier and pay a penalty making a total of $s = 20\$$ to supplier per rotten apple box.

If we consider the fixed costs to stay unchanged ($C_{fix,SI} = 0$), then Equation (2) will show the supplier's profit when using Smart Items as follows:

$$\prod_{perShipment}^{SI,supplier} = 6,170.40\$ - 2,158.00\$ - 72.00\$ + 80.00\$ - C_{fix} = 4,020.40\$ - C_{fix}.$$

$$(8)$$

The supplier's profit will increase by 803.60\$ per shipping. The one time investment of 5,000\$ for buying 100 Smart Items amortizes after 7 shipments. Now let us see how the use of Smart Items influences the shipper's business. The shipper charges the customer $p_{transport} = 4\$$ for each box shipped. His costs are $c_{transport} = 2\$$. Through Equation (3) we get the shipper's profit for each delivery with a simple Smart Items technology, such as barcode:

$$\prod_{perShipment}^{barcode,shipper} = 800\$ - C_{fix} \qquad (9)$$

When using advanced Smart Items, the supplier will charge $p_{special} = 0.50\$$ extra since he also has to sustain a Smart Items infrastructure. But he will also have to refund the supplier and pay a penalty making a total of $s = 20\$$ for damaged products. The shipper's profit calculated through Equation (4) is

$$\prod_{perShipment}^{SI,shipper} = 900.00\$ - 80.00\$ - C_{fix} = 820 - C_{fix}. \qquad (10)$$

The shipper's profit will increase by 20\$. Even though he is responsible for the goods he damages during transportation, he will also be earning more money.

Now let us consider how the use of Smart Items influences the customer. We expect him to profit from the smaller amount of rotten apples, even though he will be paying higher transport costs. When using barcodes, we get $m' = (17.14 + 4.00) * 400 = 8,456.00$ and $u(\psi', \omega') = (1 - 0.20)^2 = 0.64$. And the use of Smart Items results in $m^* = (17.14 + 4.50) * 400 = 8,656.00$ and $u(\psi^*, \omega^*) =$

$(1 - 0.10)^2 = 0.81$. We assume the following model structure for the customer's utility function: $\widetilde{u}(x) = 1 - e^{-kx}$. This utility denotes the normalized value of an additional (financial) effort x for the customer. The scalar k expresses the slope of the value of the additional effort. In this example, an additional financial effort of $m^* - m' = 200\$$ for more powerful Smart Items leads to 10% less rotten apples. This makes the delivery more valuable for the customer. However, this has to be compared with the value of an additional investment of 200\$ per shipment. Inserting the above values into Equation (6) and assuming $k = 0.9\%$ results in the following equation

$$\widetilde{u}(m^* - m') = 0.16 < u(\psi^*, \omega^*) - u(\psi', \omega') = 0.81 - 0.64 = 0.17, \qquad (11)$$

The right side of the inequality evaluates to 0.17 and denotes how much more the delivery becomes valuable for the customer. This is due to the reduction of the amount of rotten apples. The customer's additional financial effort has a value of 0.16 according to his utility function. The inequality evaluates to true. The delivery becomes more valuable than the additional effort spent by customer. Hence, the use of more powerful Smart Items pays off.

5   Guidelines for Parameter Estimation

One of the cornerstones of our model is the use of simple abstract parameters that estimate certain values within one type of supply chain. This allows us to compare various types of supply chains, e.g. traditional vs. Smart Items supported supply chains. The major problem for doing so is how to obtain these parameters in a practicable way.

We proposed to estimate the parameters using a black-box approach, as we see it difficult to measure detailed values or to uncover complex interplay within a supply chain. This approach is less sophisticated than a full-blown analysis and may be more error prone. On the other hand, the proposed method is faster and can be carried out at lower costs. Furthermore, it can be applied to supply chains where it is impossible to retrieve a detailed understanding, e.g. because not all participants in the supply chain are willing to participate. The model we present here requires only three stakeholders to work together: The supplier, the customer and the shipper. In the simplest form of the model, we consider only one instance of these stakeholders within the process.

Our proposal for the black-box oriented approach is to estimate parameters based on small trial-runs of the technology. Here, the technology is brought into the supply chain for selected items only, and parameters are continuously measured. In a first run, parameters will not be used for improving the process, but used for quantification of the factors ω and ψ. Additionally, from the calculation model of the supply chains other parameters are derived (p_{good}, $c_{production}$, q_{sales}, c_{retour}, s, C_{fix}, $c_{capacity}$). In a second run, Smart Items technology is used to additionally quantify parameters ν, ω, ϕ and ψ. From the cost calculation for the introduction of the Smart Items technology we finally project the total cost of a full-blown application of technology, and their parameter $C_{fix,SI}$, $c_{operation}$, c_{SI}, c_{GSM}, plus additional known and estimated parameters (F, W).

6   Discussion

The derived cost model is mainly linear. This may be considered as an over-simplification. However, an iterative approach for the parameter estimation could compensate this and reflect a close to the real-world model. If one of the parameters changes, we will initiate a re-investigation of the other parameters according to the method described in section 5. If any two or more parameters depend on each other, this re-investigation will figure out a new parameter set. This accounts for the non-linearity in real-world processes. One has to be aware that this approach increases significantly the effort to work with the Smart Items cost model.

Another point of discussion is the usage of the Cobb-Douglas function introduced in section 3.3. This function structure is neither derived, nor does it have its fundament in a larger theory of logistics processes. However, it has attractive mathematical features and introduces a non-linear behavior which is inherent in real-world processes, but on the other side very hard to model. In our opinion, the non-linearity accounts for the effects that some ranges of parameters have less influence on the overall result than others. In our example a decreasing ratio of loss and defective goods will contribute to the overall utility. The utility gets largest, when approaching zero-loss. However, this is quite hard as the non-linear slope of the Cobb-Douglas function illustrates.

Related to Cobb-Douglas is the customer utility $\tilde{u}$. It is difficult to determine and may involve many parameters which may require a broader study. The selection of the function and its parametrization may partially depend on psychological factors, e.g. previous experiences in a business domain or personal risk assessment. The utility function is very open to an adaptation according to the specific needs of a concrete domain.

Another point of criticism is the simplifications in the model. We assumed an ideal environment, where the supplier only produces one type of good and Smart Items operate error-free. However, experiences from field trials involving Smart Items, e.g. CoBIs [2], revealed a variety of errors. For logistics applications, the most crucial are RF shielding, i.e. the Smart Items cannot communicate to each other anymore, and the power supply of the electronics. Latter adds significantly to the operation costs.

A deeper investigation on the effects of our design decisions is clearly a task for future work.

7   Related Work

The research on RFID and related technologies for supply chains of specific business and market contexts is well established. In many cases the research is driven by applications or scenarios where technological solutions for specific market segments (e.g. grocery stores) are developed or evaluated [15][16][17]. Examples can be found in various areas, e.g. livestock tracking [18], military [19] and evaluations of pilot projects of retailers such as Gillette [20], Tesco, Wal-Mart [21],

Metro AG [22], and Smart Packaging of Hewlett-Packard [23]. The main discussions of RFID driven applications is currently appearing in whitepapers of technology consultants or magazines (e.g. RFID Journal, Information Week or Infoworld.com) and are facing the challenges of poor forecasting accuracy, low effectiveness and responsiveness, high inventory, high returns processing cost and the presence of counterfeit products in their value chain [24].

Many researchers concentrate on technical aspects of RFID applications which are emphasized in several engineering and computer science publications outlining the system architecture and circuit design. The replacement of barcodes to automatically tag and inventory goods in real-time situations, including the whole process chain is just seen as the beginning. The real benefits come from high level uses like theft and loss prevention, reduced turnaround times, avoidance of unnecessary handling and streamlined inventories [25][26]. The main focus of many consulting-oriented and management-related publications is the integration of new technology in ERP systems to provide managerial insights. They offer an in-depth technological overview of state-of-the-art developments and outline different aspects of e-business and supply chain management [27][28][29]. From the general technological and integration approaches analytic models have been derived to show the benefits and costs resulting from the usage of the RFID in supply chains. In [30] item-level RFID usage for decentralized supply chains is discussed by means of two scenarios for one manufacturer and retailer. Within these particular scenarios they capture important benefits reflecting real-world cost considerations in a model based on RFID.

8 Conclusion and Outlook

The proposed cost model for supply chain management is a first step towards estimating the benefits of introducing Smart Items into a logistic process. We presented separate cost models for supplier, shipper and customer. This allows for split benefit calculation, which is often required in supply chain management processes where mixed calculation is not possible or not attractive.

The proposed model can be used to maximize profit according to different types of technologies for Smart Items. It also incorporates different granularities of technology applications. We have shown in this paper, that there are three classes of technology to be distinguished: the use of barcode, the use of RFID-tags and the use of (smart) sensor systems and networks. Each of these options require a set of parameters to calculate their costs. To simplify estimation and calculation of these parameters we introduced guidelines to increase practical applicability of our model.

Our ongoing and future research has two directions. Firstly, we try to evaluate the model on further trial runs and collect experiences regarding the applicability of the guidelines and the cost model. Secondly, we seek to identify parameters that can be used for standard settings, and different technology options. This requires to define standard procedures for various types of supply chain applications, and to perform test runs on the same process using various technology options. Although we have experienced, that this will be very restricted to the specific application case, we envision to commence such parameter and data collection based on our approach on a case by case basis.

Acknowledgments

The work presented in this paper was partially funded by the EC through the project RELATE (contract no. 4270), by the Ministry of Economic Affairs of the Netherlands through the project Smart Surroundings (contract no. 03060) and by the German Ministry for Education and Research (BMBF) through the project LoCostix.

References

1. Fleisch, E., Christ, O., Dierkes, M.: Die betriebswirtschaftliche Vision des Internets der Dinge. In: Das Internet der Dinge, pp. 3–37. Friedemann Mattern (2005)
2. Decker, C., Riedel, T., Beigl, M., sa de Souza, L.M., Spiess, P., Mueller, J., Haller, S.: Collaborative Business Items. In: 3rd IET International Conference on Intelligent Environments (2007)
3. Johnson, M.E., Gozycki, M.: Woolworths "Chips" Away at Inventory Shrinkage through RFID Initiative. Stanford Global Supply Chain Management Forum (2004)
4. Decker, C., van Dinther, C., Müller, J., Schleyer, M., Peev, E.: Collaborative Smart Items. In: Workshop for Context-aware and Ubiquitous Applications in Logistics (UbiLog) (2007)
5. Assam Agricultural University, D.o.R.: Post harvest practices & loss assessment of some commercial horticultural crops of assam. Crop / Enterprise Guides (2005)
6. Sznajder, M.: Towards Sustainable Food Chain - Concept and Challenges (Case study - the dairy food chain in Poland) (2006)
7. Pfohl, H.J.: Logistiksysteme: Betriebswirtschaftliche Grundlagen. Springer, Berlin, Heidelberg (2004)
8. Finkenzeller, K.: RFID Handbook: Fundamentals and Applications in Contactless Smart Cards and Identification (2003)
9. Decker, C., Kubach, U., Beigl, M.: Revealing the Retail Black Box by Interaction Sensing (2003)
10. Decker, C., Beigl, M., Krohn, A., Robinson, P., Kubach, U.: eSeal - A System for Enhanced Electronic Assertion of Authenticity and Integrity. In: Ferscha, A., Mattern, F. (eds.) PERVASIVE 2004. LNCS, vol. 3001, pp. 254–268. Springer, Heidelberg (2004)
11. Schmidt, A., Thede, A., Merz, C.: Integration of Goods Delivery Supervision into E-commerce Supply Chain. In: Fiege, L., Mühl, G., Wilhelm, U.G. (eds.) WELCOM 2001. LNCS, vol. 2232, Springer, Heidelberg (2001)
12. Cobb, C.W., Douglas, P.: A Theory of Production. American Economic Review 18 (Supplement) (1928)
13. National Agricultural Statistics Service: Agricultural Prices. Report by Agricultural Statistics Board, U.S. Department of Agriculture (2007), http://usda.mannlib.cornell.edu/usda/current/AgriPric/ AgriPric-08-31-2007.pdf

14. U.S. Department of Labor: Producer price indexes (2007),
    http://www.bls.gov/ppi/ppitable06.pdf
15. Prater, E., Frazier, G.V.: Future impacts of RFID on e-supply chains in grocery retailing. Whitepaper (2005)
16. Loebbecke, C.: RFID Technology and Applications in the Retail Supply Chain: The Early Metro Group Pilot (2005)
17. Thiesse, F., Fleisch, E.: Zum Einsatz von RFID in der Filiallogistik eines Einzelhändlers: Ergebnisse einer Simulationsstudie. Wirtschaftsinformatik (1), 71–88 (2007)
18. Beigel, M.: Taming the Beast: The History of RFID. In: Invited Presentation Smart Labels, Cambridge, MA, USA (2003)
19. DeLong, B.: How the US Military is achieving total asset visibility and more using RFID and MEMS. In: Invited presentation. Smart labels USA, Cambridge, Massachusetts (2003)
20. Wolfe, E., Alling, P., Schwefel, H., Brown, S.: Supply chain technology-track(ing) to the future. Bear Stearns Equity Research Report, Bear Stearns (2003)
21. Romanow, K., Lundstrom, S.: RFID in 2005: The what is more important than the when with Wal-Mart edict. AMR Research Alert, AMR Research (August 2003),
    http://www.amrresearch.com
22. Roberti, M.: Metro opens Store of the Future (2003),
    http://www.rfidjournal.com
23. Hewlett-Packard: The Future of the Supply Chain and the Importance of Smart Packaging in Containing Costs. PISEC (Product & Image Security) (2004)
24. Cognizant Technology Solutions: RFID Solutions in LifeSciences Industry. Whitepaper (2004)
25. Glidden, R., Bockorick, C., Cooper, S., Dioiio, C., Dressler, D., Gutnik, V., Hagen, C., Hara, D., Hass, T., Humes, T., Hyde, J., Oliver, R., Onen, O., Pesavento, A., Sundstrom, K., Thomas, M.: Design of ultra-low-cost UHF RFID tags for supply chain applications. IEEE Communications Magazine 42(8) (2004)
26. Want, R.: An introduction to RFID technology. IEEE Pervasive Computing 5(1), 25–33 (2006)
27. Cook, C.: Practical Performance Expectations for Smart Packaging. Texas Instruments Radio Frequency Identification Systems (2006)
28. Harrop, P., Das, R.: RFID Forecasts, Players and Opportunities (2006),
    http://www.idtechex.com
29. Asif, Z.M.M.: Integrating the Supply Chain with RFID: An Technical and Business Analysis. Communications of the Association for Information Systems (2005)
30. Gary, M., Gaukler, R.W., Seifert, W.H.H.: Item-Level RFID in the Retail Supply Chain. Production and Operations Management 16, 65–76 (2007)

Generalized Handling of User-Specific Data in Networked RFID

Kosuke Osaka[1], Jin Mitsugi[1], Osamu Nakamura[2], and Jun Murai[2]

[1] Keio University / Graduate School of Media and Governance
[2] Keio University / Department of Environment and Information studies
5322 Endo, Fujisawa, Kanagawa, 252-8520 Japan
{osaka,mitsugi,osamu,jun}@sfc.wide.ad.jp

Abstract. As RFID technology has been widely adopted, it has been acknowledged that user-specific data carrying RFID tags can be conveniently used. Since the typical user-specific data is sensor data, some existing studies have proposed new RFID architectures for sensor data handling. However, there are industrial demands to include non-sensor data, such as sales records, repair history, etc., as well as sensor data in the purview of our study. To realize handling of such general user-specific data, we need to satisfy requirements such as the minimum set of data semantics required to access the user-specific data, flexible user-specific data memory schema determination and partial collection of user-specific data. For this purpose, we designed the role of "session manager", which communicates with associate applications and notifies an interrogator what to do upon retrieval of a unique identification number. The session manager with the structured user-specific data can provide flexible collection and processing of user-specific data in the networked RFID. Since the fundamental drawback of the session manager might be the lookup overhead, we examined the overhead with a simulation based on experimental data. It was revealed that the overhead is equivalent to and in some cases even better than that of the existing method, which relies on structural tag memory.

Keywords: RFID middleware, Sensor tag, Battery Assisted Passive tag (BAP), User data handling, User data collection performance, Tag memory schema resolving, Data semantic resolving.

1 Introduction

In past years Radio Frequency Identification (RFID) technology has been designed and realized for asset identification in the supply chain. Recently, however, RFID systems have spread to wider applications. As RFID technology has become more widely adopted, it has been acknowledged that user-specific data in tag memory can be conveniently used in conjunction with globally unique identification numbers such as EPC [1].

An example of user-specific data can be found in life-cycle management of consumer electronics. Major consumer electronics makers have been conducting research on and pilot testing how to manage the whole life cycle of consumer electronics using RFID technology. They recommend storing a variety of data such as the manufacturing-specific code, shipping and sales dates, repair history and other information in a consumer item's tag memory [2].

C. Floerkemeier et al. (Eds.): IOT 2008, LNCS 4952, pp. 173–183, 2008.

Another example of user-specific data can be found in the sensor-integrated RFID system, which records environmental information such as temperature, humidity and/or acceleration information, as well as evidence of tampering, into the existing RFID system. Using sensor devices with high-performance RFID tags (sensor tag) would enable manufacturers to realize value-added asset identification. For example, the smart seal is a typical sensor-integrated RFID system [3] for tamper protection. Sensor-enabled cold-chain management is another exemplary usage. A sensor-integrated RFID system can realize not only traceability but management of the proper temperature for the merchandise, which may justify an additional tariff.

It is, thus, expected that a myriad of user-specific data together with unique IDs will be adopted in industrial RFID systems. A networked RFID system which accommodates user-specific data in a general manner, therefore, needs to be established.

The de facto networked RFID standard, EPCglobal network, essentially handles events which comprise tag IDs and their read times [1]. Sensor-integrated RFID networks have been studied in the literature [4][5][6][7][8][9]. The authors' research group compiled a short review on sensor-integrated RFID and related wireless sensor network technologies [10]. International standardization bodies have been working to associate sensor data in networked RFID systems [11][12]. While sensor data can be structured as in a plug-and-play sensor [13], the life-cycle information in consumer electronics, for example, might not always fit this sensor data structure. Since the emerging technology on battery-assisted passive and active tags may house a large logging memory, 4Mbit for example [14], partial collections of tag data need to be accommodated because of the relatively slow air communication speed [15]. Naruse has proposed a network-based tag memory schema resolver for sensor-integrated networked RFID, which only can apply to a fixed memory length [4]. In this paper we propose a network RFID system which provides flexible association with user-specific data in tag memory. The fundamental idea is a combination of structured tag memory and lookup of a registry in a network, which provides memory schema information to interrogators.

The paper is organized as follows. In Section 2, we investigate the architecture requirements to establish a networked RFID system which can handle a variety of user-specific data efficiently. The qualitative benefit of the proposal is also stated. Section 3 examines the registry lookup overhead of the proposed method compared with those of previous works [11][12]. A conclusion is presented in Section 4.

2   The Proposed Networked RFID System Description

In this section we first consider the networked RFID requirements for generalized handling and processing of a user-specific data collection. Secondly, a networked RFID system that meets those requirements is proposed.

2.1   Requirements

General user-specific data can be accommodated
User-specific data may be sensor data or another type of record such as sales and repair history. Different from sensor data, which may be well structured as in [13], the

structure of the sales and repair history likely is diverse. All or a part of the data may also need to be encrypted or may need to be human-readable depending on the usage. As such, general user-specific data can be collected with the minimum set of semantics in the data.

Resolving tag memory schema
An interrogator needs to resolve the tag memory schema, which depends on the type of user-specific data, and also the length of the memory block before collection. The procedure for handling user-specific data may not enforce any change in the procedure to inventory ID tags[1].

Resolving data semantic
In an RFID application layer, user-specific data collected from RFID tags need to be incorporated with data semantics, in order to generate more meaningful events in a business context. For example, the raw temperature data "0" will need to be translated into a physical quantity (e.g. 0 degree Celsius or 0 Kelvin) with calibration.

Collection of dynamically changing data
User-specific data can be classified into static and dynamic types. In this paper, "static" user-specific data means that the data length and data types are consistent throughout the operations. "Dynamic" user-specific data, on the other hand, has variable data length. The typical dynamic data is data logging.

Partial collection of user-specific data
The most rudimentary method for collecting user-specific data is to read all the tag data with every interrogator. This may cause significant reading performance degradation particularly when the target tag is equipped with a number of sensors or has a large memory for data logging. There may be a situation in which the owner of the user-specific data demands to control which part of the data can be read depending on the interrogator. Thus, an interrogator needs to be controlled such that only the group of user-specific data which is requested from applications is collected from a tag.

Minimum collection overhead
User-specific data collection needs to be time-wise efficient.

2.2 Proposal Description

The proposed networked RFID system is shown in Figure 1.

The essential entity in the proposal is the session manager, which communicates with the associate applications and notifies the interrogator of the specific memory address and its range for user-specific data collection.

---

[1] ID tags denote RF tags which do not have user-specific data or handling capability in this paper.

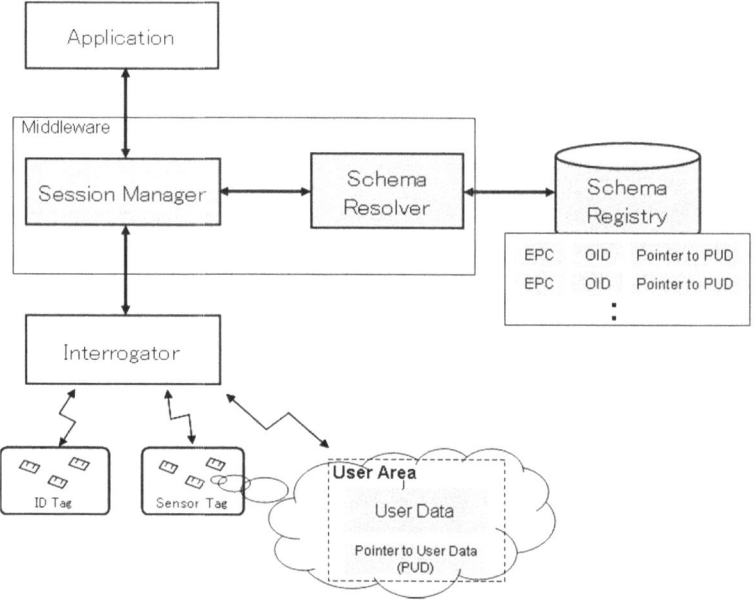

Fig. 1. Principal entities of the proposal

The typical procedure to collect user-specific data is shown in Figure 2.

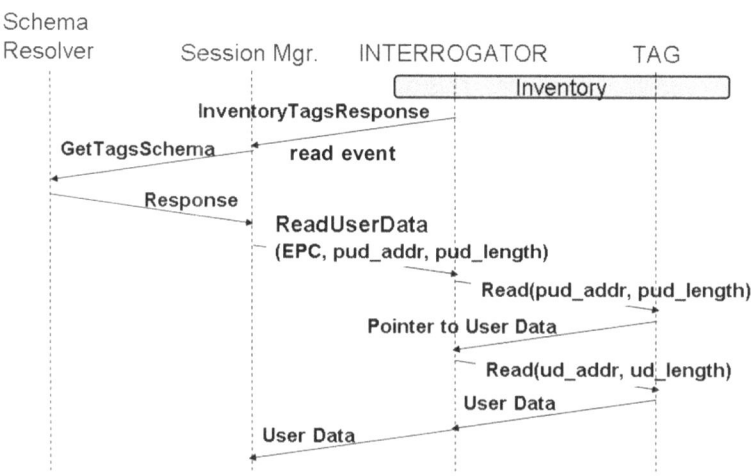

Fig. 2. User-specific data collection

The interrogator collects the user-specific data in the following procedure.

1) The interrogator collects globally unique IDs such as EPC in Tag (UII) by tag inventory.

2) Upon receipt of the unique ID, the interrogator generates a read event to place an inquiry for further action to the associate session manager.

3)   The session manager examines the unique ID in the read event and looks up the tag memory schema, which contains OID (Object ID) and Pointer to PUD (Pointer to User Data), via the schema resolver.

4)   The interrogator may collect a part of the tag memory (PUD) to determine the memory range to be collected.

5)   The interrogator collects the body of user-specific data and delivers it to associate applications through the session manager.

For the purpose of comparison, the user-specific data collection in [11][12] can be summarized as follows.

1)   The interrogator collects a globally unique ID in Tag (UII) by tag inventory and identifies the sensor data capability either by user memory indicator (UMI) or XPC (extended PC bit) or XI (extended PC bit indicator).

2)   The interrogator accesses a particular bit of area in the Tag (TID) and collects meta-data named PSAM (pointer to Sensor Address Map) by the next tag inventory.

3)   Upon receipt of the PSAM, the interrogator collects the NoS (Number of Sensors), which indicates how many sensor devices are attached.

4)   The interrogator collects the SAM-Entries (User) that contain sensor record addresses.

5)   From NoS and SAM-Entries[2] information, the interrogator collects an SID (Sensor Identifier) (User) and then judges which sensor record needs to be acquired by an associated application.

6)   The interrogator collects the body of user-specific data (User) and delivers it to the application.

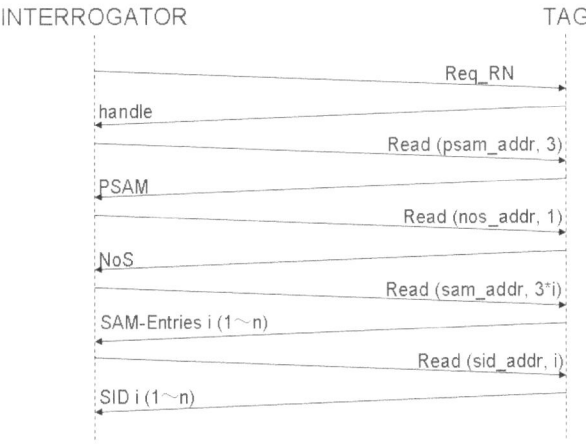

Fig. 3. Transaction for resolving memory schema on tag (fixed pattern)

Figure 4 shows the tag memory map example in [12].

---

[2] A SAM-Entry and SID are prepared in each sensor device. If a sensor tag contains two sensor devices, it has two SAM-Entries and SIDs in its memory.

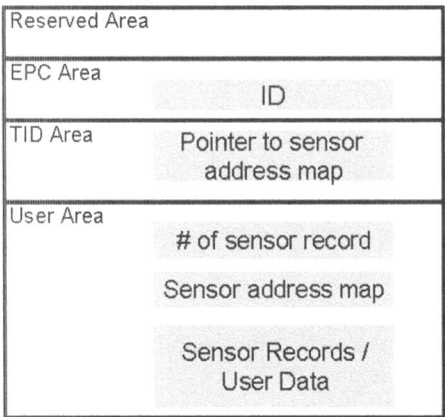

Fig. 4. RFID tag memory map example in ISO 24753

It should be stated that, in the existing method shown in Figure 3, the interrogator needs to collect all the data from the tag unless the interrogator has knowledge of which sensor data needs to be collected. Similarly, in a write operation without an inquiry to the session manager, the interrogator can only write the predefined pattern of data to the tag, or cannot determine whether it is requested to write a datum or collect a datum. In this regard, the proposed method (Figure 2) and the existing method (Figure 3) can be complementary. If we have a network connection, the interrogator may place an inquiry depending on the UIM or XPC to a session manager to find any updated request for user-specific data collection and writing. If there is no updated instruction, the interrogator collects or writes in the predefined pattern, which may be to collect all the data or no data.

A qualitative comparison between the proposed and the existing methods is shown in Table 1.

Table 1. Comparison table with related studies and our proposed model

item	ISO Working Draft	Proposed Model
General user-specific data can be accommodated	Sensor data centric	Combination of registry and structured address map on tag
Resolving tag memory schema	Sensor address map on tag	Combination of registry and structured address map on tag
Resolving data semantic	Canonical spec. sheet data in tag	Combination of registry and structured address map on tag
Collection of dynamically changing data	Sensor address map involves range	Combination of registry and structured address map on tag
Partial collection of user-specific data	Not in the charter	Combination of registry and structured address map on tag

3   Registry Lookup Overhead Estimation

Since the proposed method is a combination of a structured memory map in the tag and the network session manager, it is important to quantitatively evaluate the turnaround time for placing an inquiry to the session manager, which is referred to as the registry lookup overhead in this paper. We have already measured the registry lookup time in [2] in accordance with the number of user-specific data. In that experiment, a data base server which is operating the schema registry is set up in the same network segment as the session manager. But we don't have a reference turnaround time to compare with since we don't have an interrogator that can be instructed to collect user-specific data in accordance with the procedure in Figure 3. For this, we measured the actual user-specific data collection time using a commercial interrogator and a tag assuming we have already resolved the memory schema before data collection. The turnaround time involving both the processing and transmission time between the interrogator and the tag was analyzed in detail and computationally re-assembled to estimate the collection time using the procedure in Figure 3. It should be noted that the collection of a body of user-specific data was out of the scope for this evaluation since it is irrelevant how the memory map is resolved.

3.1   Experiment Setup Description

The experimental setup to measure the turnaround time is shown in Figure 5. A commercial interrogator (R/W) is connected to a battery-assisted passive (BAP) tag [14] by a cable, and user-specific data is collected using an ISO/IEC 18000-6 Type C [15] command. The turnaround time was measured by monitoring the demodulated signal and the backscattering signal at the tag using an oscilloscope (OSC). Figure 6 shows a screen shot of the signals in the oscilloscope when user data (30 words) are retrieved using a commercial interrogator. The configuration of the communication between the interrogator and the tag is summarized in Table 2.

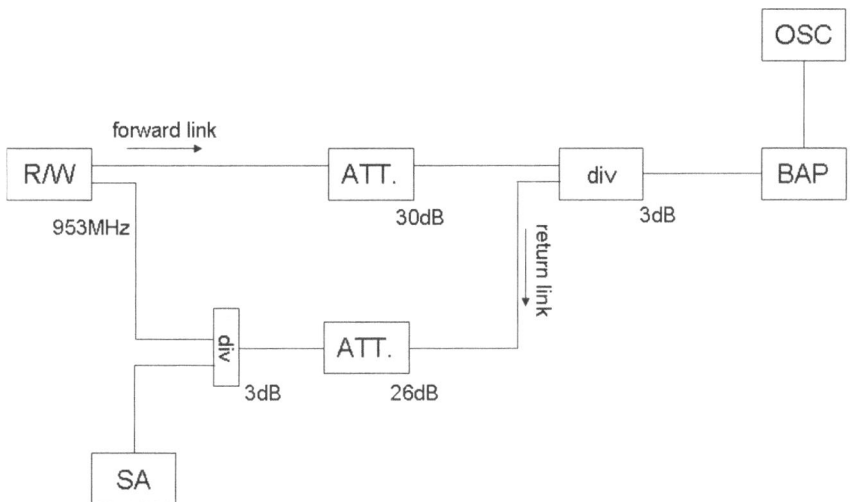

Fig. 5. Experimental setup

Table 2. Experimental configuration

Item	value
Air protocol	ISO18000-6 Type C
communication speed	Forward link: 40kbps Return link: 40kbps
Frequency	953MHz
Band width	200KHz
R/W output PWR	30dBm

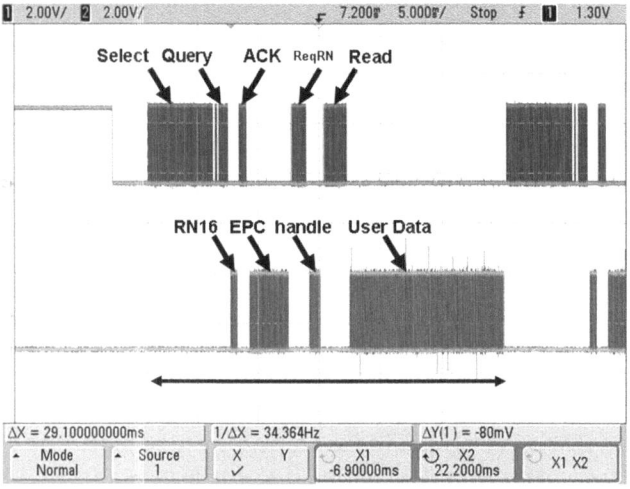

Fig. 6. Screenshot of Oscilloscope

3.2 Estimation Result

With the above experiment environment, we estimated the turnaround time for the protocol shown in Figure 7. The numbers of SAM-Entries and SIDs depend on the number of sensor devices installed in a tag. If a certain tag contains numerous sensor

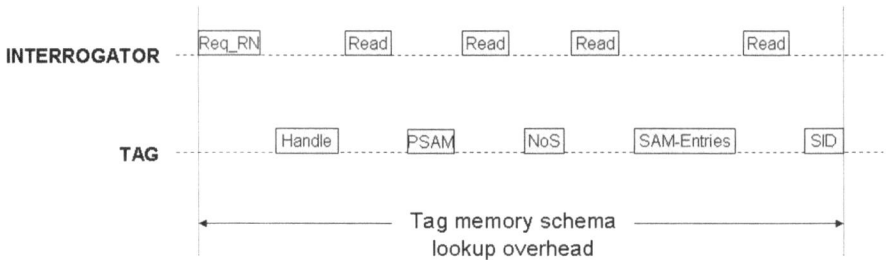

Fig. 7. Air protocol sequence

devices, the length of the data that should be read by the interrogator becomes longer. The data length of one SAM-Entry and one SID are 3 words and 1 word, respectively, for a total of 4 words (1 word = 16 bits).

Figure 8 shows the result of tag memory schema lookup overhead in the two methods. The legends "proposed" and "existing" in Figure 8 represent the procedures in Figure 2 and Figure 3, respectively. The horizontal axis denotes the number of sensor devices installed on one tag and the vertical axis is the time (msec) required to collect that amount of user-specific data. The lookup overhead of the proposal was found to be around several 10 msec, which is almost equivalent to that of the existing method and even shorter when the number of sensor devices installed on a tag is increased. This is because we set up the registry to be network-wise close to the interrogator and the communication protocol is relatively slow, resulting in longer time for plural sensor data collection.

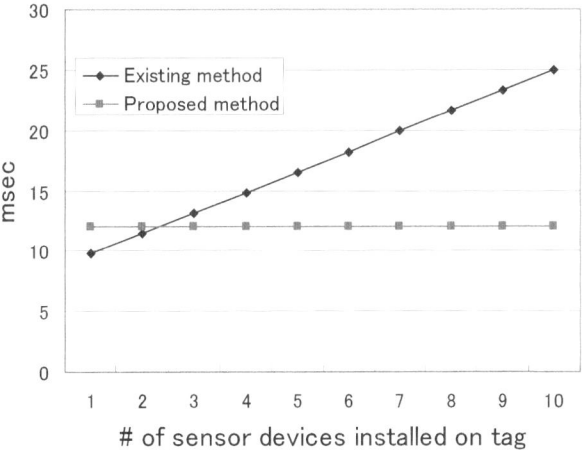

Fig. 8. Tag memory schema look up overhead

4 Conclusion

As the industrial adoption of networked RFID expands, the importance of user-specific data has been widely acknowledged. Typical user-specific data is sensor data, which can be well structured. When we consider also handling non-sensor user-specific data, such as sales records and repair history, a generic method for handling user-specific data is needed. The requirements for general handling of user-specific data in the networked RFID are the identification of a memory schema for user-specific data, accommodation of dynamically changing data length and partial collection of the user-specific data. These requirements can be satisfied when we implement the role of "session manager", which essentially instructs the interrogator what to do upon retrieval of a unique identification number. The session manager may instruct the tag to retrieve user-specific data in the specified memory range or may instruct a write operation. In order to accomplish such tasks, the session manager needs to connect with associate applications and the tag memory schema resolver to determine the instruction. The session manager role can be complementary to that of structured tag memory, in which all the information required to collect user-specific data is stored in tag memory, tailored to sensor data. Despite the qualitative advantage of having a session manager, the mandated registry lookup by the session manager may degrade the reading performance. Our simulation revealed, however, that the registry lookup is equivalent to or even faster than that of structured tag memory because the interaction between tag and interrogator in the latter is relatively slow and a number of data needs to be exchanged to determine the range of tag memory to be collected.

References

[1] EPCglobal Architecture Review Committee: EPC Tag Data Standard (TDS) Standard (2006)

[2] Yoshimura, T.: RFID: Expectation and Requirement from Consumer Electronics Industry. In: APMC 2006 workshop on Emerging Technologies and Applications of RFID, pp. 121–124 (2006)

[3] Banks, J., Hanny, D., Pachano, M., Thompson, L.: RFID APPLIED, pp. 328–329. Wiley, Chichester (2007)

[4] Naruse, D.: Integration of Real Space Information Based on RFID Model. Master Thesis Keio University (2006)

[5] Emery, K.: Distributed Eventing Architecture: RFID and Sensors in a Supply Chain. Master Thesis, MIT, http://db.lcs.mit.edu/madden/html/theses/emery.pdf

[6] Deng, H., Varanasi, M., Swigger, K., Garcia, O., Ogan, R., Kougianos, E.: Design of Sensor-Embedded Radio Frequency Identification (SE-RFID) Systemsl. In: IEEE International Conference on Mechatronics and Automation, pp. 792–796 (2006)

[7] Floerkemeier, C., Lampe, M.: RFID middleware design - addressing application requirements and RFID constraints. In: sOc-EUSAI 2005, pp. 219–224 (2005)

[8] Ranasinghe, D.C., Leong, K.S., Ng, M.L., Engels, D.W., Cole, P.H.: A Distributed Architecture for a Ubiquitous RFID Sensing Network. In: 2005 International Conference on Intelligent Sensors, Sensor Networks and Information Processing Conference, pp. 7–12 (2005)

[9] Zhang, L., Wang, Z.: Integration of RFID into Wireless Sensor Networks: Architectures, Opportunities and Challenging Problems. In: Fifth International Conference on Grid and Cooperative Computing Workshops, GCCW 2006, pp. 463–469 (2006)

[10] Mitsugi, J., Inaba, T., Patkai, B., Theodorou, L., Lopez, T.S., Kim, D., McFarlane, D., Hada, H., Kawakita, Y., Osaka, K., Nakamura, O.: Architecture Development for Sensor Integration in the EPCglobal Network. White Paper, White Paper series (to appear in Auto-ID Lab, 2007)

[11] ISO/IEC WD 18000-6REV1: 18000-6REV1, Information technology — Radio frequency identification for item management — Part6: Parameters for air interface communications at 860MHz to 960MHz. Working draft (February 9, 2007)

[12] ISO/IEC WD 24753, Information technology — Radio frequency identification (RFID) for item management — Application protocol: encoding and processing rules for sensors and batteries. Working draft (September 21, 2006)

[13] Nist, IEEE-p1451, draft standard, http://ieee1451.nist.gov/

[14] Mitsugi, J.: Multipurpose sensor RFID tag. In: APMC 2006 workshop on Emerging Technologies and Applications of RFID, pp. 143–148 (2006)

[15] ISO/IEC 18000-6 Information technology — Radio frequency identification for item management- Part 6: Parameters for air interface communications at 860MHz to 960MHz Amendment 1: Extension with Type C and update of Types A and B (June 15, 2006)

[16] EPCglobal Architecture Review Committee, The EPCglobal Architecture Framework Final version of 1 July 2005. EPCglobal (July 2005)

A Passive UHF RFID System with Huffman Sequence Spreading Backscatter Signals

Hsin-Chin Liu[1] and Xin-Can Guo[1]

National Taiwan University of Science and Technology
Taipei, 106, Taiwan
hcliu@mail.ntust.edu.tw

Abstract. At present passive RFID standards, the tag collision problems are solved using time division multiple access technologies, including slotted ALOHA schemes and binary-tree search schemes. In order to accelerate an inventory process, a multiple access scheme that can identify more than one tag simultaneously is necessary. Considering the characteristics of a passive UHF RFID system, a novel passive tag backscatter scheme using Huffman spreading sequences is proposed, which can effectively improve an inventory process. The performances of several studied multiple access schemes are compared. Simulation results validate the performance enhancement of the proposed system.

Keywords: passive RFID, multiple access, Huffman sequences.

1 Introduction

At present passive RFID standards, the tag collision problems are solved using time division multiple access (TDMA) technologies, including slotted ALOHA schemes and binary-tree search schemes [1, 2]. Because only a single tag can be identified at one time slot, the throughput of an inventory process is hence limited. In order to accelerate the inventory process, several multiple access schemes that can identify more than one tag simultaneously have been proposed [3, 4].

In [3], similar to other code division multiple access (CDMA) systems, the proposed scheme utilizes a set of 256-bit length gold sequences to spread the tag backscatter signals, so that the RFID reader can successfully separate the signal of each individual tag. However, due to the non-zero cross-correlation of each spreading sequence, the performance of this method deteriorates when there exists power inequality amount the tag backscatter signals. In general, a passive tag merely returns its response by backscattering the incident continuous wave (CW), which is emitted from a reader. The strength of a received tag backscatter signal is determined by a variety of factors, including the power of the incident CW, the radar cross section (RCS) of the tag, the polarization of the tag antenna, the propagation loss in the tag-to-reader link, and so on. In order to mitigate the near-far problem, a sophisticated power control scheme is required. Unfortunately, the implementation of power control mechanism on each individual tag is impractical due to the limitations of the tag power and cost. Moreover, because a tag singulation must be accomplished within

C. Floerkemeier et al. (Eds.): IOT 2008, LNCS 4952, pp. 184–195, 2008.

1 ms, the protocol complexity is strictly constrained. Another approach is given in [4], which utilizes the orthogonality of sinusoidal functions to singulate individual tag by estimating their signal phases. The paper however does not disclose how the phases of backscatter signal can be estimated. Consequently, it is difficult to evaluate the feasibility of the method.

In this paper, a novel passive tag backscatter scheme using Huffman spreading sequences is proposed. The Huffman sequences are more near-far resistant and can preserve code orthogonality without precise synchronization of received signals. Consequently, it is more suitable for a passive UHF RFID system. Simulation results demonstrate that the proposed scheme can effectively speed up an inventory process than other methods do.

The reminder of this paper is organized as follows: Section 2 briefs the tag-to-reader (T-R) communication, which brings challenges of multiple access technologies in present passive RFID systems; Section 3 introduces Huffman spreading sequences used in this work; Section 4 presents the simulation results; Section 5 draws conclusions. Finally, some measurement results of link timing are presented in Appendix.

2   The Tag-to-Reader Communication of a Passive UHF RFID System

In general, for example in an EPC Gen2 system, a reader broadcasts a reader-to-tag (R-T) command, and listens for individual tag response thereafter. In order to initialize an inventory process, the reader sends a QUERY R-T command with tag clock instruction to all tags within its coverage. After receiving the QUERY command, each tag should reply its modulated backscatter signal (MBS) in its corresponding time slot. The tag-to-reader communication is asynchronous and the powers of received MBS of different tags are usually unequal as explained below.

2.1   Asynchronous T-R Communication

According to the link timing specification [2], the tag backscatter signal should begin after T_1 seconds starting from the last rising edge of R-T command as shown in Fig. 1. In [2] the parameter T_1, as shown in Table 1, is defined in the link timing parameter table with typical value as

$$T_1 = MAX\left(RTcal,\ 10T_{pri} \right), \tag{1}$$

where $RTcal$ denotes reader-to-tag calibration symbol, and T_{pri} denotes link pulse-repetition interval. Some examples of the parameter T_1 with divide ratio (DR) as 8 are given in Table 2. When multiple tags reply the same received R-T command in the same time slot (that is, a tag collision occurs), they may not be synchronized due to the deviation of T_1 parameter in each tag. Moreover, since the variation of T_1 can be nearly a tag backscatter symbol long, the orthogonality of the spread backscatter signals may be impaired if the spreading sequences require precise synchronization to maintain their mutual orthogonality. The Walsh codes, for instance, may hence be unsuitable to such a system.

Experimental measurements of three different type Gen2 tags, presented in Appendix, validate that the T_1 parameter of these tags are indeed different.

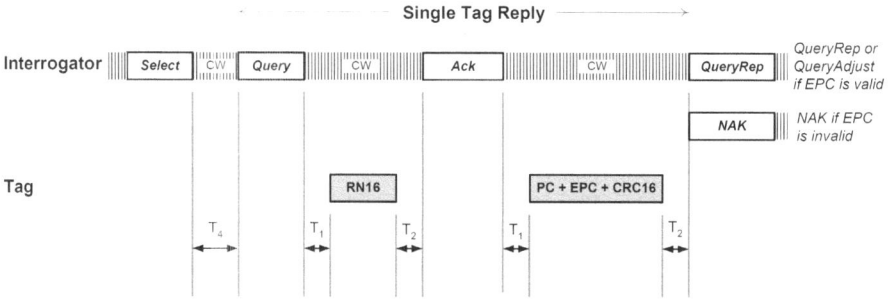

Fig. 1. The tag response time [2]

Table 1. T_1 link timing

Parameter	Minimum	Typical	Maximum
T_1	MAX(RTcal,$10T_{pri}$) $\times$ (1-FT)-2µs	MAX(RTcal,$10T_{pri}$)	MAX(RTcal,10 T_{pri}) $\times$ (1-FT)+2µs

Table 2. Some examples of T_1 link timing

DR(Divide Ratio)=8			
LF(Link Frequency, KHz)	Minimum(µs)	Typical(µs)	Maximum(µs)	Max.-Min.(µs)
40	238	250	262	24
160	56.125	62.5	68.88	12.75
320	26.13	31.25	36.38	10.25

2.2 Unequal Received MBS Power

In a free space propagation model without consideration of reader antenna pattern (assuming unit gain, omni-directional), there are two main factors that can affect the strength of received MBS: one is the distance d between the reader and the tag, and the other is the tag antenna gain G_{tag}. According to the well-known Friis free space equation, the received MBS power can be written as

$$P_r(d) = \frac{P_{Back} G_{tag} \lambda^2}{(4\pi)^2 d^2 L},$$ (2)

where P_{Back} denotes the power of tag MBS, λ is the wavelength of the reader operation frequency, and L is the system loss factor not related to propagation. Because P_{Back} is proportional to G_{tag} and reverse proportional to the squares of d, we can rewrite (2) as

$$P_r(d) \propto \frac{G_{tag}^2}{d^4},$$ (3)

where $\propto$ denotes proportional relationship.

There are a variety of tag antennas, in which the half-wavelength dipole antenna is most extensively used. The normalized antenna power gain of a half-wavelength dipole antenna is sensitive to the angle of arrival (AOA) as presented in (4).

$$G_{tag}(\theta) = \frac{\cos^2\left[\dfrac{\pi \cos\theta}{2}\right]}{\sin^2\theta},$$ (4)

where θ denotes the AOA of an incident wave with $\theta = 0$ as the direction of antenna boresight. Therefore, the powers of MBS signals from any two tags can be very different because of the different AOA of impinging CW from the reader; even they are equally separated away from the reader.

Consequently, we can conclude that the received MBS power from each tag is often unequal. This phenomenon results in the inevitable near-far problem in a commonly used direct sequence spread spectrum system. Therefore, the Gold sequences, for example, may not be suitable for the system because they are not sensitive to the near-far effect.

3   Huffman Sequence Backscatter Signals

Unlike a cellular phone, a passive tag is designed as a low cost simple device, whose operation must rely on external limited power source. The tag is incapable to perform a sophisticate power control mechanism as a cellular phone does. Hence, commonly used spreading sequences with low cross-correlation, such as gold sequences used in [3], are not suitable for a passive RFID system due to the severe near-far effect.

On the other hand, the orthogonal spreading sequences like Walsh codes can resist the near-far effect. However, the orthogonal spreading sequences require synchronization to preserve their mutual orthogonality. Unfortunately, multiple tags hardly backscatter their responses simultaneously as described in section 2.1.

In order to apply CDMA technology to the passive UHF RFID system, orthogonal spreading sequences without precise synchronization are hence desired.

Huffman [5] has defined a $N+1$ chip long complex (or real) sequence, whose autocorrelation value is zero for all shift except no shift or N chip shift.

A Huffman sequence $\begin{pmatrix} c_0 & c_1 & \cdots & c_N \end{pmatrix}$ can be represent as

$$\mathbf{c}_0 = \begin{pmatrix} c_0 & c_1 & \cdots & c_N \end{pmatrix}^T, \tag{5}$$

and the i-th shift sequence can be written as

$$\mathbf{c}_i = \begin{pmatrix} c_i & c_{i+1} & \cdots & c_N & c_0 & \cdots & c_{i-1} \end{pmatrix}^T. \tag{6}$$

Equation (6) can be obtained from (5) using

$$\mathbf{c}_i = P^i \mathbf{c}_0, \tag{7}$$

where P is a N+1 by N+1 permutation matrix as

$$P_{N+1 \times N+1} = \begin{bmatrix} 0 & 1 & 0 & \cdots & 0 \\ 0 & 0 & 1 & 0 & \cdots \\ \vdots & \ddots & \ddots & \ddots & \vdots \\ 0 & \cdots & 0 & 0 & 1 \\ 1 & 0 & 0 & \cdots & 0 \end{bmatrix}. \tag{8}$$

The normalized autocorrelation function ρ of a Huffman sequence is presented as

$$\rho_{ij} = \mathbf{c}_i^H \mathbf{c}_j = \begin{cases} 1, & i = j \\ \varepsilon, & i - j = N \\ \varepsilon^*, & i - j = -N \\ 0, & otherwise \end{cases}, \tag{9}$$

where ε^* denotes the complex conjugate of ε, and the magnitude $|\varepsilon|$ can be smaller than $1/(N+1)$.

Recently, a method to generate a real Huffman sequence (also known as a shift-orthogonal sequence) has been proposed [6] and [7]. In this work, a real Huffman sequence, shown in Table 3, is used in our simulation to verify the performance of our proposed scheme.

An MBS can be generated by varying the tag reflection coefficient Γ [8] and [9]. In order to produce a Huffman sequence spreading MBS, both magnitude and phase of the time-variant tag reflection coefficient $\Gamma(t)$ in (10) need to be changed according to the Huffman sequence.

$$\Gamma(t) = \frac{Z_L(t) - Z_A^*}{Z_L(t) + Z_A}, \tag{10}$$

where $Z_L(t)$ denotes the time-variant tag load impedance, Z_A denotes tag antenna impedance, and Z_A^* denotes complex conjugate of Z_A. Note that, a normalized Huffman sequence is used because the magnitude of $\Gamma(t)$ is always less or equal to 1.

Rearranging (10), we have the tag load impedance $Z_L(t)$ at time t as

$$Z_L(t) = \left(\frac{\Gamma(t) Z_A + Z_A^*}{1 - \Gamma(t)} \right). \tag{11}$$

Therefore when the reflection coefficient $\Gamma(t)$, which is equal to the Huffman sequence at time t, is given, the corresponding tag load impedance $Z_L(t)$ can be obtained.

The last digit of 16-bit random number (RN16) [2] in each tag is used to determine its associated Huffman spreading sequence. In this work (N=8), if the last digit of a tag RN16 is 0, the tag uses c_0 to spread its FM0 encoding backscatter signals; otherwise, it uses c_5 to spread its FM0 symbol. Note that, c_0 and c_5 are orthogonal. There is only one Huffman sequence is used, the reader design is hence simpler than other CDMA system because it only use different shift of the same code to separate the received signals.

When two tags reply in the same time slot, a tag collision occurs at present passive RFID system. Depending on the power difference of the two received MBSs, a reader of present RFID system may identify one or none of the tags. However, using Huffman sequence spreading scheme, the reader can have 50% probability to simultaneously identify both tags successfully when their spreading sequences are different. If both tags have the same last RN16 digit, the reader can still indentify both tags if the link timing parameter T_1 of the two tags separate more than 1 chip of the Huffman spreading sequence (providing that both MBSs are still orthogonal).

Table 3. A 9-chip normalized real Huffman sequence used in this work

C_0	C_1	C_2	C_3	C_4	C_5	C_6	C_7	C_8
0.4577	0.5763	0.1160	-0.2660	0.0633	0.2120	-0.3640	0.3440	-0.2733

4   Simulation Results

4.1   Simulation Setting

In this work, a 9-chip (M=9) Huffman spreading sequence is used as depicted in Table 3, and the corresponding waveforms of MBS are illustrated in Fig. 2. In the simulations, we assume that all tags are uniformly distributed within a sphere with radius 2.64 meter. An omni-directional reader with $P_{EIRP} = 1watt$ is located in the center of the sphere. The antenna of each tag is assumed as a half-wavelength dipole antenna. The orientation of each tag is arbitrary, which means that AOA of the incident CW (both azimuthal angle and elevation angle) from the reader are uniformly distributed from 0 to π ; Fig. 3 presents the sketch of simulation environment.

In order to simulate more realistic scenario, the tags are assumed EPCgloble Gen2 compatible [2]; link frequency (LF) of 40KHz and A Type A Reference Interval (Tari) of $25\mu s$ are assumed. The duration of a FM0 symbol in the given condition is also $25\mu s$. The parameter T_1 , according to the specification [2], can vary from $238\mu s$ to $262\mu s$ with typical value as $250\mu s$. It is noteworthy that the deviation of T_1 can be as large as $24\mu s$, which is nearly a FM0 symbol duration. The T_1 of each tag is assumed Gaussian distributed with mean of the typical value of T_1 , and less than 0.1% probability that the T_1 falls out of the valid duration.

The simulation block diagram is presented in Fig. 4. In the simulation channel we consider free space propagation only. The power of additive white Gaussian noise (AWGN) is normalized by the power of maximum MBS signal from a tag separated 2 meter away from the reader.

In the simulation, an 8-chip normalized Gold code spreading MBS and an 8-chip normalized Walsh code spreading MBS are compared with the proposed scheme. The normalized spreading sequences are listed in Table 4.

In order to optimize the performance of the system throughput, a modified slotted Aloha algorithm is used with the mean number of tags in a time slot as 2. The details of the cross-layer design, however, are beyond the scope of this paper.

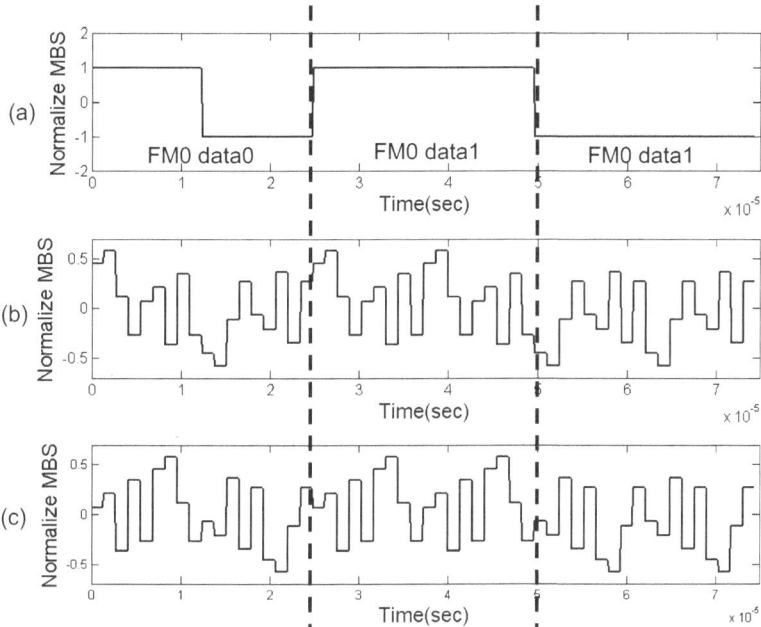

Fig. 2. (a) FM0 backscattering signal. (b) Huffman sequence spreading waveform using $\mathbf{c}_0$. (c) Huffman sequence spreading waveform using $\mathbf{c}_5$.

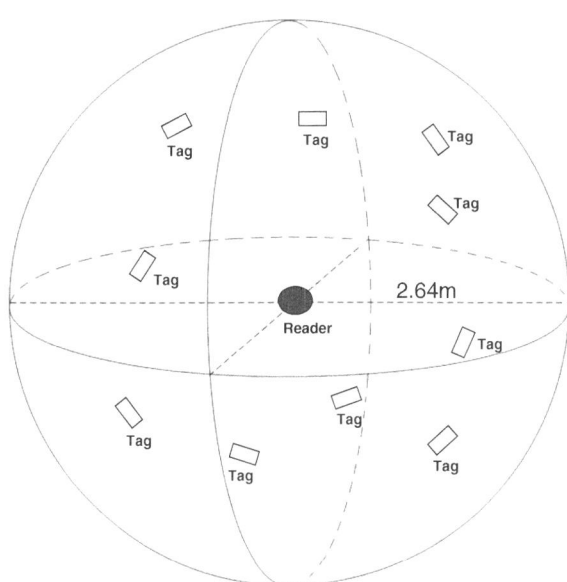

Fig. 3. The sketch of simulation environment

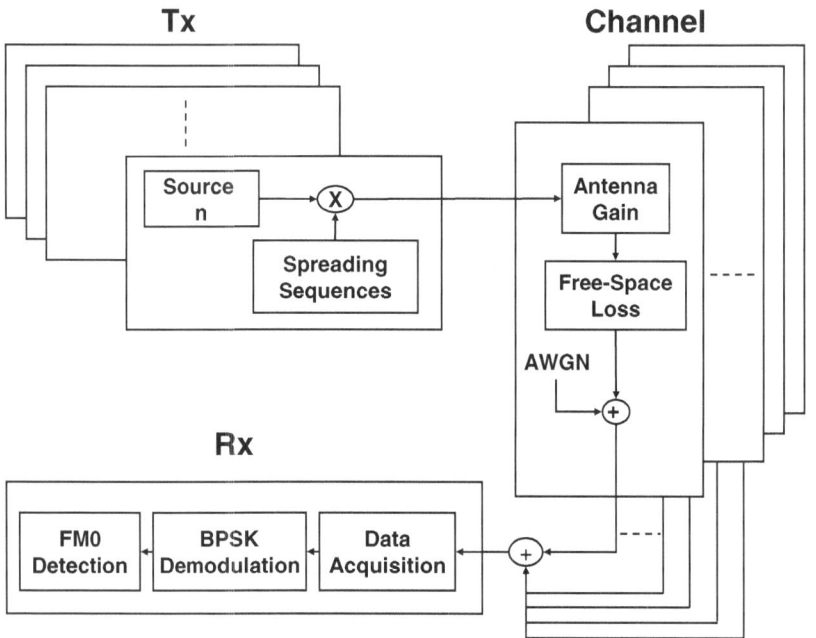

Fig. 4. The simulation system

Table 4. The spreading sequences used in the simulation

	Last bit of RN16	C_0	C_1	C_2	C_3	C_4	C_5	C_6	C_7	C_8
Gold sequences	0	1/8	-1/8	1/8	-1/8	1/8	-1/8	1/8	-1/8	NA
	1	-1/8	1/8	1/8	1/8	1/8	-1/8	-1/8	1/8	NA
Walsh Sequences	0	1/8	-1/8	1/8	-1/8	1/8	-1/8	1/8	-1/8	NA
	1	1/8	1/8	-1/8	-1/8	1/8	1/8	-1/8	-1/8	NA
Huffman sequence	0	0.4577	0.5763	0.1160	-0.2660	0.0633	0.2120	-0.3640	0.3440	-0.2733
	1	0.2120	-0.3640	0.3440	-0.2733	0.4577	0.5763	0.1160	-0.2660	0.0633

4.2 Simulation Results

The throughput of the system is defined as

$$Throughput = \frac{number\ of\ total\ tags}{number\ of\ used\ slots\ in\ an\ inventory\ process}. \quad (12)$$

The performance comparisons for different numbers of tags are listed in Fig. 5 to Fig. 7. Apparently, the proposed scheme outperforms other multiple access methods, especially in high SNR conditions. Walsh code spreading method is the second best choice. All results demonstrates that using TDMA technology only, such as EPC Gen2 slotted Aloha algorithm, results in poor performance in an inventory process.

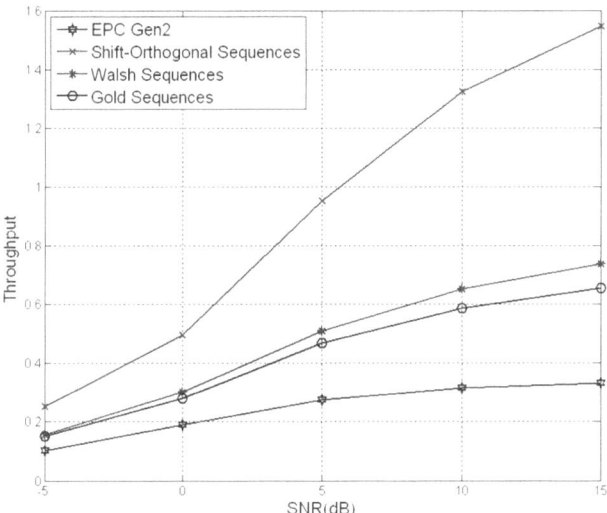

Fig. 5. An inventory of 500 tags. Note that the Shift-Orthogonal sequence denotes the proposed scheme.

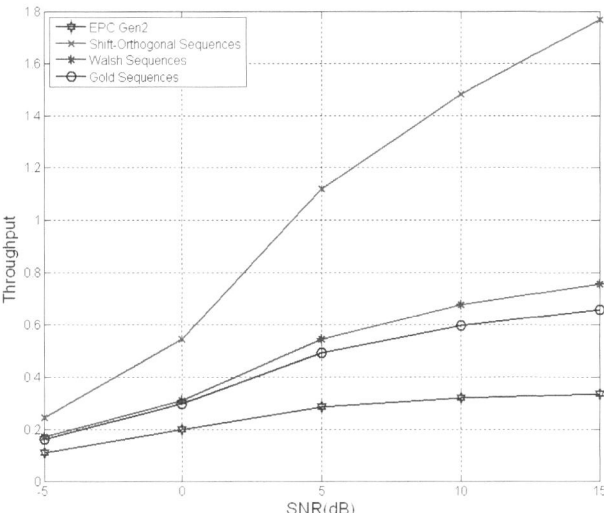

Fig. 6. An inventory of 1000 tags. Note that the Shift-Orthogonal sequence denotes the proposed scheme.

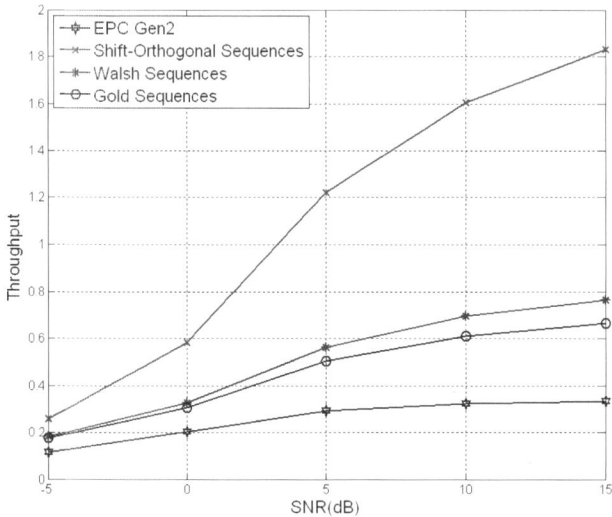

Fig. 7. An inventory of 2000 tags. Note that the Shift-Orthogonal sequence denotes the proposed scheme.

5 Conclusions

A novel passive tag backscatter scheme using Huffman spreading sequences is proposed, which can effectively improve an inventory process. The performance of several studied multiple access schemes are compared. Simulation results validate the performance enhancement of our proposed system. However the system may require a reader with high sensitivity, so that the levels of backscatter signals can be detected correctly. In addition, to implement such a system, further studies of cross-layer technologies combining TDMA (Mac layer) and CDMA (Physical layer) technologies are necessary.

Acknowledgments. This study is partially supported by the National Science Council of Taiwan under grant no. NSC 96-2221-E-011-017.

References

1. 18000-6 Part 6 – Parameters for Air Interface Communications at 860 to 960 MHz, International Organization for Standardization (2004)
2. EPC^{TM} Radio-Frequency Identity Protocols Class-1 Generation-2 UHF RFID Protocol for Communication at 860MHz-960MHz Version 1.09, http://www.epcglobalinc.com
3. Rohatgi, Anil: Implementation and Applications of an Anti-Collision Differential-Offset Spread Spectrum RFID System, Master Thesis of Science in the Electrical and Computer Engineering of Georgia Institute of Technology School (2006)
4. Pillai, V., et al.: A technique for simultaneous multiple tag identification. In: Fourth IEEE Workshop on Automatic Identification Advanced Technologies, pp. 35–38 (2005)

5. Huffman, D.A.: The generation of impluse-equivalent pulse train. IEEE Trans. Inf. Theory IT–8, S10–S16 (1962)
6. Tanada, Y.: Synthesis of a set of real-valued shift-orthogonal finite-length PN sequences. In: IEEE 4th International Symposium on Spread Spectrum Technique and Applications, pp. 58–62 (1996)
7. Tanada, Y.: Orthogonal Set of Huffman Sequences and Its Application to Suppressed-Interference Quasi-Synchronous CDMA System. IEICE Trans. Fundamentals E89–A(9) (2006)
8. Yen, C., Gutierrez, A.E., Veeramani, D., van der Weide, D.: Radar cross-section analysis of backscattering RFID tags. IEEE Antennas and Wireless Propagation Letters 6, 279–281 (2007)
9. Finkenzeller, K.: RFID Handbook, 2nd edn. Wiley, Chichester (2003)

Appendix

Some experimental results of T_1 link timing parameter of different tags are presented in this section. Table A presents the statistical results and Fig. A illustrates how the measurement is performed.

Table A. Experimental T_1 measurements using three different type Gen2 tags

Tag	Average T_1 (µs)
Alien GEN2 SQUIGGLE	244.53
Alien GEN2 OMNI-SQUIGGLE	246.09
TI Gen2	248.04

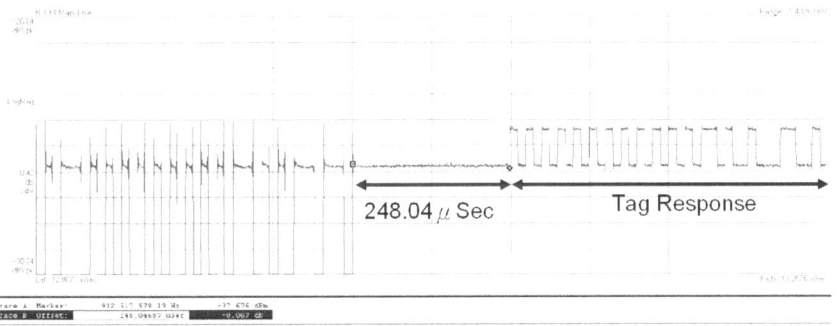

Fig. A. A snapshot of T_1 measurement of a TI Gen2 tag

Radio Frequency Identification Law Beyond 2007

Viola Schmid[*]

Technische Universität Darmstadt, Fachgebiet Öffentliches Recht,
Hochschulstr. 1, 64289 Darmstadt, Germany
schmid@jus.tu-darmstadt.de

Abstract. In 2007, RFIDs are not only a challenge for the fields of computer science and economics, but also for law. This paper develops a framework for RFID legislation and RFID lobbyism. With the coordinates "globality, verticality, ubiquity and technicity", the establishment of scenarios (RTAMP, RTAMA, EPC, AGG) and the exemplary categorization of pioneer legislature initiatives in the USA, the discussion about RFID law will be focused and qualified. The opinion presented in this article is that traditional privacy law is not sufficient to meet the challenges of RFID (Bag Paradigm).

1 RFID Law Versus Self Regulation, Self Control and Self Protection

1.1 Traditional Privacy Law (in Germany and the European Community)

Hessen – a German state – is the worldwide pioneer of privacy law. The world's first "Data Protection Act" was adopted in 1970 in Hessen. In 1983, the citizens' privacy consciousness and competence forced the highest German court – the Bundesverfassungsgericht (Federal Constitutional Court) – to render the groundbreaking "Volkszählungsurteil" (Population Census Judgment). The ratio decidendi of this decision from the ages of mainframe computers[1] amounts to:

> „A legal system in which the citizens do not know who knows what, when and on which occasion about them is inconsistent with the constitutional right to information privacy"[2] [3]

Therefore traditional privacy law has two principles: the protection of personal data of natural persons ("right to information privacy" = "Recht auf informationelle Selbstbestimmung") and the principle of transparency, which demands that the data

---

[*] The author would like to thank Ms Julia Gerhards and Ms Franziska Löw, doctoral candidates in our department, for their contributions.

[1] See, e.g., [1] p. 41 qualifying the differentiation as of no academic importance.

[2] BVerfGE 65, 1, 43: „Mit dem Recht auf informationelle Selbstbestimmung wäre […] eine Rechtsordnung nicht vereinbar, in der Bürger nicht mehr wissen können, wer, was, wann und bei welcher Gelegenheit über sie weiß".

[3] This essay uses grey shaded frames for projected law and non shaded frames for existing law.

C. Floerkemeier et al. (Eds.): IOT 2008, LNCS 4952, pp. 196–213, 2008.
© Springer-Verlag Berlin Heidelberg 2008

collection process be noticeable ("when") and calculable ("what"). US jurists, scientists and politicians can only – reflecting on their own data protection law and life experience – comment on this decision as follows: "Germany has another idea". Despite the impressive tradition of German and European privacy laws, it is a common belief that reforms are necessary. For RFID, three main criticisms about traditional privacy law are particularly relevant: firstly, German and European law concentrate on the protection of personal data of natural persons[4] and structurally disregard the sphere of information privacy of legal entities[5]. Secondly, the postulation of data calculability ("what") in the "Population Census Judgment" of 1983 is incompatible with the flood of data of "Ambient Intelligence, Nomadic, Pervasive and Ubiquitous Computing"[6]. Therefore a legal conception for "Identity Management"[5] instead of traditional "data protection and information privacy law"[7] is needed. And lastly, we require an answer to what is referred to as the "Bag Paradigm".

1.2 "Bag Paradigm" as a Challenge for Traditional Privacy Law (in the European Community)

The following "Bag Scenario" poses the question whether the traditional principle of transparency should be extended to non-personally identifiable information – what is referred to here as the "Bag Paradigm."

"Bag Scenario I.". The "Bag Scenario" can be found in a scientific study commissioned by a Panel of the European Parliament (2006/2007):

> "The Apenheul [case # 130] is a zoo specialized in all kinds of apes and monkeys. An outstanding feature of the park is the opportunity for some kinds of monkeys to move freely through the crowd of visitors. Curious as they are, the monkeys often

---

[4] European Community Law: "Personal data" shall mean any information relating to an identified or identifiable natural person ("data subject"); an identifiable person is one who can be identified, directly or indirectly, in particular by reference to an identification number or to one or more factors specific to his physical, physiological, mental, economic, cultural or social identity (Directive 95/46/EC of the European Parliament and of the Council of October 24, 1995 on the protection of individuals with regard to the processing of personal data and on the free movement of such data, Art. 2 (a)). US-Law: "Personal Information" means individually identifiable information about an individual collected online, including- (A) a first and last name; (B) a home or other physical address including street name and name of a city or town; (C) an e-mail address; (D) a telephone number, (E) a social Security Number; (F) any other identifier that the Commission determines the physical or online contacting of a specific individual; or (G) information concerning the child or the parents of that child that the website collects online from the child and combines with an identifier described in this paragraph. (Children's Online Privacy Protection Act of 1998); a new approach to defining personal data see [2].

[5] The reason being, that the constitutional right to information privacy is based upon the protection of human dignity and the right to free development of personality (Article 1 section 1 and Article 2 section 1 German Basic Law and BVerfGE 65, 1); see [3] p. 64.

[6] See [1] p. 41 qualifying the differentiation as of academic importance; Differing opinion [4] p. 416 ff. with the opinion that data protection laws in effect are flexible enough.

[7] See [6] p. 28 posing the question whether privacy is a foregone concept.

> try to open visitors' bags in hope of a free lunch. The park therefore introduced the "Monkey bag", a green bag with an extra clip lock which monkeys cannot open. The bag is obligatory, which is enforced by the receptionists providing the bag at the entrance of the park [...] Aside from this security reason for implementing the bag, the department of marketing added a marketing feature to the bag: scanning visitors movements through the park through an active RFID sewn into the bag [...] The Monkey Bag RFID has a marketing function: how do visitors move through the park and how can the flow of people be optimized. [...]"[8]

"Bag Paradigm". Clearly the "Bag Scenario" complies with traditional information privacy law because no personally identifiable information is involved. "The visitors remain unanimous, are not traced real time and do not suffer any consequences as a result of the data they provide."[9] Moreover, it is evident that such tracing should not be legally prohibited. Nevertheless future RFID law should incorporate a principle of transparency for non-personally identifiable information retrieved by RFID (the "Bag Paradigm" developed here). Consequently the marketing department of the zoo would be forced to inform the visitors about the active RFID. Nevertheless, it is then a matter of consumer choice whether they visit the zoo and a matter of provider choice whether the marketing department also offers bags without RFID. The "Bag Paradigm" poses the question whether "law for one technology" (LOT) – the RFID Law – is needed. In 2007 Germany's Green Party [8] as well as the German Parliament – the Bundestag – asked the German government to check if such a LOT [9][10][11][10] is necessary:

> „[...] In particular it has to be granted that the concerned persons are comprehensively informed about the usage, the reason for usage and the content of the RFID tag. It has to be possible to deactivate the tags used on the sale level or to delete the contained data when they are not needed anymore. Furthermore it has to be granted that data contained in RFID tags of different products are only processed in a way that prevents the profiling of secret behavior, usage or movement. The federal government is asked to report to the German parliament about a need for legislative action."[13] [11]

1.3 Law Versus Market

Self-Regulation, Self-Control and Self-Protection in the European Community. Law is the product of legislature, judiciary and administration. In 2007, RFID poses questions to the legislature and not yet to courts or agencies. The law model (legislation) is the opposite of the market model, in which providers choose self-control or self-regulation through Fair Information Practices (FIPs) [15] and customers choose self-protection through Privacy Enhancing Technologies (PETs) [6][15]. A well known [16] example for the market model is the Electronic Product

---

[8] See [7] p. 21 f.
[9] See [5] p. 16.
[10] Some opinions on RFID and Privacy Law see [12].
[11] See [14] p. 24 favoring ten different RFID fair information practices and self-regulation and privacy enhancing technologies.

Code (EPC) Guidelines by EPCglobal – "a subscriber-driven organization comprised of industry leaders and organizations focused on creating global standards for the EPCglobal Network [17]". The concept of the EPC-Guidelines is based on components like "Consumer Notice," "Consumer Education" and "Retention and IT-Security-Policy."[18] Infringements on the EPC-Guidelines only lead to a discussion of the infringement on an electronic platform. Neither penalties nor the expulsion of the infringer from the network is warranted. Therefore the EPC-Guidelines must be qualified as self-control and not as self-regulation. The EPC-Guidelines are highlighted because they can potentially provide an answer to the Bag Paradigm. Even industry spokespersons interested in global RFID marketing have understood that the customer has to be informed about the existence ("consumer notice" through special labels on tagged products) and the functionality of RFID ("consumer education"). According to the EPC-Guidelines, even RFIDs that do not contain any personally identifiable information have to be "discardable." With EPCglobal's self-control strategy, the "Bag Paradigm" could be answered as follows: Those individuals who are interested in a zoo visit have to be informed. Additionally the tags in the provided bags have to be discardable or the customers have to be offered an untagged bag. Still, EPC-Guidelines are an instrument of self-control, a strategy of the so called business ethics[12].

From the legal perspective of this essay, the EPC-Guidelines' lack of an enforcement strategy has to be seen as a disadvantage. Naturally it is the creed of a jurisprudent: only law that can be enforced by jurisdiction and administration creates certainty for users of RFID and for persons being confronted with RFID (wearable and implanted RFIDs). Consequently the Commission of the European Communities communicated in March, 2007:

> "It is clear that the application of RFID must be socially and politically acceptable, ethically admissible and legally allowable [...].[13] Privacy and security should be built into the RFID information systems before their widespread deployment (security and privacy-by-design), rather than having to deal with it afterwards.[14]"

The Commission did not answer the question: "Does this "security-and-privacy-by-design-concept" include the necessity of RFID legislation?"[23] Nevertheless the Commission asked this question in an online-consultation in 2006[15]. At least 55 % of the 2014 respondents believed that legislation is necessary and did not only want to rely on the market model with "Self Control" and "Fair Information Principles" (which 15 % preferred). From a 2007 perspective it can be summarized as follows: The Commission deals with the usage, the standardization and the interoperability (termed here as "Utilization Legislation") of RFID.[16] Concerning RFID privacy law, the European Commission only announced the establishment of an informal task force

---

[12] See [19] p. 28; other aims of business ethics are described in [20].

[13] See [21] p. 5; RFID in European Law see [22] p. 107 ff.

[14] See [21] p. 7.

[15] See [24] p. 18; [25] describing the European RFID initiatives.

[16] Commission Decision (2006/804/EC) of 11/23/2006 on harmonization of the radio spectrum for radio frequency identification devices operating in the ultra high frequency band (Art. 2 No.1).

(so called "Stakeholder Group"[17]). Still the Commission reserved the possibility for a future change of European data protection law.

RFID Legislation in the USA. Surprisingly, the worldwide pioneers of RFID law come from US state level, not from US federal level.[18] The future is forecast in the bills of state senates, houses of representatives and assemblies that are now processed in eighteen states of the USA – compare the attached table that is based on internet research in September 2007.[19] These RFID law initiatives might be the omen for change to privacy paradigms in the USA.[27][20] And this change might bridge the privacy gap between the USA and Europe. RFID in the USA is not only a matter of legislative policies, but is already a matter of law. In two states of the USA – Wisconsin (2006) and North Dakota (2007) – a chaptered referred to here as "Prohibition Legislation" exists:

> (1) No Person may require an individual to undergo the implanting of a microchip.
> (2) Any person who violates subsection 1 may be required to forfeit not more than $ 10 000[...]."[21]

These two states have assumed a positive obligation for a protection against the implantation of RFID chips. Both laws do not provide a concept of informed consent to the implantation of RFID.[30] This omission might be criticized, because embedded RFIDs could be useful in high-security environments (e.g. nuclear plants and military bases) in order to protect people and valuable gears or data.[31][22] Therefore this RFID legislation can really only be the beginning. In summary, the following agenda can be drawn:

1. Do we need law or is the market sufficient?
2. If legislation is postulated: Is traditional law sufficient or do we need new law?
3. If we decide on new law – like it is discussed in the USA right now: When do we need this law?

Technology First or Law First? There are few examples for the pre-existence of law before the technology reached the market. In Germany, such an example is the Law of Electronic Signatures that is older than a decade and that wanted to open the market for electronic signatures. RFID law could also precede the spread of RFID technology in the market – but this is exactly what Governor Arnold Schwarzenegger (California) rejected in 2006 [34][23]. He vetoed a RFID "IT Security Bill" with the words:

---

[17] See [21] p. 6: "This group will provide an open platform allowing a dialogue between consumer organizations, market actors and national and European authorities, including data protection authorities, to fully understand and take coordinated action [...]"

[18] See [26] discussing the various options for RFID Law on state and federal level.

[19] See the attached table at the end of section A.

[20] [28] p. 12 arguing the constitutionality of the tracking of persons with RFID; [29] p. 6 designing principles of fair information practice.

[21] 2005 Wisconsin Act 482 for 146.25 of the statutes (2005 Assembly Bill 290); Chapter 12.1-15 of the North Dakota Century Code (2007 Senate Bill No. 2415).

[22] See [32], p. 4, discussing the VeriChip, an implantable device carrying an unique key that hospitals could use to access medical records in emergency cases; [33] p. 8, describing the information market place for implantable and wearable chips.

[23] Arguing about a legal moratorium [35] p. 13.

"I am returning Senate Bill 768 without my signature. SB 768, which would impose technology regulations on RFID-enabled ID cards and public documents, is premature. The federal government, under the REAL ID Act, has not yet released new technology standards to improve the security of government ID cards. SB 768 may impose requirements in California that would contradict the federal mandates soon to be issued. In addition, this bill may inhibit various state agencies from procuring technology that could enhance and streamline operations, reduce expenses and improve customer service to the public and may unnecessarily restrict state agencies. In addition, I am concerned that the bills provisions are overbroad and may unduly burden the numerous beneficial new applications of contactless technology."

This veto to RFID "IT-Security-Law" arouses transatlantic interest, because the German and European legislators also deal with the IT security of identification documents at the moment. The European Regulation Law and the German Passport Act[24] is comparatively non-specific regarding security standards.[25]

"Passports and travel documents shall include a storage medium which shall contain a facial image. Member States shall also include fingerprints in interoperable formats. The data shall be secured and the storage medium shall have sufficient capacity and capability to guarantee the integrity, the authenticity and the confidentiality of the data."[26]

New RFID Legislation – A Systematic Approach. The representation of the developments in the USA, Europe and Germany requires a systematic approach.

(2)   The systematization of the challenges RFID places on law is done by coordination along the axes globality, verticality, ubiquity and technicity (Radar Chart, "Question position" of lobbyists and the legislature).
(3)   The systematization of facts about RFID using scenarios (beyond the Bag Scenario I.). The scenarios enable legal and jurisprudential discourses on the grounds of identified facts.
(4)   The systematization of the legislative initiatives and laws serves to qualitatively classify the answers to the challenges placed on RFID law ("Answer Position").

---

[24] New§ 4 Sec. 3 PassG of 07/20/2007 in action since 11/01/2007: „[…] sind der Reisepass […] mit einem elektronischen Speichermedium zu versehen, auf dem das Lichtbild, Fingerabdrücke, die Bezeichnung der erfassten Finger, die Angaben zur Qualität der Abdrücke und die in Absatz 2 Satz 2 genannten Angaben gespeichert werden. Die gespeicherten Daten sind gegen unbefugtes Auslesen, Verändern und Löschen zu sichern. […]".

[25] The „Bundesrat" has passed the Bill on 07/08/2007. Describing the IT-Security measures [36] p. 176 and [37] criticizing the security standards.

[26] Art. 1 Sec. 2 Council Regulation (EC) No 2252/2004 on standards for security features and biometrics in passports and travel documents issued by Member States, Official Journal L 385/1.

Table 1. Pending State-RFID-Legislation in the USA in 2007 (State September 2007)

State	Right-To-Know-Legislation	Prohibition Legislation	IT-Security-Legislation	Utilization-Legislation	Task-Force-Legislation
Arkansas	-	SB 195	-	SB 183	SB 846
California	SB 388	SB 29 SB 31 SB 362	SB 30	-	-
Georgia	-	HB 276	-	-	-
Massachusetts	HB 261 SB 159	-	-	-	-
Michigan	-	HB 4133 HB 5061 HB 5091	-	HR 51	-
Missouri	SB 210 SB 13	-	-	-	-
New Hampshire	HB 686		-	HB 269	-
New Jersey	AB 3996	SB 1866	AB 3015	AB 4061	-
New York	AB 222 AB 261				AB 225 SB 165
North Dakota	-	SB 2415	-	-	-
Oregon	HB 3277	-	-	-	-
Pennsylvania	HB 993	-	-	-	-
Rhode Island	-	SB 474	-	-	-
Tennessee	HB 2190	-	-	-	-
Texas	-	HB 1925	SB 2027	SB 574 HB 1308 HB 2990	-
Virginia	HB 2086	-	-	-	-
Washington		HB 1031 SB 6020		HB 1133 SB 5366	-
Wisconsin	-	AB 141 AB 488	-	-	-

2   RFID Legal Challenges in a Radar Chart

Every RFID lobbyist and legislator must first ask the question, if there should be a homogenous RFID law or if RFID law should be heterogeneous ("Question Position"). The classification of two RFID scenarios in the coordinate system with the four axes "globality, verticality, ubiquity and technicity" proves that RFID law will be differentiated and heterogeneous. The four axes can be defined briefly as follows:

1. Globality: In how many countries should RFIDs be freely produced and marketed?
2. Verticality: How long is the lifespan and/or the length of use of the tag/chip, reader and background device?
3. Ubiquity: To what extent are RFIDs part of our inner (subcutaneous use) or outer (ambient intelligence) present? To what extent are we surrounded by RFIDs – or not?
4. Technicity: Which technical qualities does the system show with respect to the processing of data, as well as the protection against unauthorized access?

Both scenarios – one EPC (Electronic Product Code II.) scenario and one RTAMP (Real-time Authentication and Monitoring of Persons III.) scenario – can be outlined as follows:

– In the EPC scenario II. obviously a huge number of products are supposed to be identified worldwide with a unique identifier number contained in a tag (this goes from clothing to household articles).
– In the RTAMP scenario III. here chosen, a chip is implanted into a person. This chip contains medical information (blood type, allergies or special diseases e.g. diabetes or epilepsy). In emergencies the chip enables doctors to access life saving information.

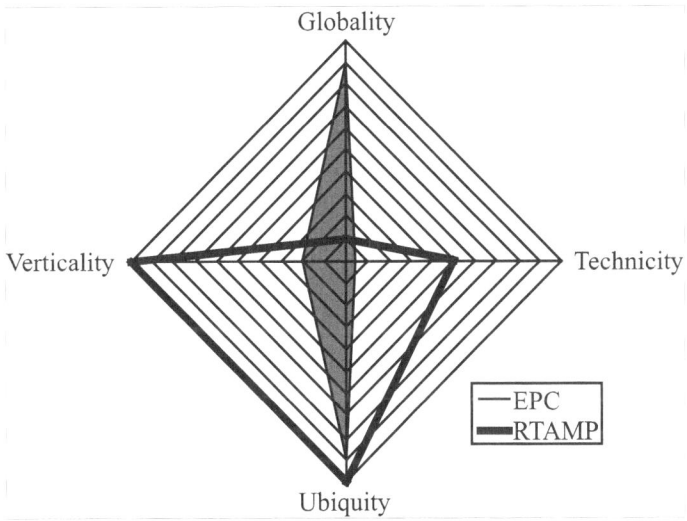

Fig. 1. Legal Challenges in a Radar Chart

2.1 Globality

EPC-Scenario II. It is indisputable that at the beginning of the third millennium economy and parallel technology – its companion or product – see a multinational or even global marketing potential. An increasing number of products will be offered to more consumers everywhere and anytime. Differing national RFID law complicates a multinational or global marketing of products and to be attached RFID technology. An obvious example is the projected "RFID Right to Know Act" of the state of New York, USA. Right now (September 2007) the New York legislators deliberate whether RFIDs have to be irrecoverably disabled after the purchase of the tagged item.[27] If not, the tags could always be reactivated.

> "Every retail mercantile establishment that offers items or packages that contain or bear radio frequency identification tags shall remove or deactivate all tags at the point of sale.[28]
> A radio frequency identification tag, once removed or deactivated, shall not be reactivated without express consent of the consumer associated with the tagged item.[29]"

The RFID marketing industry would have to diversify their marketing strategies if the New York legislative policy is going to be chaptered:

- Either the tags used in New York would have to fulfill higher standards than tags in other states where no option for deactivation or removal is demanded,
- Or the products destined for export to New York would not be allowed to carry a RFID tag.

A primary advantage of the tagging with RFID – the worldwide possible identification of a given product by one number – would then be crossed or complicated by a single state's law. In summary it can be said: the more global the manufacturing and sales strategies for RFID and the RFID-tagged products are, the more legal systems have to be synchronized. RFID is still a technology waiting for its market breakthrough. In the absence of empirical percentages, only educated guesses have been made in 2007 in this paper. Because of the high output and sales figures and the participation of many countries, the EPC scenario has a high globality value (90 %).

---

[27] The deactivation challenge is not only known in New York but also a component of the aforementioned online consultation of the European Commission [24]. As of 1984 respondents two thirds thought that RFID product tags in supermarkets should be automatically deactivated at the point of sale. Other solutions – i.e. a removable sticker attached to the product itself and a "proximity tag" with a very short reading distance – are favored by 51 % and 44 % of the respondents, respectively. One quarter of all respondents preferred the RFID tag to be part of the product's packaging.

[28] Section 219 (3) General Business Law as in New York 2007 Assembly Bill 222 ("RFID-Right-to-Know Act").

[29] Section 219 (3 A) General Business Law as in New York 2007 Assembly Bill 222 ("RFID-Right-to-Know Act").

RTAMP-Scenario III. The medical RTAMP scenario has a much lower globality value (10 %). Considering the comparatively high costs for the product and the ethical doubts, it is expected that only a little number of states will choose such a technology that – literally – "goes under one's skin". The second axis in the challenge chart is:

2.2 Verticality

EPC-Scenario II. Verticality means the potential durability of RFIDs. In the EPC scenario it is possibly the aim to make RFIDs last long enough to not only use them in the supply chain until purchase in a retail shop, but also for warranty instances or even waste management.[30] Of course, this verticality ambition is contradictory to RFID-Right-to-Know legislation (like in New York). This legislation would postulate that RFIDs are only permitted until the tagged product is purchased by a customer. If this "Right-to-Know" agenda is taken into account the EPC scenario has a low verticality value (manufacturing and sale = 20 %).

RTAMP-Scenario III. The RTAMP scenario, on the other hand, reaches the highest verticality value (100 %). Theoretically the tag accompanies the chipped person their whole life (with implantation as an infant). The life span of the wearer then is identical to the duration of the tag. The third axis is:

2.3 Ubiquity

Theoretically RFIDs could be used ubiquitously. The entire surrounding of a person (things, plants, animals) could be tagged (referred to here as "extrinsic ubiquity").

EPC-Scenario II. Because of the worldwide marketing strategies of many products, the EPC scenario also has a high (extrinsic) ubiquity value (90 %).

RTAMP-Scenario III. In contrast, the RTAMP scenario with implanted RFIDs is a scenario of inner ubiquity (referred to here as "intrinsic ubiquity"). RFIDs that "go under one's skin" accompany the wearer everywhere and have the full ubiquity value (100 %). The last axis in the radar chart is:

2.4 Technicity

In a legal consideration that is affected by privacy[31] and the principle of transparency, it is possible to differentiate RFIDs through (1) the (lacking) complexity of the tag, (2) the (lacking) complexity of the background devices and (3) the (maximum) reading range.

1. Tags can be differentiated into active and passive, rewriteable, processing (smart) and sensor provided products. Because of the privacy aspect, it is very important which IT-security standards and policies in the tag protect it from unauthorized reading attempts and searches. If the tags contain personal data – like in identification documents – these have to be protected from unauthorized reading.

---

[30] [38] p. 259 referring to [39].
[31] Differentiating some scenarios see [40] p. 9.

2. The validation of the complexity of the background devices – like the reader or other linked media (PC, laptop, cell phone, palm or Blackberry) – depends on how the RFID data are organized – especially if they are aggregated or profiled with other data. Furthermore, the communication between tag/chip and background device has to be protected against spying.[11][41]

3. The principle of transparency demands that the reading range of the RFID system is ascertained. Reading ranges from only a few centimeters (so called near-field-communication, e.g. at the purchase of tickets) [42] to several meters are possible. The principle being that the longer the reading range is, the less transparent is the sending or reading process. The shorter the reading range is, the more transparency exists. If a tag can only be read from 15 centimeters, the wearer will expectantly notice the reading process. If instead the tag can be read from 10 meters, the wearer probably would not notice. The reading/sending distance is – aside from the deactivation challenge – a very important aspect of RFID.[32]

EPC-Scenario II. Overall (1-3) the technicity value of the EPC scenario is low (5 %), because the tag is passive, not rewritable and it does not have sensors or any computing power. It has a small reading range.

RTAMP-Scenario III. The RTAMP scenario with the implanted chip gets a middle value (50 %). Such a tag contains sensitive data (§ 3 Section 9 German Data Protection Act) and therefore needs a more stringent IT-security policy that protects it from unauthorized reading or rewriting.

3 RFID Law Scenarios

Apart from the four challenge axes globality, ubiquity, verticality and technicity, a legal perspective can distinguish between four privacy orientated function scenarios.[43] Privacy – the protection of personal data – is a paradigm of traditional law and requires us to determine whether RFIDs are used to

– monitor products (*Electronic Product Code* scenario = EPC),
– monitor animals (*Real-time authentication and monitoring of animals* scenario = RTAMA)
– monitor persons (*Real-time authentication and monitoring of* persons scenario = RTAMP) or
– collect data for profiling purposes (*Aggregation* scenario = AGG).

The product scenario is called *EPC* scenario, because it characterizes a product just like a barcode.[33] Characteristically, EPC scenarios initially do not show any relation

---

[32] The online consultation of the European Commission [24] asked what the respondents thought about the maximum reading distance, which could be considered as acceptable for "proximity tags". It "received answers from 1342 respondents (60 %). About 10 % do not consider the concept of "proximity tags" as a valuable solution to preserve privacy. A variety of arguments are brought forward, e.g., "inappropriate for certain applications"; "useless because of the discrepancies between specified and real reading ranges"; "difficult to enforce", "consumers should be notified".

[33] Differing from a barcode scenario an individual product is identified.

to personal data. Still, a person who carries an RFID tagged item gives the entity which uses the RFID system the opportunity to collect their information. This is why the legislative initiative in New York[34] demands that EPC tags must be able to remove or deactivate. The verticality of this scenario is legally limited to the presence of the customer in the retail shop.

RTAMA scenarios – e.g. the tagging of dogs[35] or cattle[36] – do not show any personal relation if the tag contains only a number. In the case of more complex tags, even such RTAMA scenarios can involve personal data (e.g. if the tag contains information about the breeder or owner). From a legal perspective, questions about the sanitary harmlessness of the implantation of tags [44] and the owners' interest in keeping depreciating information uncovered have to be asked.

From a traditional privacy law point of view, *RTAMP* scenarios are the most challenging. Examples for RTAMP are the authentication of persons through identification documents that contain an RFID tag and make personal data remotely readable, as well as the monitoring of pupils, employees or elderly requiring care.

The privacy importance of all three scenarios can be qualified with the *AGG* scenario. An example of this is when the EPC scenario is combined with credit card data. The New York legislature initiative wants to prohibit exactly that:

> "No retail mercantile establishment shall combine or link a consumer's personal information with information gathered by, or contained within, a radio frequency identification tag."[37]

In summary, it can be said that the four challenge axes of the radar chart are coordinates for the actual usage of RFID. Each of the four scenarios reflects a different quality of privacy law. The greater the privacy relation is – as always in RTAMP scenarios – the more significance the "traditional law" has. The less privacy issues are affected by the application of RFIDs – as it is in the EPC scenarios or the "Bag Paradigm" – the greater is the importance of the legislature politics and the more relevant is the question about the need for new law. Also, this new law in action can be further categorized as follows:

4   Categories of Future RFID Legislation

Future RFID legislation might have five different goals:

(1)   Right-to-Know-legislation,
(2)   Prohibition-legislation,
(3)   IT-Security-legislation,
(4)   Utilization-legislation and
(5)   Task-Force-legislation.

---

[34] As in Fn. 29.
[35] E.g. in Hamburg, Germany.
[36] E.g. in Texas, USA.
[37] Section 219 (4 A.) General Business Law as in New York 2007 Assembly Bill 222 (RFID Right to Know "Act").

1. *Right–to-Know-legislation* denotes initiatives demanding that the customer be informed in RFID scenarios. Furthermore, the customers shall be able to deactivate the tags after purchase, so that after leaving the retail shop reading attacks are unsuccessful. In 2004 (108th congress), there was an unsuccessful legislative initiative with the title "Opt Out of ID Chips Act" on the federal level. RFID tagged products were supposed to be labeled.[38] On the state level, such legislative initiatives can be found in: California (2007 Senate Bill 388 "RFID Privacy"), Massachusetts (2007 Senate Bill 159 and House Bill 261 "Consumer Protection and RFID"), Missouri (2007 Senate Bill 13 and 210 "RFID Right To Know Act of 2007"), New Hampshire (2007 House Bill 686 "Regulation of tracking devices"), New Jersey (2007 Assembly Bill 3996), New York (2007 Assembly Bill 261 "Labeling of retail products"; 2007 Assembly Bill 222 "Radio Frequency Right To Know Act"), Oregon (2007 House Bill 3277), Tennessee (2007 House Bill 2190 "Labeling of Retail Products"), Virginia (2007 House Bill 2086 "Labeling") and Washington (2007 Senate Bill 6020, 2007 House Bill 1031 "Labeling").

2. *Prohibition-legislation* means initiatives that want to forbid or at least restrict the usage of RFID in certain scenarios. Prohibition-legislation is the opposite of Utilization-legislation. On the federal level, no examples for such legislation can be found. On the state level, several bills have been introduced. In California, the legislation was concerned with RTAMP scenarios (attendance and tracking of location of pupils shall be prevented by prohibiting their tagging, California 2007 SB 29 "Pupil Attendance: Electronic Monitoring"). A legislative initiative in Texas (Texas 2007 House Bill 1925 "Electronic Monitoring of Students") demands that the identification of students through RFIDs should only be used after prior consent of the guardians. Those who did not give their consent have to be offered an alternative option for identification. In Rhode Island, one initiative wants to prohibit the governmental use of RFIDs for the identification or monitoring of students, employees or contractual partners (Rhode Island 2007 Senate Bill 474 "Restriction of Radio Frequency Identification Devices"). A Michigan initiative plans to forbid the implantation and even the provision of RFIDs to humans without their prior consent (Michigan 2007 House Bill 4133 "Implanting RFID Chips into Individuals"). In the Wisconsin Assembly, an initiative has been introduced that wants to forbid the embedding of RFIDs in documents and bank notes (Wisconsin 2007 Assembly Bill 141 "Prohibition of RFID Tags in Documents"). An Arkansas Senate Bill wants to restrict the reading of RFIDs after the tagged product has been purchased and the item has been removed from the premises of the retail shop (Arkansas 2007 Senate Bill 195 "Limitation on Use of RFID Tags").

3. *IT-Security-legislation* denotes initiatives that demand certain IT security standards and aims for the protection of RFIDs from unauthorized reading and rewriting. Accordingly, initiatives in California and Washington demand that RFIDs in identification documents have to be provided with IT-security technologies (e.g. encryption, California 2007 Senate Bill 30 "Identity Information Protection Act of 2007"; Washington 2007 House Bill 1289 "RFID Chips in Driver's Licenses or Identicards").

---

[38] H.R. 4637 108th Congress, Sponsors: Klezka and Lewis.

4. *Utilization-legislation* means initiatives that want to use RFID in certain scenarios. On the federal level, RFIDs shall be embedded in identification documents – RTAMP scenario. On the state level – e.g. in Texas – RFIDs are supposed to be used for the identification of animals (Texas 2007 Senate Bill 574, 2007 House Bill 1308 "Deer Identification") – RTAMA scenarios. In California, a bill concerning the use of RFID for the access of visually impaired persons to financial services was introduced, but not further pursued (California 2005 Assembly Bill 1489). Since the touch screens of point-of-sales-systems were not manageable for visually impaired persons, RFID was supposed to grant identification. Another RTAMP scenario can be found in a legislative initiative of Texas. Hereafter inmates, employees and visitors of correctional facilities are supposed to carry a RFID tag with them (Texas 2007 House Bill 2990).
5. *Task-Force-legislation* means initiatives that demand a round table for the research of the legal and technical challenges of RFID. Characteristically Task-Force-legislation encompasses all scenarios (EPC, RTAMP, RTAMA, AGG). In the USA, no such legislation exists on the federal level. According to a hint from the internet, all that seems to exist in this context is an RFID Caucus in the US Senate that only describes an informal political board and does not have any legal accreditation. On the state level, legislative initiatives for RFID task forces exist in Arkansas (2007 Senate Bill 846) and in New York (2007 Senate Bill 165 and 2007 Assembly Bill 225).

5   Outcome and Outlook

This report concentrates on two pioneer states, despite having a global perspective.

5.1   RFID Law Pioneer: USA

The three questions posed by this report can be answered for the RFID Law Pioneer, the USA, in the following manner: the overview of legislative activities in eighteen states shows that the market is most likely insufficient and the fact that the discussion for new laws is both important and controversial (questions 1, 2). If and when (question 3) these laws will be put into action depends on the length and sustainability of these procedures. One cannot exclude the possibility that these legal initiatives "dry up" in the various houses (senate, assembly, house) or that one of the governors passes a veto. However, it should be highlighted that on the federal level a utilization legislature exists with the "Food and Drug Administration Amendments Act of 2007"[39]. Standards should be developed accordingly,

"to identify and validate effective technologies for the purpose of securing the drug supply chain against counterfeit, diverted, subpotent, substandard, adulterated, misbranded, or expired drugs". …Not later than 30 months after the date of the enactment of the Food and Drug Administration Amendments Act of 2007, the Secretary shall develop a standardized numerical identifier (which, to the extent

---

[39] Food and Drug Administration Amendments Act of 2007, Pub. L. No. 110-85 [45], Sec. 913 inserting Section 505d "Pharmaceutical Security".

practicable, shall be harmonized with international consensus standards for such an identifier) to be applied to a prescription drug at the point of manufacturing and repackaging (in which case the numerical identifier shall be linked to the numerical identifier applied at the point of manufacturing) at the package or pallet level, sufficient to facilitate the identification, validation, authentication, and tracking and tracing of the prescription drug." The standards developed under this subsection shall address promising technologies, which may include: radio frequency identification technology; nanotechnology; encryption technologies; and other track-and-trace or authentication technologies."

5.2 Data Protection Pioneer: Germany

The three questions posed in this report can be answered for the Data Protection Pioneer, Germany, in the following manner: The honest answer to the first two questions "Do we need law or is the market sufficient (1)" and "If we need law: is traditional law sufficient or do we need new law? (2)" is "It depends.". This answer is as unloved as it is old, and prepares us for the necessity of differentiation.

- An RTAMP Scenario III., in which patients are required to have a chip with medical information implanted, requires a legal foundation and that the IT-security be secured legally. This chip must be protected from both unauthorized access to the information stored on it and unauthorized overwriting of the information. According to German law ("de lege lata")[40], the government must first legally recognize this type of under-the-skin healthcare. In this respect, the answer is: yes, traditional law postulates that we need new law.[41]
- For an EPC Scenario II., which is only used for identification of objects in supply chains and in which no aggregation interest with personal data exists (AGG), in our opinion the market and its traditional market law is sufficient. All members of the chain can reach an agreement about whether or not they support the use of RFID. The answer is: no, we do not need new law. As a "caveat" we must adhere to the following: If EPC data are aggregated with personal data in such a way that localization and performance statements about employees become possible, then the area of application[42] – at least according to existing German law ("de lege late") – has been opened up (yes, there is constitutional and statutory law). The answer for the EPC-AGG scenario is: no, we do not need new law.
- The market is not sufficient for the Bag Scenario I. Laws currently in effect do not encompass this scenario. The creation of new laws ("de lege ferenda") is necessary, that allow the informed public to choose between providers with such profiling strategies and others ("Consumer Information" and "Consumer Choice"). The answer is: yes, we need new[43] law.

---

[40] E.g.: Positive obligation resulting from Art. 2 sec. 2 GG (German Constitution): Everyone has the right to life and physical integrity. The liberty of the individual is inviolable. These rights may be encroached upon only pursuant to statute.

[41] For experts: new statutory law.

[42] E.g. § 28 BDSG; § 87 sec. 1 item 6 BetrVG.

[43] E.g.: This scenario requires a new interpretation of constitutional law (Art. 1 sec. 1 GG and Art. 2 sec. 1 GG) and then the remittal of a statutory law.

The answer to the third question: "if we decide on new law – like it is discussed in the USA right now: When do we need this legislation?" is: as soon and qualified as possible. Legislative procedures attract high levels of publicity. In this way they inform the citizens, prevent technophobia and they are the way to weigh differing interests – with all skepticism and criticism – in our legal systems. Prerequisite for a qualified legislative procedure is that the participating antagonists can present their positions in a qualified manner. This requires technology to be defined, the economy to be calculated and law and politics to be interested and involved.

5.3 Ideal RFID Law

We must wait and see what the outcome of the legislative procedures in the USA is. It could be a valuable experience for the still hesitant lawmakers of the European Community and for Germany in particular. Only then will it be decided if this RFID Law will satisfy the ideal of "legal and market compatible" and "technology and market tolerable" law.

References

1. Mattern, F.: Die technische Basis für das Internet der Dinge. In: Fleisch, E., Mattern, F. (eds.) Das Internet der Dinge, pp. 39–66. Springer, Berlin Heidelberg (2005) (09/17/2007), http://www.vs.inf.ethz.ch/publ/papers/internetdinge.pdf
2. Data Protection Working Party, Opinion N° 4/2007 on the concept of personal data, WP 136, 20.06.2007 (2007) (09/17/2007), http://ec.europa.eu/justice_home/fsj/privacy/docs/wpdocs/2007/wp136_en.pdf
3. Roßnagel, A., Pfitzmann, A., Garstka, H.: Modernisierung des Datenschutzrechts (2007) (09/18/2007), http://www.verwaltung-innovativ.de/Anlage/original_549072/Modernisierung-des-Datenschutzrechts.pdf
4. Holznagel, B., Bonnekoh, M.: Rechtliche Dimensionen der Radio Frequency Identification. In: Bullinger, H., ten Hompel, M. (eds.) Internet der Dinge, pp. 365–420. Springer, Heidelberg (2007)
5. European Parliament, Scientific Technology Options Assessment STOA, RFID and Identity Management in Everyday Life (June 2007) (09/18/2007), http://www. europarl.europa.eu/stoa/publications/studies/stoa182_en.pdf
6. Langheinrich, M.: RFID und die Zukunft der Privatsphäre. In: Roßnagel, A., Sommerlatte, T., Winand, U. (eds.) Digitale Visionen – Zur Gestaltung allgegenwärtiger Informationstechnologien, pp. 43–69. Springer, Berlin (2008) (09/17/2007), http:// www.vs. inf.ethz.ch/publ/papers/langhein-rfid-de-2006.pdf
7. European Technology Assessment Group, RFID and Identity Management in Everyday Life (2006) (09/13//2007), http://www.itas.fzk.de/eng/etag/document/hoco06a.pdf
8. heise news 03/13/2007, Grüne fordern Kennzeichnungspflicht für RFID (2007) (09/18/2007), http://www.heise.de/newsticker/meldung/86662
9. Entschließung der 72. Konferenz der Datenschutzbeauftragten des Bundes und der Länder vom 26./27.10.2006 (2006) (09/13/2007), http:// www. bfdi.bund. de/cln_029/ nn_531946/ DE/Oeffentlichkeits-arbeit/Entschliessungssammlung/DSBund-Laender/72DSK-RFID,templateId=raw,property=publicationFile.pdf/72DSK-RFID.pdf

10. Positionspapier des Bundesverbandes BITKOM zum Working Paper 105 der Artikel-29-Datenschutzgruppe (2007) (09/13/2007), http://www.bitkom.de/files/documents/Position_zum_RFID_Paper_der_Art._29_Group_20050321.pdf

11. Holznagel, B., Bonnekoh, M.: RFID – Rechtliche Dimensionen der Radiofrequenz-Identifikation (2007) (09/13/2007), http://www.info-rfid.de/downloads/rfid_rechtsgutachten. pdf

12. Schmitz, P., Eckhardt, J.: Einsatz von RFID nach dem BDSG – Bedarf es einer speziellen Regulierung von RFID-Tags? In: CR 2007, pp. 171–177 (2007)

13. German Bundestag Drucksache 16/4882, http://dip21.bundestag.de/dip21/btd/16/048/1604882.pdf

14. Eschet, G.: Fips and Pets for RFID: Protecting Privacy in the web of Radio Frequency Identification. Jurimetrics Journal 45, 301–332 (2005)

15. Langheinrich, M.: Die Privatsphäre im Ubiquitous Computing – Datenschutzaspekte der RFID-Technologie. In: Fleisch, E., Mattern, F. (eds.) Das Internet der Dinge, pp. 329–362. Springer, Berlin Heidelberg (2005) (09/17/2007), http:// www.vs. inf.ethz.ch/ publ/ papers/ langhein2004rfid.pdf

16. Thiesse, F.: Die Wahrnehmung von RFID als Risiko für die informationelle Selbstbestimmung. In: Fleisch, E., Mattern, F. (eds.): Das Internet der Dinge, pp. 363--368, Springer, Berlin Heidelberg (2005) (09/17/2007),
http://www.springerlink.com/content/k88p376q9q420227/fulltext.pdf

17. (09/17/ (2007), http://www.epcglobalinc.org/about/

18. http://www.epcglobalinc.org/public/ppsc_guide/ (09/12/ (2007)

19. Brey, P.: Ethical Aspects of Information Security and Privacy. In: Petkovic, M., Jonker, W. (eds.) Security, Privacy, and Trust in Modern Data Management, pp. 21–38. Springer, Berlin Heidelberg (2007)

20. Koslowski, P., Hubig, C., Fischer, P. (eds.): Business Ethics and the Electronic Economy. Springer, Berlin Heidelberg (2004)

21. Communication from the Commission to the European Parliament, the Council, the European Economic and Social Committee and the Committee of the Regions, Radio Frequency Identification (RFID) in Europe: Steps towards a policy framework (2007) (09/18/2007), http://ec.europa.eu/information_society/policy/rfid/doc/rfid_en.pdf

22. Toutziaraki, T.: Ein winziger kleiner Chip, eine riesengroße Herausforderung für den Datenschutz, DuD 2007, pp. 107–112 (2007)

23. Article-29-Group: Working document on data protection issues related to RFID technology (10107/05/EN WP 105) (2005) (09/13/2007), http://ec.europa.eu/justice_home/fsj/privacy/docs/wpdocs/2005/wp105_en.pdf

24. Commission of the European Communities: Results Of The Public Online Consultation On Future Radio Frequency Identification Technology Policy, SEC (2007) 312 (2007) (09/13/2007), http://ec.europa.eu/information_society/policy/rfid/doc/rfidswp_en.pdf

25. Huber, A.: Radiofrequenzidentifikation - Die aktuelle Diskussion in Europa. In: MMR 2006, pp. 728–734 (2006)

26. Hildner, L.: Defusing the Threat of RFID: Protecting Consumer Privacy Through Technology-Specific Legislation at the State Level. 41 Harvard Civil Rights - Civil Liberties Law Review 133 (2006)

27. NIST Guidelines for Securing Radio Frequency Identification Systems 4-4 (2007) (09/13/2007), http://csrc.nist.gov/publications/nistpubs/800-98/SP800-98_RFID-2007.pdf

28. Dalal, R.S.: Chipping away at the Constitution: The increasing use of RFID chips could lead to an erosion of Privacy Rights, 86 Boston University Law Review 485 (2005)

29. Levary, R.R., Thompson, D., Kot, K., Brothers, J.: Radio Frequency Identification: Legal Aspects. 12 Richmond Journal of Law & Technology 6 (2005)

30. FIDIS, Future of Identity in the Information Society, D7.7:RFID, Profiling and AmI (2006) (10/12/2007), http://www.fidis.net/fileadmin/fidis/deliverables/fidis-wp7-del7.7. RFID_Profiling_AMI.pdf
31. (12/10/2007), http://rfidlawblog.mckennalong.com/archives/privacy-rfid-tags-in-humans-okay.html
32. Eden, J.M.: When Big Brother Privatizes: Commercial Surveillance, the Privacy Act of 1974, and the Feature of RFID, 2005 Duke Law & Technology Review 20 (2005)
33. Schwarzt, P.M.: Property, Privacy, and Personal Data, 117 Harvard Law Review 2055 (2004)
34. Bill No.: 2006 SB 768; Veto Date: 09/30/2006 (2006) (09/12/2007),
35. http://www.leginfo.ca.gov/pub/05-06/bill/sen/sb_0751-0800/sb_768_vt_20060930.html
36. Handler, D.: The Wild, Wild West: A Privacy Showdown on the Radio Frequency Identification (RFID) Systems Technological Frontier, 32 Western State University Law Review 199 (2005)
37. Kuegler, D., Naumann, I.: Sicherheitsmechanismen für kontaktlose Chips im deutschen Reisepass, DuD 2007, pp. 176–180 (2007)
38. FIDIS, Budapest Declaration on Machine Readable Travel Documents (MRTDs) (2007)(09/13/2007), http://www.fidis.net/press-events/press-releases/budapest-declaration/
39. Roßnagel, A., Hornung, G.: Umweltschutz versus Datenschutz? In: UPR 2007, pp. 255–260 (2007)
40. BIN IT! – The Intelligent Waste Management System (2005) (09/13/2007), http://www.embedded-wisents.org/competition/pdf/schoch.pdf
41. Arbeitskreis „Technische und organisatorische Datenschutzfragen" der Konferenz der Datenschutzbeauftragten des Bundes und der Länder: Datenschutzgerechter Einsatz von RFID (2006) (09/13/2007), http://www.datenschutz-berlin.de/to/OH-RFID.pdf
42. Holznagel, B., Bonnekoh, M.: Radio Frequency Identification – Innovation vs. Datenschutz? In: MMR 2006, pp. 17–23 (2006)
43. heise news 09/05/2007: Österreichische Fahrkarten per RFID kaufen, http://www.heise.de/newsticker/meldung/95490
44. Schmid, V.: RFID and Beyond. In: Heinrich, C. (ed.) Growing Your Business Through Real World Awareness, pp. 193–207. Wiley Publishing Inc., Indianapolis (2005)
45. heise news 09/11/2007, Krebsverdacht bei implantierten RFID-Chips (09/18/2007), http://www.heise.de/newsticker/meldung/95797
46. http://frwebgate.access.gpo.gov/cgi-bin/getdoc.cgi?dbname=110_cong_public_laws&docid=f:publ085.110.pdf (12/10/2007)

Why Marketing Short Range Devices as Active Radio Frequency Identifiers Might Backfire

Daniel Ronzani

Centre for Applied ICT, Copenhagen Business School
Howitzvej 60, 2000 Frederiksberg, Denmark
dr.inf@cbs.dk

Abstract. This paper analyses why marketing short range devices (SRD) as radio frequency identifiers (RFID) might backfire on the RFID industry. To support this claim it provides a legal use case as basis and gives an overview of selected technical parameters of RFID. Furthermore an analysis of 43 legal articles shows that legal experts perceive the technology of RFID in an undifferentiated way. Finally an analysis of 11 tag providers of so called "active RFID tags" shows that SRD are marketed as active RFID. It concludes that in order to avoid inhibiting legal consequences which might have negative effect on the RFID industry a differentiated approach regarding the functionality of short range transmitters and RFID is necessary.

Keywords: Radio frequency identification (RFID), short range device (SRD), passive tag, active tag, law, industry.

1 Introduction

This paper analyses the legal perception and technical use of so called "active" radio frequency identification (RFID) tags. It is of value not only to the legal community to clarify and differentiate the technology of RFID but also to the RFID industry in that it assumes that marketing short range transmitters as active RFID tags could be counter-productive to the RFID industry after all.

Admittedly, the understanding of RFID technology is not always clear. For instance, there is a narrow and a wide understanding of what an active RFID tag is insofar as short range transmitters are also marketed as RFID. Therefore the question about an accurate and also differentiated understanding of RFID technology in the legal discussion becomes crucial. While the RFID industry includes short range transmitters in the product line of RFID to increase sales, the legal consequences of merging the functionalities of both RFID and short range devices (SRD) might lead to a restrictive legal interpretation and understanding of RFID because the technical features of SRD are broader than those of RFID.

First, a legal use case is presented as basis for the subsequent legal and technical analysis (section 2). Second, an overview of RFID technology with regard to tag functionality, energy supply and coupling is given in section 3 to set the technical ground in the discussion of RFID and SRD. Third, an analysis of 43 journal articles

C. Floerkemeier et al. (Eds.): IOT 2008, LNCS 4952, pp. 214–229, 2008.
© Springer-Verlag Berlin Heidelberg 2008

shows the legal perception and understanding of RFID technology between 2001 and 2007 (section 4). Fourth, an industry analysis with an empirical sample of 11 providers of so called "active" RFID tags illustrates how the RFID industry is marketing SRD as RFID (section 5).

This paper concludes by suggesting that the industry's approach of marketing short range transmitters as RFID could backfire on the industry (section 6).

2 Use Case

The use case selected to support the claim of this paper is taken from the empirical sample provided in section 3.1. It illustrates why it makes a difference to discriminate RFID tag functionality.

Dalal [5] examines the use of RFID technology in various contexts and privacy invasions. The use case of this paper focuses especially on the potential use of RFID and the Fourth Amendment of the Constitution of the USA [29] which stipulates that it is "[t]he right of the people to be secure in their persons, houses, papers, and effects against unreasonable searches and seizures […]" [29]. There has been extensive ruling on this topic with the following four landmark cases on telephony, radio transmitter and thermal imaging surveillance:

1. In Olmstead v. United States [22] the plaintiff(s) was (were), among others, convicted of violating the National Prohibition Act [21] for unlawfully possessing, transporting and importing intoxicating liquors. The (divided) U.S. Supreme Court held that evidence collected by wiretapping telephone lines did not violate the Fourth Amendment because "[t]he language of the amendment cannot be extended and expanded to include telephone wires" [22]. The wiretapping had been effectuated without a physical trespass by the government, and was thus legally obtained.

2. In 1967 the U.S. Supreme Court overruled its decision in Olmstead [22]. In Katz v. United States [15] the court argued that evidence overheard by FBI agents who had attached an electronic listening and recording device to the outside of a public telephone booth from which the plaintiff had placed his calls for bets and wagers in violation of the Criminal Law was searched unconstitutionally. Whether a given area was constitutionally protected deflected attention from the problem presented, as the Fourth Amendment protected people, not places: "[…] the Fourth Amendment protects people – and not simply 'areas' – against unreasonable searches and seizures [… so] it becomes clear that the reach of that Amendment cannot turn upon the presence or absence of a physical intrusion into any given enclosure." [15]

3. In 1983 the U.S. Supreme Court ruled on a Fourth Amendment Case that monitoring the progress of a car carrying a container with a "beeper" (i.e. a battery operated radio transmitter which emits periodic signals that can be picked up by a radio receiver) did not violate the defendant's constitutional rights. In United States v. Knotts [30] the court decided that monitoring the beeper signals did not invade any legitimate expectation of the defendant's privacy because there was a diminished expectation of privacy in an automobile: "One has a lesser expectation of privacy in a motor vehicle because its function is transportation and it seldom serves as one's residence or as the repository of personal effects. A car has little capacity for escaping public scrutiny. It travels public thoroughfares where both its occupants and its contents are in plain view." [4]

4. Finally, in Kyllo v. Unites States [19], the U.S. Supreme Court ruled in a case of thermal heating surveillance that use of thermal imaging devices to gather information about heat in a house's interior is not removed from scope of Fourth Amendment search merely because the device captures only heat radiating from external surface of a house, and thus involves "off-the-wall" rather than "through-the-wall" observation. In this case agents of the United States Department of the Interior suspected that marijuana was being grown in the petitioner's home. The court argued that where "the Government uses a device that is not in general public use, to explore details of a private home that would previously have been unknowable without physical intrusion, the surveillance is a Fourth Amendment "search," [sic!] and is presumptively unreasonable without a warrant." [19]

To date there have been no Supreme Court rulings on RFID surveillance. However, spinning forth the courts' present decisions, Dalal [5] argues that among the many factors to be considered in a potential RFID ruling, RFID searches are likely to be found constitutional under the Fourth Amendment because "tracking devices in public places are not considered to violate an objective expectation of privacy". [5]

After having provided the use case in this section an overview of selected technical parameters necessary for the subsequent analysis is discussed in the next section.

3 Technical Parameters

This section covers the technical parameters of RFID technology. An overview of these parameters is important to understand the inaccuracy in legal analyses (section 4) and to better understand the industry approach (section 5).

According to Kern [16], Finkenzeller [11] and Glover [12] the most common classifications for RFID are:

Table 1. Differentiators of RFID according to Kern [16], Finkenzeller [11] and Glover [12] (adapted). Characteristics that are italicized often but not necessarily (or solely) group in the vertical.

Differentiator	Characteristics		
Frequency	*Low frequency (30 – 300 kHz)*	*High frequency (3 – 30 MHz)*	*UHF (300 MHz – 3 GHz) and Micro-ware (> 3 GHz)*
Memory and data	1-bit (no chip)	n-bit (chip with ID)	
Energy supply chip	*Passive*	*Semi-active/-passive*	*Active*
Communication	Full Duplex	Half Duplex	Sequential
Coupling	*Capacitive (electrical) coupling*	*Inductive coupling*	*Backscatter coupling*
Read range	*Close proximity: ≈ < 1cm*	*Remote (or vicinity): ≈ 1cm – 1m*	*Long Range: ≈ > 1m*
Antenna	*Coil*	*Ferrite*	*Dipole*

A more detailed explanation follows for the energy supply of RFID tags (section 3.1), and the way they broadcasts to the reader (section 3.2).

3.1 Energy Supply

Three types of transponders vary in energy supply: passive, semi-active/semi-passive, and active RFID tags. It is important to note and it is herein argued that the common division of tags by energy supply – i.e. passive tags without own energy supply and active tags with their own energy supply for the chip in the tag (for example batteries or solar cells) – has nothing to do with the transmission of data from the transponder to the reader. In either case the tag needs the energy of the reader to transmit the data.

Active transponders use their own power source only to supply the chip in the tag with energy and not to transmit the data from the transponder to the reader. The advantage of an own power supply in active tags is that all energy from the reader can be used for data transmission because the chip is already supplied with energy by a separate source (e.g. battery). This dual energy supply has positive effects on the read range because all energy derived from the reader can be used for transmission and no energy is lost for powering the chip. [11] [16] Active RFID tags – herein understood as active RFID tags in the narrow sense according to Finkenzeller [11] – do not have the capability of emitting their own high frequency signal. According to Finkenzeller [11] transponders with the capability of emitting an own frequency signal are not RFID transponders but rather SRDs. These devices emit their own high frequency electro-magnetic field without influencing the field of the reader. [11] Bensky [2] makes the same differentiation:

> "A quite different aspect of the data source is the case for RFIDs. Here the data are not available in the transmitter but are added to the RF signal in an intermediate receptor, called a transducer. [...] This transducer may be passive or active, but in any case the originally transmitted radio frequency is modified by the transducer and detected by a receiver that deciphers the data added [...] A basic difference between RFID and [transmitter-receiver] is that RFID devices are not communication devices per se but involve interrogated transponders." (Bensky [2])

Glover et al. [12] acknowledge the differentiation of power source for passive and active tags. Traditionally active tags use the internal energy source to power the chip and the reader to power communication. However, these authors opt to use the term semi-passive for tags that only use the internal power supply to feed the chip (or other devices) but not for communication. [12]

A further source of definitions for passive, semi-passive/semi-active and active RFID tags is EPC Global Inc.[1] and RFID Journal[2]. EPC Global and RFID Journal discriminate the following functionalities of the different tag types in Table 2.

---

[1] http://www.epcglobalinc.org (last visited December 9, 2007). EPC Global Inc. is a leading organisation for the industry-driven standards of the Electronic Product Code (EPC) to support RFID.

[2] http://www.rfidjournal.com (last visited December 9, 2007)

Table 2. Selected definitions of RFID functionalities

Source	Type	Tag definition and functionality [own emphasis]
EPC Global Inc.	Passive tag	"RFID tag that does not contain a power source. The tag generates a magnetic field when radio waves from a reader reach the antenna. This *magnetic field powers the tag* and enables it to send back information stored on the chip." [43]
	Semi-passive/ semi-active tag	"A class of RFID tags that contain a power source, such as a battery, to power the microchip's circuitry. *Unlike active tags, semi-passive tags do not use the battery to communicate with the reader.* Some semi-passive tags are dormant until activated by a signal from a reader. This conserves battery power and can lengthen the life of the tag." [43]
	Active tag	"A class of RFID tag that contains a power source, such as a battery, to power the microchip's circuitry. Active tags *transmit a signal to a reader* and can be read from 100 feet (35 meters) or more." [43]
RFID Journal	Passive tag	An RFID tag without its own power source and transmitter. When radio waves from the reader reach the chip's antenna, the energy is converted by the antenna into electricity that *can power up the microchip in the tag.* The tag is able to send back information stored on the chip. [...]" [44]
	Semi-passive/ semi-active tag	"Semi-*passive* tags are"[s]imilar to active tags, but the *battery is used to run the microchip's circuitry but not to broadcast a signal to the reader.* Some semi-passive tags sleep until they are woken up by a signal from the reader, which conserves battery life. Semi-passive tags can cost a dollar or more. These tags are sometimes called battery-assisted tags." [44]
	Active tag	"An RFID tag that has *a transmitter to send back information, rather than reflecting back a signal from the reader*, as a passive tag does. Most active tags use a battery to transmit a signal to a reader. However, some tags can gather energy from other sources. Active tags can be read from 300 feet (100 meters) or more [...]." [44]

EPC Global Inc. and the RFID Journal also define SRD as active RFIDs (Table 2): First, the definitions of semi-active/semi-passive tags by both EPC Global and RFID Journal stipulate that these tags use their battery to power the chip's circuitry. Second, EPC Global states that "[u]nlike active tags, semi-active tags do not use the battery to communicate with the reader". This means e contrario that these active tags use the battery power to broadcast the signal. Furthermore, EPC Global's definition of active tags states that they transmit a signal to the reader. Third, the RFID Journal defines the active RFID tag to include a "transmitter to send back information, rather than reflecting". These definitions of semi-active/semi-passive tags and active tags also clearly show that (i) active tags as understood by Finkenzeller [11] and Kern [16] are referred to as semi-active/semi-passive tags by EPCglobal and RFID Journal (and Glover [12] respectively); and that (ii) active tags as understood by EPCglobal and RFID Journal (and Glover [12] respectively) are referred to as SRDs by Finkenzeller [11] and Bensky [2].

Based on the differentiation between these three different RFID tag types, the next subsection explains how passive and active RFID tags as understood in the narrow sense by Finkenzeller [11] and Kern [16] derive the necessary power for transmission of the data.

3.2 Coupling

The following subsection introduces the technology of coupling. It gives a brief technical overview on how energy is derived from radio waves of a reader to power an RFID tag. This understanding is necessary to differentiate between energy supply and data transfer to the reader as discussed in the industry review (section 5.2).

Coupling is the mechanism by which a transponder circuit and a reader circuit influence one another for energy supply of the transponder as well as for data transfer to the reader. There are three main coupling modes: inductive coupling, capacitive coupling and backscatter coupling.

First, transponders of inductive coupling systems are mostly only used in passive tags. The reader must provide the required energy for both the data signal as well as for the operation of the chip. The inductively coupled transponder usually comprises a chip and an antenna coil. The reader's antenna generates an electromagnetic field. When a tag is within the interrogation zone of a reader the tag's antenna generates voltage by electromagnetic induction which is rectified and serves as power supply for the chip. The data transfer back to the reader works by load modulation: When a resonant transponder is within the range of the electromagnetic field it absorbs and reduces the energy of the reader's magnetic field which can be represented as change of impedance. Switching on and off a load resistor by the transponder can also be detected by the reader. The course of this change allows the interpretation of a signal (data transfer). [11] [16]

Second, capacitive coupling systems use plate capacitors for the transfer of power from the reader to the transponder. The reader comprises an electrode (e.g. metal plate). Through the very precise placement of the transponder on the reader a functional set-up similar to a transformer is generated. If high-frequency voltage is applied to this electrically conductive area of the reader, a high frequency field is generated. Electric voltage is generated between the transponder electrodes if the transponder is placed within the electrical field of the reader. This electrical voltage supplies the transponder with power. Similar to load modulation of inductive coupling, the read range of a reader is dampened when an electrically coupled tag is placed within the resonant circuit. This allows switching on and off of the modulation resistor (data transfer). [11]

Last, long distance backscatter systems are often active (in the narrow sense) or semi-passive (in the wide sense) tags, i.e. they are supported by an additional energy source for the chip within the transponder. The source energy for the transponder emitted by the reader is partly reflected by the transponder and sent back to the reader. Backscatter coupling is based on the principle of radar technique that electromagnetic waves are reflected by objects with dimensions larger than half the length of a wave. Also in this coupling mode a load resistor is switched on and off in time to transmit data from the transponder to the reader, thereby modulating the amplitude of the reflected power (modulated backscatter). [11] [16]

4 Legal Analysis

The two previous sections have introduced the use case and provided the technical background on energy supply and coupling of RFID. This section presents an empirical sample of legal journals and reviews the technical understanding by legal experts. It forms the legal basis of the claim that marketing SRD as RFID might backfire on the RFID industry.

4.1 Research Method and Legal Empirical Sample

The legal articles that form the basis for the empirical sample have been searched and selected in the WestLaw[3] and LexisNexis[4] databases. Both databases are leading providers of comprehensive legal information and business solutions to professionals. Two queries were conducted in each legal database. The parameter set in both databases for retrieval used the following keywords: radio frequency ident* (as group word and truncated using a wild card to allow variances such as ident*ification* or ident*ifiers*) and RFID (the acronym for radio frequency identification). Within Westlaw and LexisNexis the search was conducted in the database for combined journals and law reviews. Apart from searching for articles containing the keywords listed above, a filter was set by searching articles written in English and limited to the regions North America and Europe.

As both legal databases have similar but not identical content some results overlapped. After manually sifting out the duplicates of both databases, 141 legal journal articles, reviews and reports dating from 1997 to several months into 2007 remained (gross sample). Almost 80 per cent of the selected articles date back to the years 2005 and 2006. From the total 141 retrieved articles some 98 articles were excluded because a full text analysis showed neither their main nor side content relates specifically to the technology of RFID. The 98 excluded articles in many cases only mention RFID as an example in one sentence, part of a sentence or footnote. These 98 articles do not have any in-depth relation to RFID, explanation or analysis of RFID technology. A total of 43 articles were selected for this analysis (net sample). These selected articles have either a (sub-) chapter dedicated to RFID technology or comprise RFID as secondary content in either main text or footnotes with technological and/or legal analysis of RFID (as compared to the excluded 98 articles).

The year 2001 has 1, the year 2004 has 3, the years 2005 and 2006 each have 17 and 2007 (not full year) has 5 in-scope articles. Most in-scope articles are found in legal journals, reviews and reports for Information and Communication Technology law (42%). General legal journals and legal policy journals account for another third (30%). The remaining selected articles relating to RFID (28%) are distributed among administrative/public law, finance law, food and drugs reports, human rights law, intellectual property law, lawyer associations, and procurement law.

---

[3] http://www.westlaw.com (accessible only with license; last visited May 18, 2007)
[4] http://www.lexisnexis.com (accessible only with license; last visited May 18, 2007)

In summary, the large part of the 43 legal articles researched for this analysis has a similar structure. The articles mostly comprise (i) an introduction, followed by (ii) *a technical description*, (iii) a legal analysis and lastly (iv) a conclusion. Other articles explain the technical advancements of RFID only in footnotes. Within the technical description many authors recall the history of RFID as far back as World War 2, differentiate between passive and active tags, and provide a few examples of RFID implementation.

It is recognised that there are several limitations to the proposed methodology. First, the search is limited to two legal databases. Second, English language articles are relied upon exclusively which introduce a certain bias regarding countries, especially in Europe. Lastly, no quality assessments regarding the type of law journal is made. Despite these possible sources of error, it is suggested that analysing these 43 articles is a reasonable method for conducting the proposed analysis of the technical RFID understanding legal articles.

4.2 Legal Review

This section reviews the technical RFID understanding and perception of legal experts based on the legal empirical sample.

Of the 43 in-scope legal journal articles a little less than 50 per cent differentiate between passive, semi-passive/semi-active tags and active tags: Two articles use the functionality for tags in the narrow sense according to Finkenzeller [11] and Kern [16], thirteen articles refer to the functionality in the wide sense according to Glover [12], and five articles are inexplicit in the definition of the active tag (i.e. narrow or wide sense).

Both Landau [20] and Smith [23] mention the necessity of the active tag first being activated to transmit the data to the reader regardless of whether they have a battery or not. This makes them active according to Finkenzeller [16] or semi-active/semi-passive according to Glover [12]. Unclear remain the technical statements of Brito [3], Thompson [28], Eng [9], Kobelev [17], and Eleftheriou [8]. These five authors mention (active) tags with batteries or tags that are self-powered, but do not explain whether such energy is also used to transmit the data or whether transmission power is generated by the reader.

Stark [25], Delaney [6], Terry [27], Eschet [10], Asamoah [1], Herbert [14], and Smith [24] not only refer to the necessity of a battery (or other power supply) in an active tag, but more importantly consider such power supply essential for the transmittal of the data to the reader. Other authors like Eden [7], Willingham [31], and Stein [26] especially emphasise the lack of need for an active tag to be within activation range of a reader. Such tags continuously transmit their data. The tag range referred to by Eden, Willingham and Stein is only exceeded by tags as mentioned by Handler [13] and Koops [18] with transmitters capable of sending the signal over up to several kilometers. The tags as referred to by these ten authors are active tags by definition of Glover [12] and SRD as understood by Finkenzeller [11] and Bensky [2].

With passive tags there is no doctrinal or industry driven differentiation similar to the one with active and semi-passive/semi-active tags. In principle all reviewed authors agree that passive tags do not have an internal power supply and transform the (electromagnetic) waves of the reader by induction. Many reviewed authors explicitly mention the lack of battery supply and/or the energy powering by the reader. Thompson et al. [28] also refer to the virtually unlimited operational life of a passive tag while Smith [24] inaccurately states that a passive tag cannot be turned off. Indeed the opposite applies: a passive tag is always off and needs external manipulation to be "switched on" by influence from the reader. If at all, it would be more accurate to envision the metaphor of active tags (as defined by Finkenzeller) or semi-passive tags (as defined by Glover) being "woken up".

5 Industry Analysis

This section analyses the RFID industry by first providing an empirical sample of the active tag providers and then by reviewing the active RFID tag providers' marketing strategy.

5.1 Research Method and Industrial Empirical Sample

Similar to the legal analysis an empirical sample is drawn for the industry analysis. The Buyer's Guide 2007 online database[5] of the RFID Journal has been searched for industry suppliers of active RFID tags. The RFID Journal claims to be the only independent media company devoted solely to RFID and its many business applications. Its mission is to help companies use RFID technology to improve the way they do business.

Two queries were conducted for this empirical sample of industry suppliers of active RFID tags. The search parameters were set to search the RFID Journal Buyer's Guide 2007 database by type of technology (e.g. passive or active tag). In addition, a geographical filter was set to limit the search to the U.S. and to Europe respectively.

The database search for the U.S. market provided 104 hits; the database search for the European market provided 72 hits. A manual comparison of these 176 query hits resulted in an overlap of 64 resources (i.e. RFID providers based in both the U.S.A. and in Europe). Subsequently, the product range of these 64 RFID providers was analysed (gross sample). They offer among others tag hard- and software, readers, printers, and services such as consulting and system integrations. To qualify in the empirical sample of this analysis the provider must supply active RFID tags and issue a tag datasheet (PDF or html format) for evaluation and verification of the tag parameters. A total of 16 providers meet these selection criteria: AAiD Security Solutions[6], AeroScout[7], Axcess[8], Deister Electronic[9], Ekahau[10], Identec Solutions[11],

---

[5] http://www.rfidjournal.com/article/findvendor (last visited December 9, 2007)

[6] http://www.autoaccessid.com (last visited December 9, 2007)

[7] http://www.aeroscout.com (last visited December 9, 2007)

[8] http://www.axcessinc.com (last visited December 9, 2007)

Multispectral Solutions[12], RF Code[13], RFID Inc.[14], RFind[15], Savi[16], Smartcode[17], Synometrix[18], Tagmaster[19], Ubisense[20], and Wherenet[21].

It is also acknowledged in this empirical sample that there are several limitations to the proposed methodology. First, the search is limited to the online database of the Buyer's Guide 2007 offered by the RFID Journal. Second, the search is geographically limited to fit the legal empirical sample (North America and Europe). This excludes providers in other regions of the world like South America, Africa and Asia Pacific. Lastly, only tag providers with a datasheet for evaluation of the technology are included in the sample. Despite these possible biases it is suggested that analysing the RFID tags of these providers is a reasonable method for conducting the proposed analysis of the marketing approach of the RFID industry.

5.2 Industry Review

The selected technical parameters in the previous section 3.1 have shown that technically there is a difference between RFID and SRD. While Finkenzeller [11] and Bensky [2] argue in favour of a strict differentiation of these technologies, Glover et al. [12] concede this distinction as accurate but opt to lump active RFID and SRD together under the term active RFID. These authors leave the traditional path of distinguishing the energy for the chip and energy for broadcasting, and use the term active RFID as synonym for SRD. The industry also seems to follow the wider functionality of active RFID tags.

From the total of 16 active tag providers selected in the empirical sample (footnotes 5 through 20) five are eliminated from the count. Although they meet the selection criteria ("active" RFID tag and online datasheet) it is not clear from the datasheet description whether they broadcast automatically to the reader or not. From the remaining eleven active tag datasheets (net sample) eight refer explicitly to the marketed tag as "active tag" (or similar), while all eleven datasheets include the feature of self-dynamic signal transmission to the reader, i.e. these tags beacon or blink periodically. The RFID tags offered by these providers include the following features as outlined in Table 3. Only one example is presented per tag provider even if its product line includes more than only "active" tag.

---

9 http://www.deister.com (last visited December 9, 2007)

10 http://www.ekahau.com (last visited December 9, 2007)

11 http://www.identecsolutions.com (last visited December 9, 2007)

12 http://www.multispectral.com (last visited December 9, 2007)

13 http://www.rfcode.com (last visited December 9, 2007)

14 http://www.rfidinc.com (last visited December 9, 2007)

15 http://www.rfind.com (last visited December 9, 2007)

16 http://www.savi.com (last visited December 9, 2007)

17 http://www.smartcodecorp.com (last visited December 9, 2007)

18 http://www.synometrix.com (last visited December 9, 2007)

19 http://www.tagmaster.com (last visited December 9, 2007)

20 http://www.ubisense.net (last visited December 9, 2007)

21 http://www.wherenet.com (last visited December 9, 2007)

Table 3. Selected short range transmitters with features [own emphasis]

Tag name	Tag feature as stated in the online datasheets
AAiD AA-T800	"The AutoAccess AA-T800 Long Range Tags are designed for high value asset identifcation, real-time loss prevention, inventory management and tracking applications. [It features] low power consumption. Tag life is estimated at 5-7 years when transmitting *at 1.5 second intervals.*" [32]
Aeroscout T3 Tag	"The AeroScout T3 Tag is the most advanced Wi-Fi based Active RFID tag on the market, from the market leader in the WI-FI-based Active RFID industry. The T3 Tag is a small, battery-powered wireless device for accurately locating and tracking any asset or person. [The *t]ransmission interval [is] programmable [from] 128 msec to 3.5 hours.*" [33]
AXCESS ActiveTag Container Tag	"The AXCESS Container Tag provides a low cost solution for improving cargo container security, efficiency of movement and tracking capabilities while ensuring the integrity of the cargo within shipping containers. It uses the AXCESS ActiveTag™ RFID technology. [...] Under normal conditions the container tag *will 'beacon'* to the AXCESS system, letting the system know the tag is still in place." [34]
Multispectral Model Sapphire Vision	"Multispectral Solutions' Sapphire VISION puts this unique technology to work for active RFID applications [with a t]ag battery life in excess of 5 years (*at one update per second*)." [35]
RF Code M100 Active RFID Tag	"RF Code designs and manufactures active Radio Frequency Identification (RFID) monitoring systems that utilize beacon tags that *periodically broadcast* their status using encoded radio transmissions. [...] Every tag broadcasts its unique ID and a status message at a periodic rate (that is programmed at the factory). [...] Motion activated tags can be programmed to operate at *2 beacon rates*: slow when the tag is stationary, and faster when the motion sensor is activated (to provide immediate notification when objects are moving)." [36]
RFID Inc. EXT1 Personnel Tag	"Our Extend-a-Read product is based on 433 MHz active (battery powered) anti-collision (read many simultaneously) Tags. Tags simply *emit a data signal every 1.8 to 2.2 seconds* which is picked up by the Reader." [37]
RFind active RFID Talon Tag	"The RFind active RFID Talon Tag is a core component of our 3-element 915MHz RTLS architecture. [...] Battery Lifetime: 5 years @ *400 communication events per day.*" [38]
SaviTag ST-656	"The SaviTag™ ST-656 is an innovative, data rich, active RFID tag for ISO containers, enabling shippers, carriers and logistics service providers to monitor their shipments in real-time as they move through the global supply chain. [One of the key features is the] UHF transmitter to transmit alarms, *beacon* and Savi Reader Interrogation Responses." [39]
SynoTag SMPTK-002	"Read Write Active RFID Tag with LED. [This] tag transmits *signal to reader every 300ms - 500 ms.*" [40]

Table3. (*continued*)

Ubisense Compact Tag	"The Compact Tag is a small, rugged device that, when attached to assets and vehicles, allows them to be dynamically located to a precision of 15cm in 3D. [Power supply & battery life last o]ver 5 years at a continuous 5 second *beacon rate*." [41]
WhereTag III	"The WhereTag III is […] a small device that can be attached to assets of many kinds, [and i]t is used to manage those assets by allowing them to be identified and located by the system. The WhereTag III *'blinks' an RF transmission* at pre-programmed rates ranging from 5 seconds to one hour between blinks [with a] User Configurable Blink Rate of 5 sec to 1 hr." [42]

It is here argued that the RFID tags marketed by these suppliers are SRD. While all eleven tags contain a battery equally to the active RFID tags in the narrow sense referred to by Finkenzeller [11], Kern [16] and Bensky [2], the RFID tags by the providers listed in Table 3 continually and indiscriminately broadcast a signal to the environment. They blink at different intervals with beacon rates ranging from a few milliseconds to several hours. They have battery lifetimes of up to several years. To such extent they need neither the energy supply from the reader (section 3.1) nor do they broadcast by coupling (section 3.2). The active tags in the wide sense as listed in Table 3 have an independent energy supply and transmitter for broadcasting.

6   Discussion

Following the technical and legal outlines this section discusses the arguments supporting the claim why marketing SRD as (active) RFID tags might backfire on the RFID industry. It also sheds light why this strategy is unlikely to backfire on the SRD industry.

6.1   Backfire on the RFID Industry

Neither Dalal [5] nor various other legal authors accurately differentiate the types of RFID tags. It is argued in this paper that it will make a difference in the outcome of a court decision and in policy making whether the surveillance is with a passive RFID tag, an active RFID tag, or with a short range transmitter. By the definitions used in this analysis (details especially in section 3.1), the radio transmitting device used in United States v. Knotts [30] is a short range transmitter, not an RFID. To such extent the findings of SRD should not be and as argued herein are not applicable to active RFID in the narrow sense (see section 3).

People carrying RFID tags will generally fall within the protection of the Fourth Amendment as the Fourth Amendment protects people, not places. [15] RFID tags in the narrow sense as understood and advocated in this paper need power from the reader to transmit the information back to the reader. While people in public will generally expect other people to see where they go [30], it must remain a persons right of privacy (not) to disseminate information generally contained in an RFID tag, i.e. information that is not generally accessible by other people. As both passive and active RFID tags in the narrow sense need the energy of a reader to transmit data back to the reader, broadcasting the data is not possible without an antecedent manipulation by the reader. By contrast SRDs continually and indiscriminately transmit information contained in them in intervals to the environment. In the case of a SRD the person (deliberately) carrying the SRD will expect the environment to pick up the transmitted data and "searching" such data will thus not be unconstitutional under the Fourth Amendment.

On the one hand, eleven tag manufacturers and suppliers are marketing their short range transmitters as active RFID tags. The tags of these suppliers do not transduce the radio frequency of the reader. They have their own energy supply for broadcasting and have a transmitter to broadcast a signal in intervals indiscriminately to the environment. On the other hand, the legal community does not differentiate accurately between the different tag functionalities. If at all, it relies on Glover's understanding of functionality for passive, semi-passive/semi-active and active tags. This means that the legal community includes the self-dynamic functionalities of short range transmitters in their understanding and analysis of RFID. The legal analysis should differentiate between the tag functionalities but it does not. It is here argued that legally it makes a difference if the tag needs the reader's radio frequency for energy supply and broadcasting, or not.

In line with this argumentation the claim can be made that if the RFID industry keeps marketing their short range transmitters as RFID, the legal community might continue including such broad and self-dynamic device functionalities in its legal interpretation and analysis of RFID. The inclusion of broad SRD functionalities by the legal community in their interpretation, policy and decision making might lead to restrictive interpretation, use and limited legal acceptance of RFID. Why? Because if monitoring a beeper that broadcasts its signals in public is *not* unconstitutional under the Fourth Amendment and the legal community perceives SRD and RFID to be the same technology due to the marketing endeavours of the industry, then the privacy advocates might join forces to legally stop the implementation and deployment of RFID in order not to run the risk of having constitutional surveillance of RFID tags in the narrow sense without a warrant. Hence, the marketing strategy of riding on the trend wave of RFID might backfire on the RFID industry as the industry will need to follow (more) restrictive law and case decisions. The result might be restrictive implementation and deployment possibilities and therefore limited device and service sales.

6.2 Backfire on the SRD Industry?

Why should the lack of differentiation between RFID and SRD be a problem for the RFID industry and not for the SRD industry? Could the SRD industry not also suffer from the joint marketing of both technologies as RFID? Could marketing SRD as active RFID backfire on the SRD industry?

It has been argued that reading an (active) RFID in the narrow sense is more intrusive as compared to an SRD because it requires an antecedent manipulation by the reader to trigger the data broadcast. Consequently it could be argued that marketing SRD as RFID will have negative effect on the SRD industry (and not the other way around as stated in the previous section) because the more restrictive legal interpretation of RFID in the narrow sense could spill over to SRD and make the surveillance of SRD broadcasts unconstitutional without a warrant.

The following arguments disapprove such assumption: SRD is being marketed as RFID, not vice versa. Section 5.2 lists short range transmitters that are promoted as active RFID. The indiscriminate broadcasting of SRD merges into the RFID technology in the narrow sense, not vice versa. RFID is in focus, SRD is out of perception. What remains is the notion that short range transmitters are active RFID tags.

From a legal perspective the analysis in section 4.2 reveals that the majority of authors in the investigated legal articles use Glover's and not e.g. Finkenzeller's understanding of active RFID tags (in the narrow sense). Hence, they perceive the technology exactly as it has been promoted by the industry. So the legal community transposes the constitutional surveillance of a beeper as ruled in United States v. Knotts [30] into the RFID field and not vice versa.

For these reasons it is not anticipated that marketing SRD as active RFID will backfire on the SRD industry.

7 Conclusion

As has been consistently argued in this paper, marketing SRD as active RFID might backfire on the RFID industry. This leads to the following two conclusions:

1. The industry needs to clarify its terminology for SRD and RFID. The term SRD seems deceptive anyhow since it infers that the range is even shorter than with RFID (whereas in reality it is much longer). Furthermore the current marketing strategy of marketing SRD as active RFID (section 5.2) might need to be reconsidered.
2. The legal community needs to better differentiate both RFID and SRD technology. In order accurately analyse the technical and make distinguished legal recommendations and regulations the legal community must better understand the underlying technology.

References

1. Asamoah, A.K.: Not as easy as it appear: Using radio frequency identification technology to fulfill the prescription drug marketing act's elusive pedigree requirement. Food and Drug Law Journal 61, 385 (2006)
2. Bensky, A. (ed.): Short-range wireless communication, 2nd edn(ed.). Elsevier, Amsterdam (2004)
3. Brito, J.: Relax don't do it: Why RFID privacy concerns are exaggerated and legislation is premature. UCLA Journal of Law and Technology 5 (2004)
4. Cardwell v. Lewis, 417 U.S. 583 (1974)
5. Dalal, R.S.: Chipping away at the constitution: The increasing use of RFID chips could lead to an erosion of privacy rights. Boston University Law Review 86, 485 (2006)
6. Delaney, K.: RFID: Privacy year in review: America's privacy laws fall short with RFID regulation. A Journal of Law and Policy for the Information Society 1, 543 (2005)

7. Eden, J.M.: When big brother privatizes: Commercial surveillance, the privacy act of 1974, and the future of RFID. Duke Law & Technology Review 20 (2005)
8. Eleftheriou, D., Berliri, M., Coraggio, G.: Data protection and E-commerce in the United States and the European Union. International Lawyer 40, 393 (2006)
9. Eng, G.: Technology trends affecting the practice of law. Los Angeles Lawyer 79, 28-APR (2005)
10. Eschet, G.: FIPS and PETS for RFID: Protecting privacy in the web of radio. Jurimetrics Journal 45, 301 (2005)
11. Finkenzeller, K.: RFID Handbuch (4. Auflage). Hanser, München (2006)
12. Glover, B., Bhatt, H.: RFID essentials, 1st edn (ed.) O'Reilly, Beijing (2006)
13. Handler, D.: The wild, wild west: A privacy showdown on the radio frequency identification (RFID) systems technological frontier. Western State University Law Review 32, 199 (2005)
14. Herbert, W.A.: No direction home: Will the law keep pace with human tracking technology to protect individual privacy and stop geoslavery? A Journal of Law and Policy for the Information Society 2, 409 (2006)
15. Katz v. United States, 389 U.S. 347 (1967)
16. Kern, C.: Anwendung von RFID-systemen. Springer, Berlin (2006)
17. Kobelev, O.: Big brother on a tiny chip: Ushering in the age of global surveillance through the use of radio frequency identification technology and the need for legislative response. North Carolina Journal of Law & Technology 6, 325 (2005)
18. Koops, B., Leenes, R.: 'Code' and the slow erosion of privacy. Michigan Telecommunications and Technology Law Review 12, 115 (2005)
19. Kyllo v. Unites States, 533 U.S. 29 (2001)
20. Landau, S.: Digital age communications law reform: National security on the line. Journal on Telecommunications & High Technology Law 4, 409 (2006)
21. U.S. National Prohibition Act, ch. 85, 41 Stat. 305 (1919)
22. Olmstead v. United States, 277 U.S. 438 (1928)
23. Smith, S.L.: Symposium review: RFID and other embedded technologies: Who owns the data? Santa Clara Computer and High Technology Law Journal 22, 695 (2006)
24. Smith, S.L.: Gone in a blink: The overlooked privacy problems caused by contactless payment systems. Marquette Intellectual Property Law Review 11, 213 (2007)
25. Stark, S.C., Nagle, E.P.: Full speed ahead with DOD identification requirements: Next stop, radio frequency identification. Procurement Lawyer 40-Fall, 11 (2004)
26. Stein, S.G.: Where will consumers find privacy protection from RFID?: A case for federal legislation. Duke Law & Technology Review 2007, 3 (2007)
27. Terry, N.P.: Assessing the technical, conceptual, and legal frameworks for patient safety information. Widener Law Review 12, 133 (2005)
28. Thompson, D., Kot, K., Brothers, J.: Radio frequency identification: Legal aspects. Richmond Journal of Law and Technology 12, 6 (2005)
29. U.S. Const. Amend. IV
30. United States v. Knotts, 460 U.S. 276 (1983)
31. Willingham, K.M.: Scanning legislative efforts: Current RFID legislation suffers from misguided fears. North Carolina Banking Institute 11, 313 (2007)
32. AAiD: (last visited December 9, 2007), http://www.autoaccessid.com/Files/File/Cutsheets/AA-T800.pdf
33. AeroScout: (last visited December 9, 2007), http:// www.aeroscout. com/data/ uploads/ AeroScout%20T3%20Tag%20Data%20Sheet.pdf

34. Axcess: (last visited December 9, 2007), http://www.axcessinc.com/products/docs/containertag.pdf
35. Multispectral: (last visited December 9, 2007), http://www.multispectral.com/pdf/Sapphire_VISION.pdf
36. R.F Code: (last visited December 9, 2007),
    http://www.rfcode.com/images/media_kit_docs/m100_tech_spec_sheet.pdf
37. RFID Inc.: (last visited December 9, 2007), http://www.rfidinc.com/products1.html
38. RFind: (last visited December 9, 2007),
    http://www.rfind.com/pdf/RFind%20Tags%20Gateway%20_V1.5_%2005-01-07.pdf
39. Savi: (last visited December 9, 2007), http://www.savi.com/products/SaviTag_656.pdf
40. Synometrix: (last visited December 9, 2007), http://www.synometrix.com/SYNOMETRIX_SMPTK-002_active_RFID_tag_specification.pdf
41. Ubisense: (last visited December 9, 2007),
    http://www.ubisense.net/media/pdf/Ubisense%20Compact%20Tag%20EN%20V1.0.pdf
42. Wherenet: (last visited December 9, 2007),
    http://www.wherenet.com/pdf/products/WhereTagIII.6.14.07.pdf
43. EPC global Inc., Glossary: (last visited December 9, 2007), http:// www. epcglobalinc.org/what/cookbook/chapter5/039–GlossaryV.6.0May2005FINAL.pdf
44. RFID Journal: (last visited December 9, 2007),
    http://www.rfidjournal.com/article/glossary/3

Object Recognition for the Internet of Things

Till Quack[1], Herbert Bay[1], and Luc Van Gool[2]

[1] ETH Zurich, Switzerland
{quack,bay}@vision.ee.ethz.ch
[2] KU Leuven, Belgium
luc.vangool@esat.kuleuven.be

Abstract. We present a system which allows to request information on physical objects by taking a picture of them. This way, using a mobile phone with integrated camera, users can interact with objects or "things" in a very simple manner. A further advantage is that the objects themselves don't have to be tagged with any kind of markers. At the core of our system lies an object recognition method, which identifies an object from a query image through multiple recognition stages, including local visual features, global geometry, and optionally also metadata such as GPS location. We present two applications for our system, namely a slide tagging application for presentation screens in smart meeting rooms and a cityguide on a mobile phone. Both systems are fully functional, including an application on the mobile phone, which allows simplest point-and-shoot interaction with objects. Experiments evaluate the performance of our approach in both application scenarios and show good recognition results under challenging conditions.

1 Introduction

Extending the Internet to physical objects - the Internet of Things - promises humans to live in a smart, highly networked world, which allows for a wide range of interactions with this environment. One of the most convenient interactions is the request of information about physical objects. For this purpose several methods are currently being discussed. Most of them rely on some kind of unique marker integrated in or attached to the object. Some of these markers can be analyzed using different kinds of wireless near field communication (for instance RFID tags [24] or Bluetooth beacons [11]), others are visual markers and can be analyzed using cameras, for instance standard 1D-barcodes [2] or their modern counterparts, the 2D codes [21].

A second development concerns the input devices for interaction with physical objects. In recent years mobile phones have become sophisticated multimedia computers that can be used as flexible interaction devices with the user's environment. Besides the obvious telephone capabilities, current devices offer integrated cameras and a wide range of additional communication channels such as Bluetooth, WLAN or access to the Internet. People are used to the device they own and usually carry it with them all day. Furthermore, with the phone-number,

C. Floerkemeier et al. (Eds.): IOT 2008, LNCS 4952, pp. 230–246, 2008.

a device is already tied to a specific person. Thus it is only natural to use the mobile phone as a personal input device for the Internet of things.

Indeed, some of the technologies mentioned above have already been integrated in mobile phones, for instance barcode readers or RFID readers. The ultimate system, however, would not rely on markers to recognize the object, but rather identify it by its looks, i.e. using visual object recognition from a mobile phone's camera image. Since the large majority of mobile phones contain an integrated camera, a significant user base can be addressed at once. With such a system, snapping a picture of an object would be sufficient to request all the desired information on it. While this vision is far from being reality for arbitrary types of objects, recent advances in the computer vision field have led to methods which allow to recognize certain types of objects very reliably and "hyperlink" them to digital information.

Using object recognition methods to hyperlink physical objects with the digital world brings several advantages. For instance, certain types of objects are not well suited to attach markers. This includes tourist sights, which are often large buildings and a marker might only be attached at one or few locations at the building, an experiment which has been attempted with the Semapedia project [1]. Furthermore, a user might want to request information from a distance, for instance for a church tower which is up to several hundred meters away. But even if the object is close, markers can be impractical. A barcode or RFID attached to the label of an object displayed in the museum would be difficult to access if the room is very crowded. Taking a picture of the item can be done from any position where it is visible. Furthermore, consistent tagging the objects is often difficult to achieve. One example are outdoor advertising posters. If a poster company wanted to "hyperlink" all their poster locations, they would have to install an RFID or bluetooth beacon in each advertising panel or attach a barcode to each of them, which requires a standardized system and results in costs for installation and maintenance. Another field of application are presentation screens in smart meeting rooms or information screens in public areas. The content displayed on the screen is constantly changing and it would be a involved process to add markers to all displayed content.

Using object recognition to interact with these objects requires only a database of images. That being said, object recognition does not come without restrictions, either. For instance, it is currently (and maybe always) impossible to discriminate highly similar objects, such as two slightly different versions of the same product in a store. Furthermore, efficient indexing and searching visual features for millions or billions of items is still a largely unsolved problem.

In this paper we present a method and system enabling the Internet of Things using object recognition for certain types of objects or "things". At the core of our server-side system lies a retrieval engine which indexes objects using scale invariant visual features. Users can take a picture of an object of interest, which is sent to the retrieval engine. The corresponding object is recognized and and an associated action is executed, e.g. a web-site about the object is opened. The

---

[1] http://www.semapedia.org

Fig. 1. The user "tags" a presented slide using our mobile application by taking a picture (left), which is automatically transmitted to the server and recognized (middle), a response is given in an automatically opened WAP browser (right).

system is completed with a client-side application which can be installed on a mobile handset and allows true point-and-shoot interaction with a single click.

We present two fully functional applications, which demonstrate the flexibility of the suggested approach. The first one is slide tagging in smart meeting rooms. Users have the ability to "click" on slides or sections of slides that are being presented to record them for their notes or add tags. The second application is a cityguide on the mobile phone. Users have the possibility to take a picture of a sight, send it to a recognition service, and receive the corresponding Wikipedia article as an answer. For this application, the search space is limited by integrating location information, namely cell-tower ids or GPS.

Both systems are experimentally evaluated in different dimensions, including different phone models with different camera qualities, for the trade-offs using different kinds of search space restriction (geographic location etc.), and with and without projective geometry verification stage.

The remainder of this paper is organized as follows: we start with an overview over related work in section 2. The main body of the paper is built around the two applications presented, namely hyperlinked slides for interactive meeting rooms in section 3 and hyperlinked buildings for a cityguide in section 4. Each of these sections discusses method and implementation, followed by an experimental evaluation of the respective system. Finally, conclusions and outlook are presented in section 5.

2   Related Work

Our method can be related to other works in several aspects. One aspect covers work related to our smart meeting room application, for instance the use of camera-equipped mobile phones as an interaction device for large screens. Here, Ballagas et al. have suggested a system [4] which allows users to select objects on large displays using the mobile phone. However, their method relies on additional 2D barcodes to determine the position of the camera and is meant to use the mobile phone like a computer mouse in order to drag and drop elements on the screen. Very recently, in [7] a system similar to ours has been proposed for recognizing icons on displays. While the screens are conceptually similar to the ones used in meeting rooms, we are not aware of any other work that has proposed using camera-equipped mobile phones for tagging or retrieval of slides in smart meeting rooms. The most similar works in that respect deal with slide retrieval from stationary devices. For instance, Vinciarelli et al. have proposed a system [23] which applies optical character recognition (OCR) to slides captured from the presentation beamer. Retrieval and browsing is done with the extracted text, i.e. the method cannot deal with illustrations or pictures in the slides. SlideFinder [18] is a system which extracts text and image data from the original slide data. Image retrieval is based on global color histograms and thus limited to recognize graphical elements or to some extent the global layout of the slide. Using only the stored original presentation files instead of using the captured image data does not allow to synchronize the slides to other meeting data such as recorded speech or video. Both systems are only meant for query-by-keyword retrieval and browsing from a desktop PC. While our system could also be used for off-line retrieval with query-by-example, we focus on tagging from mobile phones. This requires the identification of the correct slide reliably from varying viewpoints, which would not be possible with the cited approaches.

Another aspect that relates to our work are guiding applications on mobile devices. Bay et al. have suggested a museum guide on a tablet PC [5]. The system showed good performance in recognizing 3D exhibition objects using scale invariant local features. However, in their system the whole database resisted on the client device, which is generally not possible for smaller devices such as mobile phones and larger databases. A similar system on a mobile phone, but with somewhat simpler object recognition is the one proposed in [12]. The suggested recognition relies on simple color histograms, which turns out not to be very robust to lighting changes in museum environments. Discriminating instances of the objects in our applications, namely slides or outdoor images of touristic sights, is even less reliable with global color histograms.

The work most similar to our city guide application is maybe [20]. Similar to the cityguide application presented in this paper, the authors also suggest a cityguide on a mobile phone using local features. However, their focus is on improving recognition capabilities using informative and compact iSift features instead of SIFT features. Our work differs significantly in several points: we use multiple view geometry to improve recognition, we rely on SURF features (which are also more compact and faster than SIFT features), and we also investigate

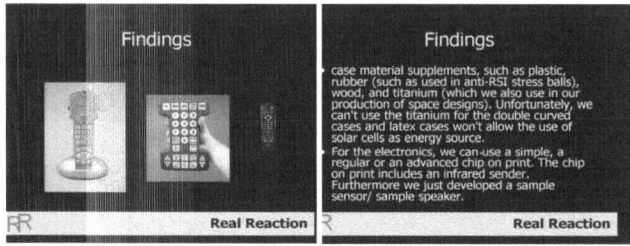

Fig. 2. Typical presentation slides from the AMI corpus [8] database

numerically the effects of restriction by GPS or cell ids on the recognition rate and matching speed. That is, instead of improving the features themselves, we add a global geometry filter as a final verification stage to the recognition system. Finally, the test databases we propose contains images taken from viewpoints with much larger variation than the databases used in [20].

The main contributions of this work are thus: a full object recognition system pipeline, including a server side recognition server and a client side software for single-click interaction with the environment; a complete object recognition pipeline for the Internet of Things, which starts with local feature correspondences, verification with projective geometry, and search space restriction by multimodal constraints, such as GPS location; the implementation evaluation for two sample applications, namely slide tagging and bookmarking in smart meeting rooms, as well as a cityguide application for the mobile phone; last but not least, for both cases the evaluation on challenging test datasets.

3   Hyperlinked Slides: Interactive Meeting Rooms

Today's meeting rooms are being equipped with an increasing number of electronic capturing devices, which allow recording of meetings across modalities [1,3]. They often include audio recording, video recording, whiteboard capturing and, last but not least, framegrabbing from the slide projector. These installations are usually deployed to facilitate two tasks: allowing off-line retrieval and browsing in the recorded meeting corpus and turning the meeting rooms into smart interactive environments. In the work at hand, we focus on the captured presentation slides which are a central part of today's presentations. As shown in figure 2, the slides usually contain the speaker's main statements in written form, accompanied by illustrations and pictures, which facilitate understanding and memorizing the presentation. Indeed, the slides can be seen as the "glue" between all the recorded modalities. Thus, they make a natural entry point to a database of recorded presentations.

A typical usage scenario for our system is as follows: Using the integrated camera of her mobile phone, an attendee to a meeting takes a picture of a slide which is of interest to her. The picture is transmitted to a recognition server over a mobile Internet connection (UMTS, GPRS etc.). On the server, features are extracted from the picture and matched to the database of captured slides. The correct slide is recognized, added to the users personal "bookmarks", and she receives a confirmation in a WAP browser on her mobile phone. Note that the messaging from the phone can be done using standard MMS or using a custom client-side application which we programmed in C++ on the Symbian platform. Figure 1 shows screenshots of our mobile application for a typical usage scenario.

Back at her PC, the user has access to all her bookmarked slides at any time, using a web-frontend which allows easy browsing of the slides she bookmarked. From each bookmarked slide she has the possibility to open a meeting browser which plays the other modalities, such as video and audio recordings, starting at the timepoint the slide was displayed. By photographing only a section of a slide, the user has also the possibility to highlight certain elements (both text or figures) - in other words, the mobile phone becomes a digital marker tool.

Please note that one could assume that a very simple slide bookmarking method could be designed, which relies only on timestamping. The client-side would simply transmit the current time, which would be synchronized with the timestamped slides. Our system does not only allow more flexible applications (the beforementioned "hightlighting" of slide elements) but is robust towards synchronization errors in time. In fact, using a "soft" time restriction of some minutes up to even several hours, would make our system more scalable and unite the best of both worlds.

The basic functionality of the proposed slide recognition system on the server is as follows: for incoming queries, scale invariant local features are extracted. For each feature a nearest neighbor search in the reference database of slides is executed. The resulting putative matches are verified using projective geometry constraints. The next two subsections describe these steps in more detail.

3.1 Slide Capturing and Feature Extraction

We start from a collection of presentation slides which are stored as images. This output can be easily obtained using a screen capture mechanism connected to the presentation beamer. From the image files, we extract scale invariant features around localized interest points. In recent years significant progress has been made in this field and has led to a diverse set of feature extraction and description methods [16,6,17], which have been successfully applied in domains such as video retrieval [22], object class recognition [15] etc. It turns out that such local features cannot only be used to describe and match objects and scenery, but work also reliably for text such as license plates [9]. Thus, this class of features is a good choice for description of the slide content which contains both text and visual data such as pictures and charts. Furthermore, as opposed to global features proposed in [18,12] they also allow the user to photograph specific sections or elements of a slide as a query to our system. In our implementation we use the publicly available SURF [6] detector and descriptor combination. This choice was motivated by the fast computation times and competitive recognition performance shown in [6]. The output of the SURF detector consists of 64-dimensional feature vector for each detected interest point in an image.

3.2   Slide Recognition System

The slide recognition approach consists of two steps: feature matching and global geometric verification. For the feature matching we compare the feature vectors from the query image to those of the images in the database. More precisely, for each 64-dimensional query vector, we calculate the Euclidean distance to the database vectors. A match is declared if the distance to the nearest neighbor is smaller than 0.7 times the distance to the second nearest neighbor. This matching strategy was successfully applied in [16,6,5,17].

Finding the best result could now be done by just selecting the query-database pair, which receives the highest number of matches. However, without verification of the geometric arrangement of the matched interest points, the wrong query-database pair may be selected. This is particularly true in our case, where we have a high number of matches stemming from letters in text parts of the slides. These matches are all "correct" on the feature level, but only their consistent arrangement to full letters and words is correct on the semantic level.

To solve this problem, we resort to projective geometry. Since the objects (the slides) in the database are planar, we can rely on a 2D homography mapping [13] from the query image to a selected candidate from the database in order to verify the suggested matching. That is, the set of point correspondences between the matched interest points from query image $\mathbf{x}_i^q$ and database image $\mathbf{x}_i^d$ must fulfill

$$H\mathbf{x}_i^q = \mathbf{x}_i^d \ i \in 1 \dots 4 \tag{1}$$

where H is the $3x3$ homography matrix whose 8 degrees of freedom can be solved with four point correspondences $i \in 1 \dots 4$. To be robust against the beforementioned outliers we estimate H using RANSAC [10]. The quality of several estimated models is measured by the number of inliers, where an inlier is defined by a threshold on the residual error. The residual error for the model are determined by the distance of the true points from the points generated by the estimated H. The result of such a geometric verification with a homography is shown in Figure 6.

3.3   Experiments

For our experiments we used data from the AMI meeting room corpus [8]. This set contains the images of slides which have been collected over a extended period using a screen-capture card in a PC connected to the beamer in the presentation room. Slides are captured at regular time intervals and stored as JPEG files. To be able to synchronize with the other modalities (e.g. speech and video recordings), each captured slide is timestamped.

To create the ground truth data, we projected the slides obtained from the AMI corpus in our own meeting room setting and took pictures with the integrated camera of two different mobile phone models. Namely, we used a Nokia N70, which is a high-end model with a 2 megapixel camera, and a Nokia 6230, which is an older model with a low quality VGA camera. We took 61 pictures

Fig. 3. Examples of query images, from left to right: with compositions of text and image, taken from varying viewpoints, at different camera zoom levels or may contain clutter, example which select a specific region of a slide, or contain large amounts of text.

with the N70 and 44 images with the Nokia 6230 [2]. Figure 3 shows some examples of query images. The reference database consists of the AMI corpus subset for the IDIAP scenario meetings, which contains 1098 captured slide images.

We extracted SURF features from the reference slides in the database at two resolutions, 800x600 pixels and 640x480 pixels. For the 1098 slides this resulted in $1.02 * 10^6$ and $0.72 * 10^6$ features, respectively. For the SURF feature extraction we used the standard settings of the detector which we downloaded from the author's website.

The resolutions of the query images were left unchanged as received from the mobile phone camera. We ran experiments with and without homography check, and the query images were matched to the database images at both resolutions. A homography was only calculated if at least 10 features matched between two slides. If there were less matches or if no consistent homography model could be found with RANSAC, the pair was declared unmatched. If there were multiple matching slides, only the best was used to evaluate precision. Since the corpus contains some duplicate slides, a true match was declared if at least one of the duplicates was recognized.

Table 1 shows the recognition rates, for the different phone models, different resolutions and with and without homography filter. At 800x600 resolution, the homography filter gives an improvement of about 2% or 4% for each both phone type, respectively. The recognition rate with a modern phone reaches 100%, the lower quality camera in the older 6230 model results in lower recognition rates. The results for the 640x480 database confirm the results of the 800x600 case, but achieve overall lower recognition scores. This is due to the fact, that at lower resolution fewer features are extracted.

4   Hyperlinked Buildings: A Cityguide on a Mobile Phone

The second scenario we present in this paper deals with a very different kind of "things". We "hyperlink" buildings (tourist sights etc.) to digital content. Users

---

[2] The query images with groundtruth are made available for download under http://www.vision.ee.ethz.ch/datasets/.

Table 1. Summary of recognition rates for slide database

	Prec. with Geometry Filter		Prec. without Geometry Filter	
	800x600	640x480	800x600	640x480
Nokia N70	100%	98,3%	98,3%	96,7%
Nokia 6230	97,7%	93,2%	91%	86,3%

can request information using an application on their mobile phone. The interaction process, the software and user interface are very similar to the meeting room scenario. However, this time the number of objects is nearly unlimited, if the application is to be deployed on a worldwide basis. To overcome the resulting scalability problems, we restrict the search space geographically. That is, we restrict the visual search to objects in the database, which lie in the geographic surroundings of the user's position.

In the following sections we describe this approach in more detail and evaluate its performance.

4.1   Visual Data and Geographic Location

From the user perspective, the interaction process remains the same as in the meeting room scenario: by the click of a button on the mobile phone, a picture is taken and transmitted to the server. However, unlike in the meeting room application, the guide client-side application adds location information to the request. This information consists of the current position read from an integrated or external (bluetooth) GPS device and of the current celltower id the so called CGI (Cell Global Identity).

This combination of a picture and location data forms a perfect query to search for information on static, physical objects. As mentioned before, location information alone would in general not be sufficient to access the relevant information: the object of interest could be several hundred meters away (e.g. a church tower), or there could be a lot of objects of interest in the same area (e.g. the St. Mark's square in Venice is sourrounded by a large number of objects of interest). Furthermore, in urban areas with tall buildings and narrow roads, GPS data is often imprecise. On the other hand, relying on the picture only would not be feasible either: the size of the database would make real-time queries and precise results very difficult to achieve.

After the query has been processed, the user receives the requested information directly on the screen of her mobile phone. In our demo application we open a web browser with the Wikipedia page corresponding to the object. This is illustrated in Figure 4.

4.2   System Design

The cityguide system consists of a server side software and a client-side software on the mobile phone.

Fig. 4. Client software for the cityguide application: the user snaps a picture, waits a few seconds, and is redirected to the corresponding Wikipedia page

The server side elements consist of a relational database for storage of image metadata (GPS locations, cell information etc.) and information about the stored sights. We used mySQL for this purpose. The image recognition is implemented as a server in C++ which can be accessed via HTTP.

Queries from the client-software are transmitted to the server as HTTP POST requests. A middleware written in PHP and Ruby restricts the search by location if needed and passes this pre-processed query to the recognition server. The associated content for the best match is sent back to the client software and displayed in an automatically opened browser, as shown in figure 4.

Client software on the mobile phone was implemented both in Symbian C++ and Java[3]. Note that the feature extraction of the query happens on the server side, i.e. the full query image is transmitted to the server. It is also possible to extract SURF features on the mobile phone and then transmit them as a query to the server. An implementation of this method showed, that SURF feature extraction on the phone is currently too slow: our un-optimized version in Symbian C++ on a Nokia 6630 required about 10 seconds to calculate the query features. In contrast, on a modern PC SURF feature extraction takes a few hundred ms [6]. Since the SURF features are not much more compact than the original image (several hundred 64 dimensional feature vectors per image), the main advantages of feature extraction on the phone would be increased privacy (only features transmitted instead of image) and the possibility to give a user

---

[3] Unfortunately, only the Symbian version allows access to the celltower ids.

instant feedback if a query image contained too few features, for instance due to blur, lack of texture, or low contrast due to back light.

Alternatively our system can also be accessed using the Multimedia Message Service MMS. A picture is transmitted to the server by sending it as an MMS message to an e-mail address. The response (Wikipedia URL) is returned as an SMS message.

4.3   Object Recognition Method

The data from the client-side application are transmitted to the recognition server, where a visual search restricted by the transmitted location data is initiated. If GPS data is used, all database objects in a preset radius are searched (different radii are evaluated in the experimental section of this paper). If only cell-tower information is used, the search is restricted to the objects annotated with the same CGI string.

The object recognition approach is very similar to the method discussed for the meeting room slides. That is, putative matches between pairs of query and databases images are found by nearest neighbor search for their SURF [6] descriptors. These putative matches are validated with a geometry filter. However, since we deal with 3-dimensional objects in the cityguide application, the precise model is now the $3x3$ Fundamental matrix F instead of the Homography matrix H [13]. The Fundamental matrix maps points in one image to epipolar lines another view. Residual errors for the models are thus determined by the distance of the true points from the epipolar lines generated by the estimated F [13].

From a practitioners point of view, for objects such as buildings which consist basically of multiple planes (facades) one can approximate the results by using a homography nevertheless, which requires less point correspondences. The estimation of the model from putative point correspondences can be done with RANSAC [10] in both cases.

Note that the model is particularly important to filter out false positive recognitions: Especially on structures on buildings, there are a lot of repeated patterns which match between different buildings. Only their correct arrangement in space or the image plane, respectively allow for a robust decision if an object was truly detected. Simply setting a threshold on the number matches is dangerous particularly, since discriminating a false positive recognition (e.g. a query image of an building which is not even in the database) from a query with few matches due to challenging conditions (e.g. image taken from a distance) is infeasible.

4.4   Experiments

To evaluate the proposed method, we collected a database of 147 photos covering 9 touristic sights and their locations. The 147 images cover the 9 objects from multiple sides, at least 3 per object. The database images were taken with a regular point-and shoot camera. To determine their GPS location and CGIs (cell tower ids) we developed a tracker application in Symbian C++ which runs on a mobile phone and stores the current GPS data (as obtained from an external bluetooth GPS device) and CGI cell information at regular time intervals. This log is synchronized by timestamps with the database photos.

We collected another 126 test (query) images, taken with different mobile phones (Nokia N70 and Nokia 6280, both with 2 Megapixel camera) at different days and times of day, by different users and from random viewpoints. Of the 126 query images 91 contain objects in the database and 35 contain images of other buildings or background (also annotated with GPS and cellid). This is an important to test the system with negative queries, an experiment which has been neglected in several other works. Compared to the MPG-20 database [4] we have fewer object but from multiple sides (in total about 30 unique representations), more challenging viewpoints for each side (distance up to 500 meters), full annotation with both GPS data and celltower ids, and more than 4 times as many query images. The database with all annotations (GPS, cellids, objects Wikipedia pages etc.) is available for download under [5]. Both database and query images were re-scaled to $500x375$ pixels. (Sample images from the database are visible in Figure 7 and are discussed a few paragraphs below).

Note that the CGI (Cell Global Identity) depends on the network operator, since each operator defines its own set of cell ids. If the operator does not release the locations of the cells (which is common practice in many countries for privacy reasons), we have to find a mapping between the cellids of different operators. We achieved such an experimental mapping by using our tracker application: tracks obtained with SIM cards of different mobile network operators were synchronized by their GPS locations: if GPS points were closer than 50m a correspondence between the respective cell-ids was established. This mapping is far from complete, but it simulates an approach which is currently followed by several initiatives on the Web.

We present experiments for three scenarios: linear search over the whole database without location restriction, restriction by GPS with different search radii, and restriction by cellid. For all cases we compare the trade-off between search time and recognition rate. A pair of images was considered matched, if at least

Table 2. Summary of recognition rates for cityguide

	Prec. with Geometry Filter		Prec. without Geometry Filter	
	Rec. rate	Avg. Matching Time	Rec. rate	Avg. Matching Time
Full database linear	88%	5.43s	67.4%	2.75s
GPS 300m Radius	89.6%	3.15s	76.1%	1.62s
Cell id	74.6%	2.78s	73%	1.34s

20 features matched. From the images which fullfiled this criterion the one with the most matches was returned as a response. Table 2 summarizes the results.

[4] http://dib.joanneum.at/cape/MPG-20/

[5] http://www.vision.ee.ethz.ch/datasets/

For the baseline, linear search over the whole database without geometry filter we achieve 67.4% recognition rate. This value is outperformed by over 20% with the introduction of the geometry filter, resulting in 88% recognition rate. This is due to the removal of false positive matches. However, the improved precision comes at a price in speed.

Restricting search by GPS position with a radius of 300 meters is about 40% faster while increasing precision slightly for the case with geometry filter and more substantially for the case without filter. Restriction by celltower CGI is slightly faster but significantly worse in precision. This seems mostly due to the fact, that our CGI correspondences for different operators might be incomplete. For a real world application where an operator would hopefully contribute the cell-id information or a search radius bound by GPS coordinates we would thus expect better results.

Overall the best results are achieved with GPS and a rather large radius of several hundred meters. In figure 5 we plot the precision versus time for different radii. At 100 meters we retrieve most of the of the objects correctly, but only between 300 and 500 meters we achieve the same recognition rates as for linear search, however at significantly higher speed. In fact, this speed-up over linear search will obviously be even larger, the more items are in the database. The recognition times can be further sped up with a suitable indexing structure such as [14,19]. We have compared several methods, however the results are preliminary and beyond the scope of this paper.

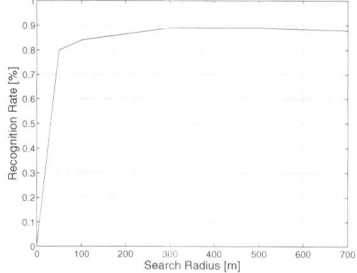 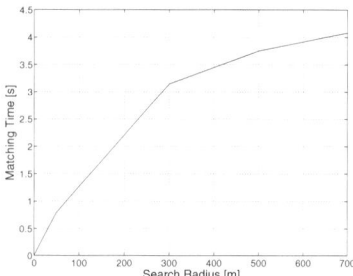

Fig. 5. Recognition rate (left) and matching time (right) depending on radius around query location

Visual results are shown in Figure 7. Section (a) shows query images in the left column and best matching database images for each query in the right column. Note the distance of the query image to the database image in the first row and the zoom and low contrast of the query in the second row. Section (b) contains a query image at the top and the best database match at the bottom. Besides the viewpoint change and occlusion through the lamp and railing, note that query and database image have very different clouds and lighting since they were taken several weeks apart. Section (c) shows an other query database pair, this time for a facade with strong cropping and change of angle. The last image in section (d)

Fig. 6. Geometric verification with a homography. Top rows: matches for a query image with the correct database image. Top left: before homography filter, top right: after homography filter. As the match between the slides is correct most of the putative feature matches survive the homography filter. At the bottom rows we match the same image to a false database image. As can be seen at the bottom left, a lot of false putative matches would arise without geometric verification, in extreme cases their count can be similar to or higher than for the correct image pair. At the bottom right all the false matches are removed, only features from the (correctly) matching frame survive and the discriminance to the correct pair is drastically increased.

Fig. 7. Result images for the city-guide application, see text for details

contains a typical "negative" query image, which should not return any matching object.

The results highlight the qualities of the suggested approach: the geometry filter improves recognition rates drastically. Restricting search to a geographic radius of a few hundred meters increases speed significantly even in our test database and will be essential for large-scale real world applications. At the same time, the results show that relying only on GPS information (objects up to several dozen meters away) would not be suitable for a real-world guiding application. Being able to "select" the objects with their mobile phone brings significant usability benefits to the user.

5   Conclusions and Outlook

We have presented an approach for object recognition for the Internet of Things, which allows users to request information on objects by taking a picture of them. We have implemented and demonstrated a full system and evaluated its capabilities in two challenging scenarios: slide tagging and bookmarking from screens in smart meeting rooms and a cityguide on a mobile phone. For both applications a server side object recognition system executes the following pipeline: local features are extracted from an incoming image. The features are matched to a database, where the search space is optionally restricted by metadata delivered with the request, for instance by geographic location from GPS coordinates or celltower ids. The resulting candidate matches are verified with a global geometry filter. The system is completed with a client-side software, which transmits query image and metadata such as GPS locations to the server with a single click.

We have demonstrated the flexibility of the suggested approach with an experimental evaluation for both sample applications. To that end, the system was evaluated on two very challenging test datasets. Building on local features and boosting the recognition rate with a geometry filter we achieved very high recognition rates. This approach worked well for both matching of slides with large amounts of text and images of tourist sights from strongly varying viewpoints which underlines the flexibility of the proposed approach. For the especially challenging cityguide application we could find a good balance between performance and recognition rate by restricting the search space using GPS location information.

The results showed, that the Internet of Things by object recognition can be realized already today for certain types of objects. In fact, the system can be seen as a visual search engine for the Internet of Things. Relying just on an image sent from a mobile phone, the system can be easily adopted by both end-users and system providers. With the advance of computer vision methods, we expect a wealth of additional possibilities in the coming years.

Acknowledgements. We acknowledge support by the Swiss project IM2 as well as the "Schweizerische Volkswirtschaftsstiftung". We also thank Fabio Magagna for his help with the implementation of the client-side software.

References

1. Abowd, G.: Classroom 2000: An experiment with the instrumentation of a living educational environment. In: IBM Systems Journal (1999)
2. Adelmann, R., Langheinrich, M., Floerkemeier, C.: A toolkit for bar-code-recognition and -resolving on camera phones – jump starting the internet of things. In: Workshop Mobile and Embedded Interactive Systems (MEIS 2006) at Informatik (2006)
3. Amir, A., Ashour, G., Srinivasan, S.: Toward automatic real time preparation of online video proceedings for conference talks and presentations. In: Hawaii Int. Conf. on System Sciences (2001)
4. Ballagas, R., Rohs, M., Sheridan, J.G.: Mobile phones as pointing devices. In: PERMID 2005 (2005)

5. Bay, H., Fasel, B., Van Gool, L.: Interactive museum guide: Fast and robust recognition of museum objects. In: Proc. Intern. Workshop on Mobile Vision (2006)
6. Bay, H., Tuytelaars, T., Van Gool, L.: Surf: Speeded up robust features. In: Leonardis, A., Bischof, H., Pinz, A. (eds.) ECCV 2006. LNCS, vol. 3951, pp. 404–417. Springer, Heidelberg (2006)
7. Boring, S., Altendorfer, M., Broll, G., Hilliges, O., Butz, A.: Shoot & copy: Phonecam-based information transfer from public displays onto mobile phones. In: International Conference on Mobile Technology, Applications and Systems (2007)
8. Carletta, J., et al. (17 authors): The ami meeting corpus: A pre-announcement. In: Renals, S., Bengio, S. (eds.) MLMI 2005. LNCS, vol. 3869, pp. 28–39. Springer, Heidelberg (2006)
9. Donoser, M., Bischof, H.: Efficient maximally stable extremal region (mser) tracking. In: IEEE Conf. on Computer Vision and Pattern Recognition (2006)
10. Fischler, M.A., Bolles, R.C.: Random sample consensus: A paradigm for model fitting with applications to image analysis and automated cartography. In: Comm. of the ACM (1981)
11. Fuhrmann, T., Harbaum, T.: Using bluetooth for informationally enhanced environments. In: Proceedings of the IADIS International Conference e-Society 2003 (2003)
12. Fockler, P., Zeidler, T., Bimber, O.: Phoneguide: Museum guidance supported by on-device object recognition on mobile phones. Research Report 54.74 54.72, Bauhaus-University Weimar (2005)
13. Hartley, R.I., Zisserman, A.: Multiple View Geometry in Computer Vision, 2nd edn. Cambridge University Press, Cambridge (2004)
14. Indyk, P., Motwani, R.: Approximate nearest neighbors: Towards removing the curse of dimensionality. In: STOC 1998: Proceedings of the thirtieth annual ACM symposium on Theory of computing (1998)
15. Leibe, B., Seemann, E., Schiele, B.: Pedestrian detection in crowded scenes. In: IEEE Conf. on Computer Vision and Pattern Recognition (2005)
16. Lowe, D.: Distinctive image features from scale-invariant keypoints. Intern. Journ. of Computer Vision (2003)
17. Mikolajczyk, K., Schmid, C.: A performance evaluation of local descriptors. IEEE Trans. PAMI 27(10), 1615–1630 (2005)
18. Niblack, W.: Slidefinder: A tool for browsing presentation graphics using content-based retrieval. In: CBAIVL 1999 (1999)
19. Nister, D., Stewenius, H.: Scalable recognition with a vocabulary tree. In: CVPR 2006 (2006)
20. Paletta, L., Fritz, G., Seifert, C., Luley, P., Almer, A.: A mobile vision service for multimedia tourist applications in urban environments. In: IEEE Intelligent Transportation Systems Conference, ITSC (2006)
21. Rohs, M., Gfeller, B.: Using camera-equipped mobile phones for interacting with real-world objects. In: Ferscha, A., Hoertner, H., Kotsis, G. (eds.) Advances in Pervasive Computing, Austrian Computer Society (OCG) (2004)
22. Sivic, J., Zisserman, A.: Video google: A text retrieval approach to object matching in videos. In: Intern. Conf. on Computer Vision (2005)
23. Vinciarelli, A., Odobez, J.-M.: Application of information retrieval technologies to presentation slides. IEEE Transactions on Multimedia (2006)
24. Want, R.: Rfid - a key to automating everything. Scientific American (2004)

The Digital Sommelier: Interacting with Intelligent Products

Michael Schmitz[1,2], Jörg Baus[1,2], and Robert Dörr[1]

[1] DFG - Collaborative Research Centre
Resource-adaptive cognitive processes
Saarland University
PO-Box 15 11 50, Germany
{schmitz,baus,rdoerr}@cs.uni-sb.de
[2] DFKI GmbH
Stuhlsatzenhausweg 3 (Building D3 2)
66123 Saarbrücken, Germany
{michael.schmitz,joerg.baus}@dfki.de

Abstract. We present the *Digital Sommelier*, an interactive wine shopping assistant that provides an intuitive multi-modal interface to general product information as well as to particular attributes of a certain product, such as its current temperature. Wine bottles sense their state via attached wireless sensors and detect user interaction over RFID and acceleration sensors. Visitors can inquire information either through physical interaction with products or a natural language interface. We describe a framework and toolkit for efficient prototyping of sensor based applications as the foundation for the integration of different sensor networks utilized by the sommelier. We further introduce our concept of talking products, an anthropomorphic interaction pattern that allows customers to directly talk to products with personalities.

1 Introduction

Emerging sensor technologies that go beyond RFID recognition are becoming cheaper and smaller, paving their way into objects of everyday life. The potential ubiquity of wireless sensors will particularly affect future generations of products as elements of production, processing and delivery chains, enabling them to sense and record their state and communicate with their environment. The prospective immediate benefits for manufacture and retail are an essential factor that is increasing the chances for research work in this field to get adopted by industry - just as RFID technology has already made the step into the commercial sector, as it provides economic values by facilitating process automatization such as product tracking or inventory stocktaking. The end consumer does not directly benefit from this technology, as long as the introduction of RFID technology does not transparently result into price reductions. Sensoric and communication technology that continously sense and observe the state of a product open up possibilities for new services that also benefit end consumers: One simple example is quality assurance of sensitive food such as fresh fish, by monitoring

C. Floerkemeier et al. (Eds.): IOT 2008, LNCS 4952, pp. 247–262, 2008.

the continous maintenance of the cold chain. By providing information about the temperature history to end customers, their trust into the product could be increased. Temperature is only one product attribute that could provide useful information to the customer: We envision that products will be able to record a complete digital diary with all relevant information, particularly regarding their quality and operation. Being able to communicate with their environment it is further important to allow humans to access such a product memory.

We developed a demonstration application, the *Digital Sommelier* (DigiSom), that provides a simple interface for customers to retrieve general features of the product (wine and champaign bottles) as well as specific state attributes: the current temperature and whether the bottle has been recently shaken. Our aim was to support novice users in their shopping activities with an intuitive interface. The system recognizes user interactions based on RFID and sensor technology, and provides a natural language interface combined with visual output of information. We also integrated our concept of anthropomorphic products into this scenario: Each type of wine is associated with a distinct personality, which is reflected by prosodic attributes of the voice and by different phrasing styles. The implementation of the sensor integration is based on a framework that we have designed for rapid development of sensor based applications. In this work we present first results of our group in this field: The framework for sensor based applications, the anthropomorphic interaction pattern and the DigiSom, an application integrating both concepts in a demonstration system that has been exhibited on various occasions.

The remainder of this paper is organized as follows. First, we describe the architecture of our framework for sensor based development in chapter 2. Chapter 3 covers the conceptual and technical background of the talking products paradigm. An overview of the DigiSom installation is given in chapter 4 and chapter 5 concludes this paper with a brief discussion.

2   A Framework for Sensor Based Prototyping

Embedding sensing, computing and communication capabilities into physical artifacts is the essential building block for the creation of intelligent objects. Some simple examples can already be found in commercial products: Apples iPhone for instance uses an acceleration sensor for flipping images, a proximity sensor for switching off the display when speaking and a light sensor for adjusting brightness. The latter technique is also used in some desktop monitors. Such standalone applications use their collected data only for one distinct and straightforward purpose and do not communicate it to other devices. In a next step we envision networks of things, where all sorts of information about objects are accessible and exchangeable between artifacts and/or the environment.

Research in the field of wireless sensor networks has brought up commercial off-the-shelf systems, which provide sensor nodes with integrated sensing, computing and communication capabilities. Some of them can be configured only by means of basic parameters, others have a programmable microcontroller and can be flashed with custom firmware. Such devices use their computing resources for more sophisticated tasks like preprocessing sensor readings or routing of messages through the network - even applications running completely inside the network are possible.

Advances in miniaturization allow such devices to be included in a wide range of objects. In our intelligent environment [1], we use wireless sensor nodes to monitor general environmental parameters like temperature or light intensity, as well as to observe the state of objects (i.e. products).

2.1   Sensor Platforms

No single sensor platform can meet all the requirements of different applications. Therefore we work with various wireless sensor platforms, namely the μPart, the MICAz and Scatterweb, while using the latter system only for environmental observations, such as measuring the temperature in a room or detecting the presence of people. The μPart platform has been initially developed at Karlsruhe University [2]. It is produced and sold by the spin off Particle Computer GmbH. μPart nodes provide sensors for light, movement and temperature. They can be configured to set the sampling rate and other basic parameters through light sensors, but firmware modifications are not possible this way. The MICAz platform is a Zigbee compliant implementation of the Mica Mote platform [3], developed at Berkeley University in cooperation with Intel. MICAz nodes are sold by Crossbow Inc. and are shipped without integrated sensors, but they can be stacked with various sensor boards. We use them mainly in connection with the MTS300 board, which provides sensors for light, temperature and noise as well as a dual-axis accelerometer and a dual-axis magnetometer. The ATmega128L microcontroller on the nodes provides computing resources to run custom firmware from the integrated 128 KBytes of flash memory.

2.2   Framework Requirements

So far applications have to access sensor data directly through hardware dependent libraries. In many cases hardware details influence the entire software design, which complicates software maintenance especially when hardware is exchanged or updated. This particularly poses a burden in research environments, where different device platforms are interchanged and used for different applications, reusability of software modules is very restricted.

To enhance rapid development of sensor based applications, a solution had to be found, which provides a structure for data processing through reusable software components. From previous experience with sensor applications we have extracted the following requirements for such a framework:

1. Sensor data should be accessible through a high level API instead of raw byte access.
2. Data processing should be as simple and efficient as possible to cope with high data rates from a large number of sensors simultanously.

3. All sensor specific hardware and software details should be hidden.
4. A large number of data sources should be supported, including not only wireless sensor networks but also other types of input (e.g. reading from files, databases or other network sources).
5. There should be prebuilt components for common application tasks as well as for development related issues (e.g. data inspection, debugging).
6. The implementation should be open for extension, allowing further development and adaption of new sensor hardware.

2.3 Processing Sensor Data Streams

The design of the framework is based on the concept of stream processing [4] (see [5] for an overview). It uses a data driven approach where information is contained in possibly infinite streams of data entities. Each entity can be seen as an event and in our terms, each raw sensor reading is such an event. In our framework, events are produced, consumed and processed by three kinds of modules: *Input modules* encapsulate data sources such as sensor networks or files, and provide a steady flow of information into the framework. *Output modules* are for example used for displaying data, connecting to other applications or triggering actions based on events. Inbetween input and output optional *filtering modules* process the data streams to extract and prepare relevant and meaningful information from incoming sensor data. The data flow is implicitly uni-directional - if for any reason a bi-directional flow is required, it must be explicitly constructed.

The sensor based component of a new application is built by choosing one or more modules of a set of modules provided by the framework and connecting their in- and outputs appropriately to construct a desired function. Although new custom modules can be easily implemented and integrated, the set of prebuilt modules covers many common data processing tasks, such that a wide range of simple applications can be developed quickly with the given components.

As sensor nodes define their own message format, the encapsulating input modules are customized to suit the underlying sensor platform, and have to be extended, whenever new platforms have to be integrated. Such input modules provide flexibility and connectivity towards the framework, a back channel to inject messages into a sensor network is not implemented yet, since the main focus of the framework is to simplify data extraction and processing from sensors.

There are already several tools available to support event processing for intelligent environments. We are using the EventHeap [6] to share data between applications, but sending each sensor reading directly to the EventHeap is certainly not a good idea, since a single reading does not contain much information. The raw sensor readings have to be preprocessed and analyzed in order to generate higher level events that are meaningful enough to be sent to the actual application. The Context Toolkit [7] provides a well structured framework for working with context information. Its Generator-Widgets encapsulate context information gained from sensor readings, but a concept to extract high level information from raw sensor data is not provided.

2.4   Implementation

We decided to implement the framework in Java mainly because of two reasons: Most sensor networks already come with some Java library or classes giving a good starting point. Furthermore, Java is platform independent, which is particularly important when working in a heterogenous environment.

Data entities which are transmitted through the framework are represented by the *Message* class (fig. 1). Each message contains the source id, a timestamp, the data value and a type field. All fields are Strings, except of the timestamp which is of type long.

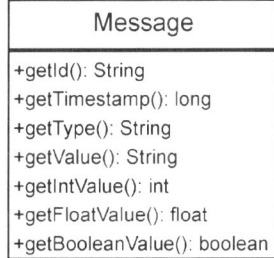

Fig. 1. The Message class (part)

As already stated, there are three types of modules, namely *Input-*, *Output-* and *Filter*-Modules. They are linked together using a design pattern from Object Oriented Programming [8] called the Listener Pattern. It allows coupling of software components and uni-directional transmission of data from a Subject (Input) to one or more Listeners (Output). Filter modules implement both, `StreamIn` and `StreamOut` interfaces and are further divided in buffered und unbuffered filters. The original pattern has been extended by a tagged version to allow Output streams being grouped together. Figure 2 shows the main interfaces.

This approach allows to seperate the application architecture from the specific implementation of modules. The architecture can be described in a model-driven fashion using block diagrams, which contain the modules as well as their connections. Figure 3 shows a diagram for a simple application based on pre-built modules. It takes sensor data from μPart nodes, filters out everything but temperature readings, and writes them to a file only when sensor value changes. The architecture can be easily implemented as shown in listing 1.1. Each box corresponds to a constructor call and each arrow corresponds to a call of the `addOutputStream()` method. More complex applications may incorporate multiple sensor networks, more sophisticated filtering and different output modalities.

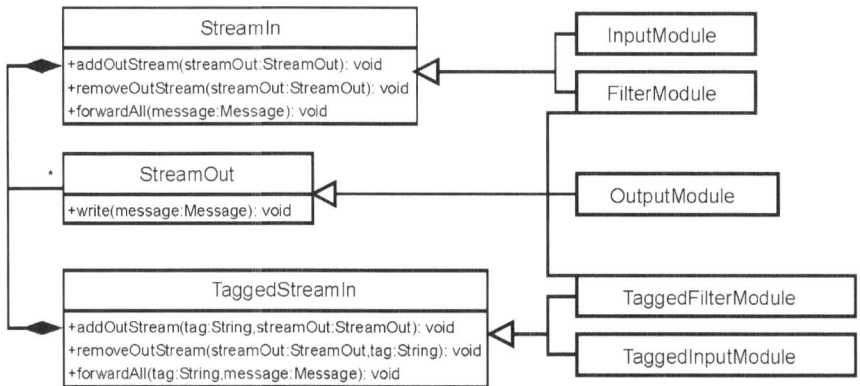

Fig. 2. Extended Listener Pattern

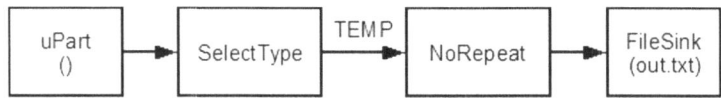

Fig. 3. Block diagramm for an application that writes sensor readings to a file

```
// construct modules
UPart input = new UPart();  // use default port
SelectType selectType = new SelectType();
NoRepeat noRepeat = new NoRepeat();
FileSink output = new FileSink(new File("out.txt"));

// link modules
input.addOutputStream(selectType);
selectType.addOutputStream("TEMP", noRepeat);
noRepeat.addOutputStream(output);

// start receiving data
input.start();
```

Listing 1.1. Implementation of the application described by figure 3

So far, there are about 50 prebuilt modules. Input can be obtained from different sensor networks, from files or from network or serial connections. Amongst others, output can be directly fused into files, databases or higher level systems like the EventHeap. For data inspection and debugging there are modules that display data either graphical of textual. Filter modules implement various computing functions like sum, variance or standard deviation of a sequence of values. There are filters that drop repeating values, split message streams by origin or type, translate certain field values of the messages and many more. The translation for example is often used to map sensor IDs to products and can be configured through an external XML file, eliminating the need for touching the binaries when changing product mappings. Modules are implemented by extending abstract base classes, which already contain functionality for establishing links between modules, so there is minimal effort needed (see listing 1.2 for an example).

```
public class EvenOddFilter extends Filter {
    protected Message filter(Message msg) {
        try {
            if(msg.getIntValue() % 2 == 0) {
                msg.setValue("Even");
            } else {
                msg.setValue("Odd");
            }
            return msg;
        } catch (NumberFormatException e) {
            return null;
        }
    }
}
```

Listing 1.2. Implementation of an example filter module

2.5 Special Purpose Modules

Not all algorithmic problems can be broken down into simple solutions. In some cases it is more reasonable to implement a complex function as a self contained module, which provides the necessary interfaces for connecting it to the framework. Robust movement analysis based on acceleration sensors is such a task, requiring sophisticated methods for information extraction. The complexity of the given data in this case requires a tailored approach, for which we generated a specialized module for our framework.

Acceleration data is sampled with a rate of 100 samples per second. In a first step the readings are preprocessed by computing several statistical values over a window of 30 samples corresponding to about 300ms. The statistical values are mean, variance, standard deviation as well as the number of local minima and maxima for each acceleration axis, forming a block of 10 characteristic values, which describe the state of acceleration for a specific time interval. These blocks are computed by our own firmware installed on the sensor boards, and are sent to the receiving PC that runs the framework and then processed by WEKA [9], first in a training phase and after that in a classification phase. In the training phase, several classes of movement states have to be specified, and incoming readings are assigned to the according movement class. Based on this training data the system can classify incoming data packages and send results further into the framework. Using our statistical approach of movement state classification, we are able to differentiate the movements regarding their intensity (no movement, slow movement, fast movement), their direction or anything else that can be extracted from the set of statistical values. The classification is done by the WEKA implementation of a Ripple-Down Rules (RDR) classifier [10], but other classifiers like C4.5 decision trees [11] have also proven to be useful. For a user of our framework, there is no need to write a single line of code during this process. A self written tool allows the training phase to be completed in a few minutes just by moving the sensor, clicking buttons and entering names for different movement classes.

2.6   Integrating Sensors into the Digital Sommelier

There are two different tasks for our sensor equipment in the DigiSom application: Collecting and storing sensor readings that define the state of wines (temperature and vibration) using the μPart nodes, and acceleration sensors on the MICAz nodes to detect, whether a user looks at the front or at the rear side of a product box. Figure 4 shows the design of the sensor processing application used for the DigiSom.

Temperature processing was a rather straightforward task, there were only two minor issues which have to be dealt with. First, the μParts do not allow to set the sampling rate of temperature and vibration sensor independent of each other. To provide adequate reaction times in detecting vibration, we had to choose a sampling interval of about one second. This results in a steady stream of equal temperature values, which can be limited by the *NoRepeat* filter module.

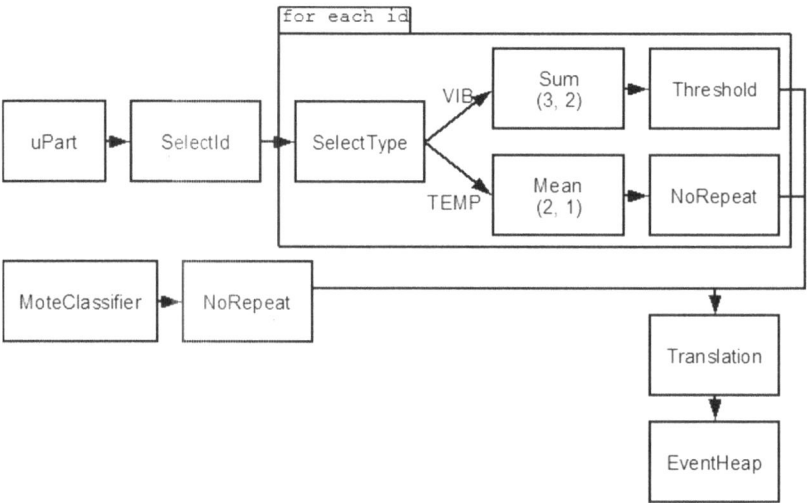

Fig. 4. Block diagramm of sensor data processing in the DigiSom

The second problem is that at the transition between two degree values, there is some interval of jittering, where sensor readings continously switch between these values. To prevent these messages from slipping through the NoRepeat filter, we compute the mean of two successive readings. This is the simplest form of a moving average represented by the module *Mean* with a buffer size of two and with an overlap between buffers of one message (see Figure 4). For vibration detection the μPart nodes use an integrated ball switch. This is great for implementing a wakeup on move function, but gives only poor results for deciding, if a bottle has been shaken or has just been moved. Vibration readings are summed up over a buffer of three messages with an overlap of two. The result is then compared to a threshold, which represents the sensitivity. To detect the orientation of the product boxes, we integrated MICAz nodes. Their two-axis accelerometer measures the acceleration caused by gravitation. A user looking at a product never holds it perfectly upright, but always in a certain angle such that she must not hold the box at eye level for a proper view. Taking this observation into account, we have trained our nodes to classify front, back and upright orientation. The classifier forwards its results into the framework, where repeating messages are dropped. After preprocessing is done, messages are translated to map sensor node IDs to poducts and published as EventHeap tuples to make them available to other parts of the DigiSom.

3   Dialogue Shell of Talking Products

One important design goal of our interactive shopping assistance is to support arbitrary users, particularly computer novices, who are not able or willing to learn the use of such a system. We therefore have to find a solution that provides a natural interaction, requiring minimal effort of a user to understand and utilize the assistance system. Nijholt et al. [12] suggest that a limited animistic design metaphor seems to be appropriate for human-environment interaction with thousands of networked smart objects. People often tend to treat objects similar to humans, according to findings of Reeves and Nass [13], which allows users to explain the behavior of a system if they lack a good functional conceptual model. In consequence, we decided to employ a natural language system, which enables the user to talk to each product.

[14] conducted a usability study of a multi-modal shopping assistant. The implemented system allows users for instance to request product information in a combination of speech and selecting gestures (i.e. taking a product out of the shelf). Findings of this study showed among others that users generally preferred direct over indirect interaction, i.e. by asking "What is your price?" instead of "What is the price of this camera?", which encouraged us to pursue this approach.

Previous studies have shown that interacting with embodied conversational agents that have consistent personalities is not only more fun but also lets users perceive such agents as more useful than agents without (consistent) personalities [15,16]. It is further shown that the speech of a consistent personality enables the listener to memorize spoken contents easier and moreover reduces the overall cognitive load [17,15]. Thus we emphasized the anthropomorphic aspect of this interaction pattern by assigning personalites to products, which are reflected by the spoken responses of a product.

Product manufacturers benefit as well, since the personalization of the product provides a new channel to communicate a brand image or distinct attributes of a certain product. A study within the context of marketing research showed that if in radio advertisements a voice fits the product, it helps the listener to remember the brand, the product and claims for that product [18].

3.1 Modelling Personality in Voices

In a first step we created voices that reflect certain personalities according to Aaker's brand personality model [19] only by adjusting prosodic parameters. We chose this model over the (rather similar) five factor model [20] commonly used in psychology, since we are applying the concept of talking objects in the shopping domain. However, both models are rather similar and to a certain extent exchangable.

We changed the four prosodic parameters pitch range (in semitones), pitch level (in Hz), tempo (as a durational factor in ms) and intensity (soft, modal or loud as in [21]) according to our literature review [22]. For example, a competent voice has a higher pitch range (8 semitones), a lower pitch level (-30%), a 30% higher tempo and a loud intesity compared to the baseline voice. In [22] we also evaluated whether it is possible to model different personalities with the same voice by adjusting these prosodic parameters, such that listeners will recognize the intended personality dimension. The study has shown that there are clear preferences for our prosody modeled speech synthesis for certain brand personality dimensions. But not all personality dimensions were perfectly perceived as intended, such that we have to amplify the effect.

Personality is certainly not only expressed in qualitative attributes of a voice, other properties of a speech dialogue are also essential, like the used vocabulary or the general discussion behaviour. For this reason we created a dialogue shell that incorporates these aspects.

3.2 Expressing Personality in Dialogues

The widely adopted personality model by Costa and McRae [20] constitutes five dimensions of human personality: Extraversion, Agreeableness, Conscientousness, Neuroticism and Openness on a scale from 0 to 100. Obviously, differentiating 100 levels in a dimension is far too much for our goals, therefore we simplified this model by discriminating 3 levels in each dimension:

- low: value between 1 and 44 (31% of population)
- average: values between 45 and 55 (38% of population)
- high: values between 56 and 100 (31% of population)

Related work, e.g. by Andre et al. [23] limited their personality modelling to only two of the five dimension - namely extraversion and agreeableness - since

these are the most important factors in interpersonal communication. Nevertheless, we discovered considerable influences of openness and conscientiousness to
speech, therefore we incorporated these two dimensions as well. The effect of the
dimension neuroticism is mainly to describe the level of susceptibility to strong
emotions, both positive and negative ones [24]. It is further shown that the level
of neuroticism is very hard to determine in an observed person [25], thus we
decided that four dimensions will suffice for our work.

We conducted an exhaustive literature review on how speech reveals different personality characteristics. Among numerous other ressources, two recent
research papers provided essential contributions to our work: Pennebaker and
Kings analysis in *Journals of Personality and Social Psychology* [26] and Nowson's *The Language of Weblogs: A Study of Genre and Individual Differences*
[27]. In both studies a large number of text blocks were examined with an application called *Linguistic Inquiry and Word Count*[1] (LIWC), which analyzes text
passages word by word, comparing them with an internal dictionary. This dictionary is divided in 70 hierarchical dimensions, including grammatical categories
(e.g. noun, verb) or affective and emotional processes. Pennebaker determined
in a study the 15 most reliable dimensions and searched for them in diary entries
of test persons with LIWC. With these results together with the given personality profiles of the probands (according to the five factor modell), he identified
correlations between the two. Nowson performed a similar study and searched
through weblogs for the same LIWC factors.

Based on these results, we provided a set of recommendations how responses
of an talking object with a given personality should be phrased. For instance,
for a high level of extraversion these recommendations are given:

- Prefered bigrams: *a bit, a couple, other than, able to, want to, looking forward*
  and similar ones.
- Frequent use of terms from a social context or describing positive emotions
- Avoidance of *maybe, perhaps* and extensive usage of numbers
- Usage of colloquial phrases, based on verbs, adverbs and pronouns
- Comparably more elaborate replies

Following these principles we implemented basic product responses (greetings,
inquiries for product attributes, farewell) for several personalities. All possible
replies of our dialogue shell are stored in one XML-file, which we named the
Anthropomorphic Fundamental Base Grammar. All entries include an associated
personality profile, for example:

```
<reply
    query=" hello "
    reply=" Hello ,  nice  to  meet  you ! "
    ag=" 1 "  co=" 2 "  ex=" 1 "  op=" 1 ">
<\reply>
```

---

[1] http://www.liwc.net/

Which means that this is the greeting of a product with average agreeableness, extraversion and openness and a high value in conscientousness. Another example:

```
<reply
    query="hello"
    reply="Hi! I'm sure I can help you! Just tell me
        what you need and I bet we can figure
        something out!"
    ag="2" co="2" ex="2" op="2">
<\reply>
```

All entries that do not regard any particular personality, should have average personality values in all dimensions.

A central product database with all products and their attributes is extended by the assigned personality profile, i.e. the values in each of the four dimensions. When the application starts up, the dialogue shell retrieves the product data of each product instance and extracts the appropriate entries from the base grammar to build the custom product grammar. If there are no entries that exactly match the given profile, the one that has the most identical values will be chosen. This dialogue shell generates a consistent speech interface to a product by knowing its attributes and a given personality profile, for instance preset by the manufacturer. With the help of such a dialogue shell we plan to fully realize our concept of products with "speaking" personalities.

4   The Digital Sommelier Installation

To test the aforementioned and other previously elaborated concepts, our group developed shopping assistants for two different shopping domains, namely a digital camera store and a grocery store including wines. One shopping assistant provides product information, product comparisons, and personalized advertisements either on tablet PC mounted on the handle of the shopping cart or wall mounted displays. In the case of the grocery scenario one instance of the shopping assistant suggests recipes based on the amount and kind of goods already placed in the shopping cart or provides information about wines and the dishes they go well with.

In a shop the user has the possibility to communicate with the store by means of multi-modal interface as described in [28] or by interacting with the offered products. Her actions in the store are recognized and monitored by a combination of *RFID*- and (in the latest version) by sensor technologies as described in this paper. The shelves in our store are equipped with RFID-readers, and every product is tagged with passive RFID-transponders, which allows for locating and identifying products. Every RFID-transponder also contains a reference to all relevant product information like its name, ingredients and nutrition facts.

In addition to that the DigiSom uses various sensors either directly attached to the wine bottles or in case of champagne to the product box, in order to gather information about the state of a product, i.e. the actual temperature of the bottle of wine and whether the bottle has been shaken or not. Furthermore, interactions with the product box can be monitored using the attached sensors, which are able to measure light, temperature, sound, and acceleration. Whenever the user takes a bottle out of the shelf the attached RFID-tag disappears from the RFID-reader field and the DigiSom receives an event from the EventHeap, which triggers the generation of the according product information page to be displayed on a nearby screen and at the same time initiates the champagne to introduce itself using the personalized voice. Furthermore, the application receives the current temperature of the champagne from the integrated sensor node, comparing the actual temperature against the proposed temperature at which the champagne should be served and generates appropriate hints, e.g. that the champagne has to be chilled before serving (see Figure 5). If the user turns the champagne to read the information on the rear side of the box, the associated sensor detects the rotation and generates the associated event, which is passed to our application. In our application such events are interpreted as an indication of the user's interest in the product and the DigiSom changes the information already displayed. In this case the application refers the user to the web page of the manufacturer, where the user can search for additional information about the product (see Figure 6).

Digitaler Sommelier

Produktinformation

Produkt:	Gosset Grande Réserve Brut
Jahrgang:	Cuvée verschiedener Jahrgänge
Alkohol:	12,00 Vol%
Preis:	30,10 EUR
Hersteller:	Gosset
Website:	www.champagne-gosset.com
Herkunftsland:	Frankreich, Champagne
Traube:	Chardonnay (46%), Pinot Noir (39 %), Pinot Meunier (15 %) und "Reserve-Weine"
Serviertemperatur:	7 °Celsius
Farbe:	Dunkles Gold-Gelb, leichte kleine Perlen
Beschreibung:	Intensiv und komplexe, aromatische Note Blumig (Lindenblüte, Narzisse) und fruchtig (Kirsche und Brombeere) Am Anfang Aroma von getrockneten Feigen und glasierten Kirschen dann geröstete Mandeln und Zwieback Langer, kräftiger Abgang
Passt zu:	Geeignet als Aperitif Champagnercreme

Tipp: Die aktuelle Temperatur des Produktes beträgt 23 °Celsius Sie sollten das Produkt vor dem Servieren kühlen!

Fig. 5. Generated page with product information including the champagne's actual temperature

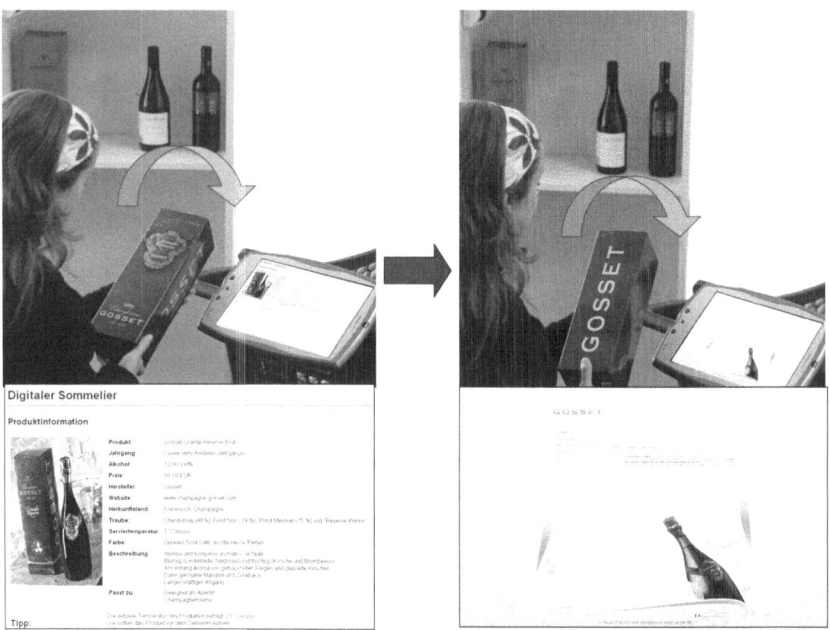

Fig. 6. A user interacting with a bottle of champagne

Through the combination of information from RFID-tags, data about the champagne's/wine's actual temperature, and whether the bottles have been shaken or not, as well as acceleration information from MICAz nodes, our Digi-Som is able to incorporate real time data about the state of products to adapt product presentations.

5   Conclusion and Future Work

In this paper we introduced the *Digital Sommelier*, an interactive shopping assistant, which uses the concept of personalized speech to present multi-modal product information. Based on the introduced framework the application combines RFID- and sensor technology to incorporate real time data about the product's state and user interactions with the product itself, in order to adapt the product presentations that will be given to the user. We believe that this technogies will pave their way into new innovative retail solutions. Furthermore, the data collected might also be used to feed and update digital product memories, that can be accessed by producers, retailers and end users. Another promising application fields for anthropomorphic, sensor-based user interfaces are interactive installations in museums or other exhibitions: The areas of education, entertainment and the combination of both provide an appropriate playground for virtual characters and tangible user interfaces, whereas in such an environment the role of integrated sensors would probably focus on detecting user interactions.

We further described a framework for sensor based prototyping, which helped us to quickly develop software components that rely on various sensor data streams. A large set of modules implemented as a Java library allows researchers and developers to assemble applications without the burden of investing much time into low level sensor stream processing. The next step will be to develop a graphical editor, for instance based on the Eclipse Graphical Editing Framework (GEF)[2], to further simplify the process of prototyping. Besides that we will continue integrating new sensor systems that will come to the market.

Our latest research on the interaction metaphor of talking products as presented in this paper will be evaluated in order to discover whether it is possible to communicate brand attributes and intentionally influence the shopping behaviour of customers. It will also be interesting to see, whether this simplified interaction pattern is more intuitive to certain customer groups than other alternatives. Pursuing the anthropomorphic approach we will try to apply an emotional model to products, such that a current emotional state will represent the state of a particular product.

Acknowledgements

This research is being supported by the German Science Foundation (DFG) in its Collaborative Research Center on Resource-Adaptive Cognitive Processes (SFB 378, Project EM 4 BAIR).

References

1. Butz, A., Krüger, A.: A mixed reality room following the generalized peephole metaphor. In: IEEE Computer Graphics & Applications, vol. 26, IEEE press, Los Alamitos (2006)
2. Beigl, M., Decker, C., Krohn, A., Riedel, T., Zimmer, T.: μParts: Low cost sensor networks at scale. In: Adjunct Demo Proceedings of Ubicomp 2005, Tokyo, Japan (2005)
3. Hill, J., Culler, D.: Mica: A wireless platform for deeply embedded networks. IEEE Micro 22, 12–24 (2002)
4. Burge, W.H.: Stream Processing Functions. IBM Journal of Research and Development 19, 12–25 (1975)
5. Stephens, R.: A survey of stream processing. Acta Informatica 34, 491–541 (1997)
6. Johanson, B., Fox, A., Hanrahan, P., Winograd, T.: The event heap: An enabling infrastructure for interactive workspaces. Technical Report CS-2000-02, Stanford University, Stanford, CA (2000)
7. Salber, D., Dey, A.K., Abowd, G.D.: The Context Toolkit: Aiding the Development of Context-Enabled Applications. In: CHI 1999: Proceedings of the SIGCHI conference on Human factors in computing systems, pp. 434–441. ACM Press, New York (1999)
8. Gamma, E., Helm, R., Johnson, R., Vlissides, J.: Design Patterns: Elements of Reusable Object-Oriented Software. Addison-Wesley Longman Publishing Co., Inc., Boston (1995)

---

[2] http://www.eclipse.org/gef

9. Witten, I.H., Frank, E.: Data Mining: Practical machine learning tools and techniques, 2nd edn. Morgan Kaufmann Publishers Inc., San Francisco (2005)
10. Comption, P., Jansen, R.: A philosophical basis for knowledge acquisition. Knowledge Acquisition 2, 241–257 (1990)
11. Quinlan, J.R.: C4.5: Programs for Machine Learning. Morgan Kaufmann Publishers Inc., San Francisco (1993)
12. Nijholt, A., Rist, T., Tuijnenbreijer, K.: Lost in Ambient Intelligence? Panel Session. In: Proceedings of CHI 2004, ACM, pp. 1725–1726 (2004)
13. Reeves, B., Nass, C.: How people treat computers, television, and new media like real people and places. CSLI Publications and Cambridge university press (1996)
14. Wasinger, R., Krüger, A., Jacobs, O.: Integrating Intra and Extra Gestures into a Mobile and Multimodal Shopping Assistant. In: Gellersen, H.-W., Want, R., Schmidt, A. (eds.) PERVASIVE 2005. LNCS, vol. 3468, pp. 297–314. Springer, Heidelberg (2005)
15. Nass, C., Isbister, K., Lee, E.J.: Truth is beauty: researching embodied conversational agents. Embodied conversational agents, 374–402 (2000)
16. Duggan, B., Deegan, M.: Considerations in the usage of text to speech (tts) in the creation of natural sounding voice enabled web systems. In: ISICT 2003: Proceedings of the 1st international symposium on Information and communication technologies, Trinity College Dublin, pp. 433–438 (2003)
17. Fiske, S., Taylor, S.: Social cognition. McGraw-Hill, New York (1991)
18. North, A., MacKenzie, L., Hargreaves, D.: The effects of musical and voice "'fit"' on responses to advertisements. Journal of Applied Social Psychology 34, 1675–1708 (2004)
19. Aaker, J.: Dimensions of brand personality. Journal of Marketing Research, 342–352 (1997)
20. McRae, R., John, O.: An introduction to the five-factor model and its applications. Journal of Personality 60, 175–215 (1992)
21. Schröder, M., Grice, M.: Expressing vocal effort in concatenative synthesis. In: Proceedings of the 15th International Conference of Phonetic Sciences, pp. 2589–2592 (2003)
22. Schmitz, M., Krüger, A., Schmidt, S.: Modelling personality in voices of talking products through prosodic parameters. In: Proceedings of the 10th International Conference on Intelligent User Interfaces, pp. 313–316 (2007)
23. André, E., Klesen, M., Gebhard, P., Steve Allen, T.R.: Integrating models of personality and emotions into lifelike characters. In: Paiva, A., Martinho, C. (eds.) Proceedings of the workshop on Affect in Interactions - Towards a new Generation of Interfaces in conjunction with the 3rd i3 Annual Conference, Siena, Italy, pp. 136–149 (1999)
24. Costa, P., McCrae, R.: The neo personality inventory manual. Psychological Assessment Resources (1985)
25. Gill, A.J., Oberlander, J., Austin, E.: The perception of e-mail personality at zero-acquaintance. Personality and Individual Differences 40, 497–507 (2006)
26. Pennebaker, J., King, L.: Linguistic styles: Language use as an individual difference. Journal of Personality and Social Psychology 77, 1296–1312 (1999)
27. Nowson, S.: The Language of Weblogs: A study of genre and individual differences. PhD thesis, University of Edinburgh. College of Science and Engineering. School of Informatics (2006)
28. Wasinger, R., Wahlster, W.: The Anthropomorphized Product Shelf: Symmetric Multimodal Interaction with Instrumented Environments. In: Aarts, E., Encarnaço, J.L. (eds.) True Visions: The Emergence of Ambient Intelligence, Springer, Heidelberg (2006)

Socially Intelligent Interfaces for Increased Energy Awareness in the Home

Jussi Karlgren, Lennart E. Fahlén, Anders Wallberg, Pär Hansson, Olov Ståhl,
Jonas Söderberg, and Karl-Petter Åkesson

Swedish Institute of Computer Science
Box 1263, 164 29 Kista, Sweden
erg@sics.se

Abstract. This paper describes how home appliances might be enhanced to improve user awareness of energy usage. Households wish to
lead comfortable and manageable lives. Balancing this reasonable desire
with the environmental and political goal of reducing electricity usage is
a challenge that we claim is best met through the design of interfaces that
allows users better control of their usage and unobtrusively informs them
of the actions of their peers. A set of design principles along these lines
is formulated in this paper. We have built a fully functional prototype
home appliance with a socially aware interface to signal the aggregate
usage of the user's peer group according to these principles, and present
the prototype in the paper.

Keywords: smart homes, domestic energy usage, physical programming, connected artifacts, distributed applications, micro level load-balancing

1   Energy Usage in the Home — An Interface Issue

Monitoring and controlling energy usage is a small component in an increasingly
intellectually demanding everyday life. Many other choices make demands on
the schedule, budget, and attention of the consumer.

However, energy usage is becoming an increasingly important facet of personal
choice: the choices made by individual consumers have consequences, both social
and environmental. In public discourse, energy usage is discussed as an arena
where individual choice makes a real and noticeable difference for the society
and the ecological system. This is a driving force and a motivating factor for the
individual consumer.

At the same time, another strand of development makes itself known in the
home and the household: the incremental digitalisation of home services. The
introduction of more and more capable entertainment, media, and gaming appliances on the one hand, and more powerful communication devices on the other,
has brought typical homes to a state where there is a high degree of computing
power but very little interconnection between appliances. It is easy to predict the
future but more difficult to get it right, and the history of "intelligent homes"

C. Floerkemeier et al. (Eds.): IOT 2008, LNCS 4952, pp. 263–275, 2008.
© Springer-Verlag Berlin Heidelberg 2008

and other related research targets is littered with mistaken projections and predictions [1,2]. However, it can safely be envisioned that the future will hold better interconnection between the various computing systems in the home, and that more digital services and appliances for the home will be introduced apace. We believe that systems for energy monitoring and control are an important vehicle for home digitalisation and that they are a natural locus for the convergence of the various functions foreseeable and present in a home.

Our research method is to instantiate various combinations of design principles into fully functional product prototypes, some of which are used to prove a point in a demonstration, others which are field tested, and yet others which are used as a basis for industrial projects. The prototypes are meant to illustrate and test the principles - not to conclusively prove or disprove them.

This paper begins by giving the points of departure for this concrete project, in terms of political and societal goals and in terms of consumer and user needs;it continues by outlining some central design principles taken into consideration during the design cycle; it then describes a concrete prototype for raising energy awareness and for providing unobtrusive mechanisms for better control of electrical energy usage in the home, instantiating the design principles in question.

2 Goals

Current societal and political attention in to a large extent focussed on questions of environmental sustainability. Chief among those question is that of energy usage and turnover. The operation, management, and maintenance of energy usage is to a large extent technological — but based on the interaction between choices made to uphold a desired lifestyle, to conserve energy, participate responsibly in attaining societal objectives, and to preserve the integrity and habitability of one's personal everyday life. To this end, our work recognises a number of background non-technological goals.

2.1 First Goal: Reducing Electricity Usage

An overriding political goal in Europe is to reduce energy usage. While energy is a raw material in several central industrial and infrastructural processes, as well as a key resource in transport and in heating and cooling indoor locations, much of energy usage is incidental to the primary task at hand – that of leading a comfortable life or performing services and producing goods in a comfortable environment. We will in this example focus on one aspect of energy usage, that of electricity usage in household environments.

Household electricity usage has risen noticeably in the most recent ten-year period [3]. There are opportunities to reduce electricity usage, inasmuch much of the increase can be attributed to unaware electricity spill, caused by standby systems or inadvertent power-up of home appliances.

The current public policy on energy matters includes the goal to reverse this trend and to considerably reduce household electricity usage within the next decade. There are studies that show that in spite of a higher density of kitchen appliances and a rapid and increasing replacement rate as regards such appliances, kitchens use less electricity today than ten years ago[4], which speaks towards the positive effect of technological advancement and towards new technology being a solution, not a hindrance to energy efficiency. Reduction in electric energy usage in households cannot be accomplished through tariff manipulation, since the politically appropriate price range of a unit of electricity cannot provide adequate market differentiation.

2.2   Second Goal: Load Shifting

The marginal cost of electricity at peak load, taking both production and distribution into account is considerable. It is desirable to reduce peak loads by shifting usage from high load to low load times over a 24-hour period or even within a shorter period. Reducing peak loads lessens the risk of overburdening the system, reduces the dependence on marginal electrity production systems – often with more noticeable environmental impact and with greater production cost per unit, and allows the power grid to be specified for a more rational capacity utilisation.

2.3   Third Goal: Preserving the Standard of Living

An immediate political concern is to accomplish a lowered electrical energy turnover and a more balanced load over time, and to do this without impacting negatively on living standards. The savings should be predominantly directed at unconscious spill and waste rather than at reducing quality of life.

In addition to this, the most immediate and pressing need of many users is expressed in terms of life management and coping – we do not wish to add burdens to the harried everyday life of consumers. Our goal is to help the individual household consumer keep electrical energy usage an appropriately small part of everyday life, afford the user higher awareness, better sense of control, without intruding on the general make-up of the home by introducing new, cumbersome, and unaesthetic devices.

3   Studies of Users

Feedback for better awareness or control of energy usage is well studied. A literature review from the University of Oxford [5] indicates savings between 5-15% from direct feedback, and also states that "... time-of-use or real-time pricing may become important as part of more sophisticated load management and as more distributed generation comes on stream.". Much of the work listed in the review concerns feedback in the form of informative billing and various types of displays (even ambient ones), with the aim of providing users with a better understanding of their energy usage. *Smart metering* is a general approach to building better and more intelligent meters to monitor electricity usage, to raise awareness among consumers, and to provide mechanisms of control either to distributors or consumers: modern electricity meters are frequently built to be sensitive to load balancing issues or tariff variation and can be controlled either through customer configuration or through distributor overrides beyond the control of the consumer [6,7,8]. Better metering has a potential impact on energy load peak reduction and the allows for the possibility for time-of-use pricing, issues which have influenced the design of the tea kettle prototype presented later in this paper.

Furthermore, initiatives such as UK's national Design Councils top policy recommendations from work with users, policy makers, and energy experts, highlight user awareness and control given by e.g. more detailed real-time monitoring of appliance energy usage, controlled through an "allowance", household "energy collaboratives" and energy trading clubs [9].

Our project shares goals with smart metering projects but focusses more on the control aspect than most other approaches — on how to allow users to influence their use of energy in the home environment.

From exploratory interview studies performed in homes and households by ourselves during the prototype design phase of our project, we have found that consumers in general are quite interested in taking control of their energy turnover: they feel they understand little of it and that the configuration and control of their home appliances are less in their hand than the consumers would like them to be. Our subjects in several of the at-home interviews we performed expressed puzzlement and lack of control in face of incomprehensible and information-dense energy bills – none wished to receive more information on the bill itself. For instance, interview subjects were not comfortably aware of the relation between billing units (kWh) and electricity usage in the household.

However, several of our subjects had instituted routines or behaviour to better understand or monitor their energy usage: making periodic notes of the electricity meter figures, admonishing family members to turn off various appliances, limiting the usage of appliances felt to be wasteful of energy, switching energy suppliers to track the least costly rates offered on the market. Many of these routines were ill-informed or ineffectual (e.g. turning off lighting, while not addressing other appliances such as home entertainment systems), and were likely to have little or no effect on the actual electricity usage or cost, but it afforded the users in question some sense of control.

Wishing to gain better control is partially a question of home economics: rising electricity bills motivates consumers to be more aware of electricity usage and to economise. However, while economy alone cannot become a strong enough motivating factor (as noted above, for reasons extraneous to the energy and environment field), consumer behaviour can be guided by other indicators. Societal pressure to behave responsibly accounts for much of consumer behaviour, e.g. in the form of willingness to participate in a more sustainable lifestyle — witness the high rate of return of drink containers in spite of the low remuneration given by the deposit system. [1] A companion motivator can be found in a general pursuit better to be able to retain the initiative and control over an increasingly complex set of systems in the household.

4   Guiding Principles

To empower users to take control of their household energy turnover, we work to harness the interest consumers show towards the electrical energy usage issue. Our conviction is that to harness this interest we need to design and develop systems that are effortless, fun, and effective to use. We wish to develop pleasurable aspects of new technology — and we do not wish to build systems to use value statements to influence or distress its users. To achieve these goals we suggest a set of guiding principles to follow. These principles are effectivisation, avoiding drastic lifestyle changes, utilizing ambient and physical interfaces, providing comparison mechanism, make systems socially aware and provide both immediate and overview feedback. In the following section we will motivate our choice of principles.

Work towards effectivisation, avoiding drastic lifestyle changes

As discussed in section 2.3, we do not wish to burden the consumer with further tasks or cognitive demands, nor to lower the standard of living in households. Rather then reducing quality of life we need to work towards more effective use of energy in general, and of electric energy in this specific case: effective in both the sense of higher efficiency but also correctly placed in time and space. In doing so we shall not enforce users to introduce drastic life style changes which would become another burden for them.

Use ambient interfaces and physical interfaces

As mentioned above an important aspect of designing artifacts for the home and the household is to avoid drastic life style changes, i.e. not disrupting the behavioural patterns of the inhabitants. Furthermore it is not desired to ruin the aesthetic qualities of the interior. Ambient interfaces [10] share the design principle to not disrupt behavioural patterns and thus the use of ambient interfaces suits very well to be employed. Furthermore utilization of physical interfaces allows us to design interaction that embed into the aesthetics of the artifact and that the actual interaction with the artifact is not drastically changed. [11]

---

[1] Close to ninety per cent of drink containers are returned by consumers in Sweden, with many European jurisdictions showing similar figures, according to the Swedish Waste Management trade association (www.avfallsverige.se). The presence of a deposit system seems to have a strong effect on the willingness to make the effort to return containers, even while the deposit usually is low and does not translate into adequate remuneration for the time and effort spent.

Use comparison mechanisms

To reduce energy usage we must slowly change the user's behaviours. As a basis for any system for behavioural modification, we must provide a benchmark or measure with which to compare individual (in our case, typically aggregated by household) performance. We will want to provide comparison mechanisms to give the consumer a sense of change, both as regards overall lifestyle and for individual actions which influence energy turnover in the home. In our current set of prototypes, we explore comparison over time ("Am I doing better than last week") and over peer groups (using a recommendation system framework [12,13]).

Build socially aware systems

Energy usage and related behaviour, as intimated above, is not only an individual question. Use aggregated over a large number of households will have greater impact both on general usage levels and on load balance than will individual usage profiles; in addition, the effects of social pressure, both to conform and to lead, can be harnessed to make a difference. Therefore it is important to create designs that operationalise some of the social context that normally is less visible in household situations. If this is done in a manner which avoids pitting users against each other in individual contests of appropriate behaviour, and which does not succumb to pointing fingers at those who do not conform to norms, such a social aware system will provide users to not only compare or learn from their own usage but others as well.

Immediate feedback and overview feedback

While the learning which is required by the consumer is on a fairly high cognitive level, and is intended to increase awareness and contribute to a sense of empowerment and control, it should be based on standard learning mechanisms. It is well established that in any learning situation, the timing of the stimulus must be in relevant juxtaposition with the contigent action — the highlevel behaviour or individual actions the consumer is engaged in. [14] The stimuli or signals to provide as feedback to the user must be *appropriate* and have *informational value* — in our case, they must not be overloaded with other information and not clash with other signals or messages the user may be interested in, and it must also have relevant cognitive content so as not to deteriorate into mere background noise. Specifically, we will keep separate the immediate feedback necessary to learn from actions from the overview sense necessary to be aware of lifestyle effects.

Plan and build for added benefits

The economical benefits provided through higher energy awareness is not enough to catch the eye of consumers, but must include other benefits and thus be designed as part of a larger platform. One possible added benefit is to provide added control and to empower the indvidual consumer and emergent groups of consumers. A possible integration of energy issues with e.g. safety monitors, time management services, and communication systems might be one solution.

5   Prototype Example – The Socially Aware Tea Kettle

Electric stand-alone tea kettles are the preferred device for heating water if the alternative is using a pot on an electric stove: tea kettles are much more energy efficient. To contribute to the overriding goal of reducing electricity usage it is thus useful to encourage households to move from stove-top saucepans to stand-alone kettles.

However, kettles occasion usage peaks: they use power up to 2 kW and are among those home appliances which require the highest current when switched on. This is observable by the consumer, in that switching on a tea kettle for many consumers causes a minor but visible brownout in the home, dimming lights momentarily. In addition, kettle usage patterns are cyclic. People have more or less the same diurnal behaviour patterns and tend to heat water for hot beverages and for other cooking purposes at about the same times.

The tea kettle is thus a useful and illustrative example of household electricity appliance: it has a high wattage and its usage is non-random with similar usage patterns across households.

As noted above, reducing peak loads by shifting usage is a desirable goal. To this end, to illustrate the possibilities inherent in aggregating households into cooperating pools of users for load-balancing purposes, we have built a socially aware tea kettle, suitable for pooling usage across several households. Pooling usage and reducing peak loads allows an aggregation of households the potential to negotiate lower average rates (possibly offset by higher peak load rates at the margin). In keeping with the guiding principles given above, we do not wish to burden the consumer with calculations or other cognitively demanding operations at the point in time where their focus is on producing hot beverages. Instead of providing numerical or graphical information, deflecting the attention of the consumer from tea to tariffs and time schedules, we provide an enhanced tool whose primary purpose remains heating water.

The underlying assumption is that if a number of households with suitably dissimilar habits are pooled, their power requirement can be balanced over a time period, if less time-critical energy requirements at times can be shifted from the immediate moment they are ordered to some less loaded point in time. A typical example of pooling currently under experimental deployment in households is delaying hot-water cisterns from immediately replacing the hot water used by morning baths and showers. If hot-water cisterns are programmed to heat the water when the overall network load is low, the aggregate cost of heating water can be kept down [7,8]. Figure 1 illustrates how the power usage of various electrical appliances (examplied by tea kettles) can be spread out in time to acheive a better load balance if the households are pooled.

Fig. 1. Pooling usage to achieve load balance

Similarly, we posit that heating water is only partially time-critical: when someone orders up a pot of boiling water, quite frequently the hot water can be delivered within a flexible time window rather than immediately. If a kettle can be made aware of the current and near-future load among a number of pooled households, it can inform its user of when within the next few minutes a kettleful water can be heated at lowest strain to the delivery grid (and, presumably, at lowest cost, if price is used as an additional signal to consumers).

Our example fully-functional kettle is mounted on a rotating base, and provides the user with rotational force feed-back to indicate booked capacity in a window over the next few minutes. A picture of the current prototype is shown in Figure 2. This allows the user to rotate the kettle to get a sense of when the load is light (less force is needed) or heavy (the kettle resists being rotated into that position). The user is always able to override the recommended time slot, and select immediate delivery or non-peak time.

5.1   Technology

The wooden cabinet on top of which the tea kettle is placed contains various hardware devices that are necessary for its operation. At the heart of this setup is a Java microcontroller responsible for Internet communication as well as measurement and control functionality. Using a clamp meter connected through i2c interface, the microcontroller monitors electric current used by the kettle. An i2c connected 240V power switch is incorporated to allow the processor to start and stop the tea kettle at designated times. The prototype also uses a modified force feedback steering wheel, modified from an off-the-shelf car game controller,

Fig. 2. Current kettle prototype. A production unit will have the same form factor as a standard off the shelf tea kettle.

connected to the base of the kettle to yield the friction-based rotation of the kettle. In addition, there is an array of LEDs positioned around the base of the kettle used to visually convey information such as active bookings.

The microcontroller runs a kettle software process which controls the hardware devices inside the wooden cabinet (forced feedback device, power switch, etc.), and also communicates with a household booking service over IP. The booking service is connected to all household electrical appliances, and monitors their current as well as predicted future power usage. The booking service is in turn connected and reports to a pooling service, which supervises the activity over many households. The pooling service is thus able to construct an aggregated "profile" of the predicted near-future power consumption of all connected households. This profile is then continuously communicated back to the booking services, and finally the individual appliances (e.g., the tea kettle prototype), where it is used to convey operating conditions that the user may wish to consider. The profile may be biased by any changes in the power suppliers future cost levels or by maximum power usage limits that may be stipulated in contracts between the power supplier and the pooled households. In the tea kettle prototype the profile is used to control the force feed-back mechanism acting on the kettle. The friction at a certain angle of rotation represents the predicted load a number of minutes into the future. The system architecture is illustrated in Figure 3. The kettle software control process, as well as the booking and pooling services, are developed using PART [15]. PART is a light-weight middleware,

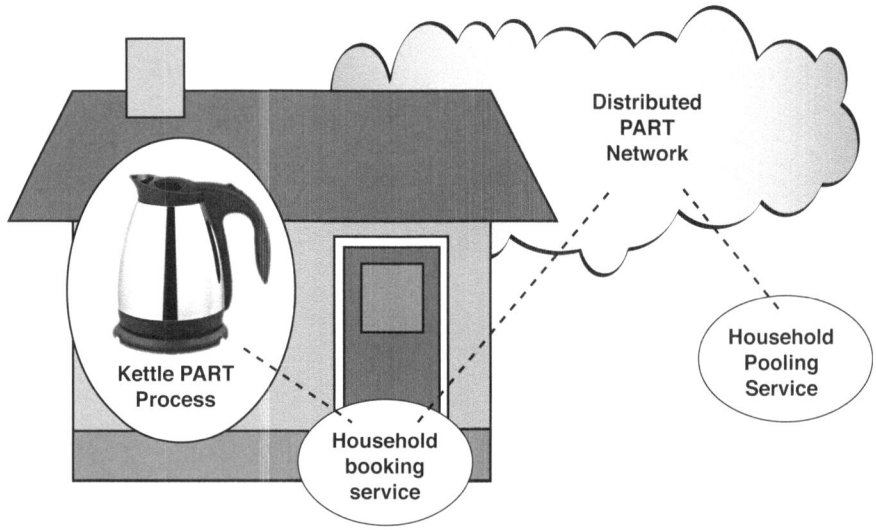

Fig. 3. System overview

written in Java, for developing pervasive applications that run on a range of different devices. PART is available as open source, under BSD license, and can be downloaded at http://part.sf.net.

The kettle can be rotated approximately 360 degrees, which currently is set to correspond to fifteen minutes of time. At one end of the interval (starting from degree 0), the friction feedback to the user indicates bookings and thus the predicted load in the immediate near future, while at the other end (around degree 360), the user can sense the predicted load fifteen minutes further along in time. When the user has found a suitable position, the kettle is activated by pressing the ordinary "on" button. This event is caught by the kettle controller hardware, which will temporarily switch off the kettle's power. The power will then be switched back on by the controller at the time corresponding to the kettle rotation, which will cause the water to start heating.

The reason why the rotation interval is fairly short (minutes rather than hours) is that we believe that roughly fifteen minutes is as long as tea and coffee drinkers can be expected to wait for their hot water. The assumption is that users want their hot water now, but that they might be prepared to wait a while in order to getter a better price, or to help spread the load among their peers. We have considered it unlikely that the user would want to schedule the heating of water for cooking or hot beverages hours into the future. One implication of this is that we also believe that once the user starts to turn the kettle, the chances are high that a booking will actually be made (the user will want the hot water now!), even if the load turns out to be high in the entire fifteen minute interval. This has allowed us to keep the interface fairly simple, based more or less entirely on the force feedback mechanism, and avoiding displays, buttons, etc, which might would have been required to support a more complex booking procedure (e.g., allowing the user to set constraints on price, time, load, etc.) All of these assumptions have to be confirmed via user studies, which are planned for the near future.

5.2   The Home Configuration Panel

Once the previous prototype is in place, the integration of various home services will hinge on getting an overview of the home configuration. How this can be achieved without overwhelming the user is a daunting design task, and this interface is only in planning stages as of yet – but it will involve a careful combination of well designed controls with defaults, examples, and preferential selections in face of difficult choice. [16,17]

We foresee in this case an integration of energy issues with e.g. safety monitors, time management services, and communication systems.

5.3   Community Tool

Similarly to the home configuration panel, an overview and control system is needed for the pooled community household power requirements. This could involve social awareness mechanisms as well as an opportunity to test different business models towards the provider. Current implementation consists of a graph visualizing web browser applet communicating to the rest of the system via a servlet.

6   Conclusions

This prototype is designed to illustrate three strands of thought. It works towards an increased *social awareness* in the household; it is based on an informed design of *immediate feedback* of aggregate information; it is an example of *unobtrusive design* to fit in the household without marring its home-like qualities, and as a platform and base for further design work, it combines our design principles in a fully functional prototype.

The kettle is an instantiation of our design principles, but by no means the only possible one, and is not intended to be deployed in the consumer market as is – it serves as an illustration of how social factors are or can be a parameter in our behaviour even within our homes. The underlying principle of an internet of household appliances can be instantiated in many ways, of which this is but one: energy usage is not a prototypically social activity, but in aggregating the behaviour of many peers into a pool of usage it becomes more social than before. The point of this prototype is that with a tangible yet unobtrusive interface we are able to provide users with a sense of social context and superimpose a behavioural overlay over individual action to aggregate usage patterns into a whole which is more functional from a societal perspective. We will be exploring this avenue of social technology in further prototypes, with other design realisations of the same principles.

Our tea kettle prototype illustrates the importance of load balancing and of user control. However, it also raises the requirement of providing users with immediate feedback directly connected to action – a well-established principle in the study of learning. As further developments of that principle we are field testing variants of electricity meters which provide both overview information of aggregated weekly usage and in other instances reacts to user action by giving auditory and visible feedback when an appliance is connected to the net.

Our tea kettle prototype illustrates the possibility of designing an interface which both provides immediate lean-forward information for the action the user is about to engage in, but without requiring cognitive effort – the information is encoded in rotation and visual feedback which needs little or no parsing. The attention of the user is not deflected from the primary goal at hand; the tool remains a tool designed for its primary function.

Acknowledgements

The research reported in this paper is part of and sponsored by the Swedish Energy Agency research program "IT, design and energy", project ERG, project number 30150-1. The ERG project adresses consumer awareness, understanding, and control of the energy usage of tomorrow's household.

References

1. Harper, R.: Inside the smart home: Ideas, possibilities and methods. In: Harper, R. (ed.) Inside the Smart Home: Interdisciplinary perspectives on the design and shaping of domestic computing, pp. 1–7. Springer, London (2003)
2. Cook, D.J., Das, S.K.: How smart are our environments? an updated look at the state of the art. Pervasive Mobile Computing 3, 53–73 (2007)
3. Energy Indicators. Annual Report from The Swedish Energy Agency (2005)
4. Bladh, M.: Snålare apparater men ökad elanvändning. In: Vardagens elvanor. pp. 18–20. (in Swedish)
5. Darby, S.: The Effectiveness of Feedback on Energy Consumption. University of Oxford, Oxford (April 2006)
6. Owen, G., Ward, J.: Smart meters: commercial, policy and regulatory drivers. Sustainability First, London (2006)
7. Abaravicius, J., Pyrko, J.: Load management from an environmental perspective. Energy & Environment 17 (2006)
8. Abaravicius, J., Sernhed, K., Pyrko, J.: More or less about data-analyzing load demand in residential houses. In: ACEEE 2006 Summer Study, Pacific Grove, California (2007)
9. Design Council Policy Paper: Designing for a changing climate. The UK Design Council (December 2007)
10. Wisneski, C., Ishii, H., Dahley, A., Gorbet, M., Brave, S., Ullmer, B., Yarin, P.: Ambient Displays: Turning Architectural Space into an Interface between People and Digital Information. In: Streitz, N.A., Konomi, S., Burkhardt, H.-J. (eds.) CoBuild 1998. LNCS, vol. 1370, Springer, Heidelberg (1998)

11. Ullmer, B., Ishii, H.: Emerging frameworks for tangible user interfaces. IBM Systems Journal, 915–931 (2000)
12. Karlgren, J.: Newsgroup clustering based on user behavior — a recommendation algebra. In: TR T94004, SICS, Stockholm, Sweden (February 1994)
13. Höök, K., Munro, A., Benyon, D.: Designing Information Spaces: The Social Navigation Approach. Springer, Heidelberg (2002)
14. Kirsch, I., Lynn, S.J., Vigorito, M., Miller, R.R.: The role of cognition in classical and operant conditioning. Journal of Clinical Psychology 60, 369–392 (2004)
15. PART Manual: PART-Pervasive Applications Runtime (November 2006) (retrieved September 19 (2007)
16. Humble, J., Crabtree, A., Hemmings, T., Åkesson, K.P., Koleva, B., Rodden, T., Hansson, P.: Playing with the bits - user-configuration of ubiquitous domestic environments. In: Dey, A.K., Schmidt, A., McCarthy, J.F. (eds.) UbiComp 2003. LNCS, vol. 2864, Springer, Heidelberg (2003)
17. Rodden, T., Crabtree, A., Hemmings, T., Koleva, B., Humble, J., Åkesson, K.P., Hansson, P.: Assembling connected cooperative residential domains. In: Streitz, N.A., Kameas, A.D., Mavrommati, I. (eds.) The Disappearing Computer. LNCS, vol. 4500, pp. 120–142. Springer, Heidelberg (2007)

Connect with Things through Instant Messaging

Jongmyung Choi[1] and Chae-Woo Yoo[2]

[1] Dept. of Computer Engineering, Mokpo National University,
61 Cheonggye, Muan-gun, Jeonnam 534-729, Korea
jmchoi@mokpo.ac.kr
[2] School of Computing, Soongsil University,
Sangdodong, Dongjakgu, Seoul 156-743, Korea
cwyoo@ssu.ac.kr

Abstract. As society and technology advance, we need more frequent communication not only with coworkers but also with "things." In this paper, we introduce a new communication paradigm and a prototype system that implements the paradigm. It is a communication façade that supports human-human, human-thing, and thing-thing communication using instant messaging. With this system, we can add any things worth talking with to our buddy list, and communicate with it through its instant messaging agent that encapsulates its services and status. We adopt a pipe & filter architecture pattern for the agent due to the requirements of extensibility. For efficient cooperation, presence information is also needed, and we suggest a statechart based presence model of things.

Keywords. Instant Messaging, Human-Thing Communication, Thing-Thing Communication.

1 Introduction

As society and technology become more complicated, we need more frequent human-human, human-thing, and thing-thing communication and cooperation. Human-human communication and cooperation have progressed using the Web and instant messaging. The Web allows people to share knowledge asynchronously and to build social networks and collective intelligence over the Internet, and instant messaging [1,2,4,7] makes it possible for people at a distance to communicate and cooperate in real time.

Compared to human-human cooperation, there are significant difficulties in human-thing and thing-thing communication and cooperation. First, there are no distinguishing technologies or paradigms for these kinds of communication and cooperation. Second, there are no standards for human-thing and thing-thing interaction. Third, there are no killer applications that support human-thing and thing-thing communication and cooperation.

Many researchers have been working to address these problems, and some have tried to extend instant messaging to connect people with devices and applications [3,10,11,12,14]. Because this approach is simple and has many advantages, we adopt it in this paper, and extend the concept of the entity in communication to include any entity worth being added, and the agent for the entity is also added. Therefore we can

C. Floerkemeier et al. (Eds.): IOT 2008, LNCS 4952, pp. 276–288, 2008.
© Springer-Verlag Berlin Heidelberg 2008

add any things to our buddy list of instant messaging, and communicate with them. For example, we can add a conference room to our buddy list, and we can talk to the room to reserve it for our project meeting. Furthermore, I can ask my printer to print my project report, and my printer can talk to me when it runs out of paper.

In this paper, we propose the concept of unified instant messaging as a communication façade, and introduce some scenarios for human-thing and thing-thing communication. For these communications, we introduce the architecture of smart agents for things. The agent consists of client module, interpreter module, and external interface module. We also suggest a presence model for things. It can be modeled with a statechart and tree.

In this paper, our contributions are three:

- Introduce a new paradigm for human-thing and thing-thing communication and cooperation
- Propose a pipe & filter type architecture for an instant messaging agent
- Propose a presence model for things

Our work will break down the barrier between human and thing in communication cooperation. It will also increase ease of work both in the workplace and at home with the support of devices, services, and things.

The remainder of the paper is organized as follows. In Section 2, we discuss other works that are closely related to our work. Then we propose our instant messaging system that supports human-thing and thing-thing cooperation and communication in Section 3. After that, in Section 4 we introduce our prototype system. Finally, we reveal the conclusions of our work in Section 5.

2   Related Works

Most of the work on instant messaging may be classified into two groups. Projects in the first group are focused on the influence of instant messaging on our daily life. Therefore, most of the research topics address how many users are using instant messaging [1], the main features in its use [4,5], why users are using it [4], how users use it in their work place [1,4,5], how teenagers use it for social networking [6], etc. These studies are very important for identifying co-working patterns with instant messaging, identifying new requirements for the workplace, etc.

Projects in the second group focus on extending instant messaging by adding new features [8-13]. Some of these projects are similar to our work. Adams [13] introduces a mail notification application based on instant messaging. It is a kind of human-application communication. Arjun [10] introduces a usage scenario in which instant messaging allows communication between networked appliances. In addition, Simon Aurell [11] shows how instant messaging can be used for monitoring and controlling remote devices in home automation. It is a kind of human-device communication. In this work, the intelligent gateway is the main part of the system, and it communicates with devices using device drivers and some form of translators. Marcela Rodríguez [14,15,16] also introduces location-aware instant messaging that supports communication with devices. Cian Foley [12] extends the concept of communication further, and introduces instant messaging that provides person-to-person, person-to-device, and person-to-service communication. Instant messaging allows the addition of services and devices to the buddy lists, and to access them in a uniform way.

Our work is differentiated from other works in three ways. First, we extend the communication entity to anything that is worthwhile talking with. Therefore, we use instant messaging as a uniform platform for communication and cooperation with people, devices, services, virtual objects, and things. Second, we introduce an extensible architecture for communication and cooperation with things. It has filter type architecture, and developers can add more features without rebuilding the systems. Third, we propose a presence model for things based on a statechart.

3 Communication Façade

3.1 Requirements for Communication Façade

In the near future, communication and cooperation between people and things or machines will be more prevalent, and its importance will increase. In the digital appliance field, there is already an identified need for human-machine and machine-machine communication in building smart homes. For example, a smart fridge [17] connects with the Internet, and it has the capacity to check its contents and send a shopping list to the owner's cell phone through short message services (SMS) or directly to the supermarket. It means that the fridge tries to communicate with humans and with things. However, we still have problems in the interaction between human and machine or between machine and machine because we do not have a unified communication method and protocol. As a result, we have difficulties due to the complex communication paths between humans and things as illustrated in Fig. 1. Furthermore, a path may be implemented with different vendor-specific protocols.

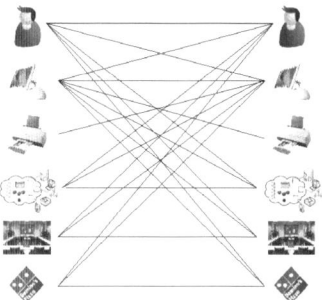

Fig. 1. Complex Communication Paths

To boost and support human-thing and thing-thing communication, we need a unified and standardized communication method. If application developers and appliance manufacturers use their own proprietary communication methods, users will experience confusion and difficulty in using the systems. Therefore, a standard way of communication is needed. For this, we have four requirements to meet. First, it must simultaneously be both synchronous and asynchronous. Because the style of the communication is real-time exchange of information or requests for services, it must be synchronous. And, sometimes it is asynchronous because it supports asynchronous communication for unavailable things.

Second, the communication method must be simple enough for a machine to understand and easy enough for users to use in comfort. A great deal of work has been done in developing convenient communication methods between humans and machines, and these methods include voice, motion, embedded menu, text, etc. Voice and motion recognition methods are very convenient for users, but they require extensive resources and there are difficulties in implementing this kind of system. On the other hand, embedded menu is physically limited, and it cannot be accessed at a distance.

Third, the communication must be extended over the network or the Internet. Then we can communicate with remote things, and things can cooperate with each other over the network. For thing-thing cooperation, the network connection is inevitable. Some intelligent agents can be in between the things and the network, and they translate the communication messages for both side of communication.

Fourth, the communication method must be easily extensible for two reasons. First, it must be able to adopt a new kind of thing or machine with the least efforts. Second, it must be able to adopt more intelligence without rebuilding the system. To support the extensibility, the communication method must be structured and encapsulated.

For a unified and standardized communication method, instant messaging is a promising technology. Fig. 2 shows that instant messaging plays the role of the communication façade for human-human, human-thing, and thing-thing communication. By using instant messaging as a gateway, we can reduce the complexity of multiple communication paths and vendor-specific protocols.

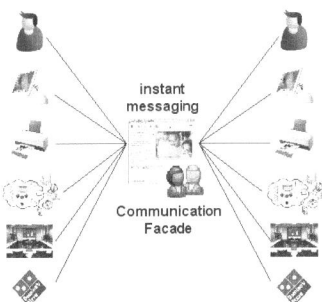

Fig. 2. Communication Facade

Instant messaging meets the four requirements mentioned above. First, it supports synchronous and asynchronous communication, and it is well suited to real time cooperation. Second, it is based on text, so it is rather simple to implement compared to voice or motion. Furthermore it is well-known to be easy to use as shown by its popularity. Third, it is connected to the Internet, and any user at a distance can communicate and cooperate with things. Fourth, instant messaging can be extended rather easily.

In instant messaging, the communication facade is possible because of virtual counterparts, called instant messaging agents of physical things. Fig. 3 illustrates the relationship of physical things with their virtual counterparts, or agents. An agent encapsulates the state and behavior of a physical thing. Fig. 3 also shows some examples of agents. With the concept of agent, we can communicate with computers, devices, services, things, or even stores in the same way as we do with our co-workers.

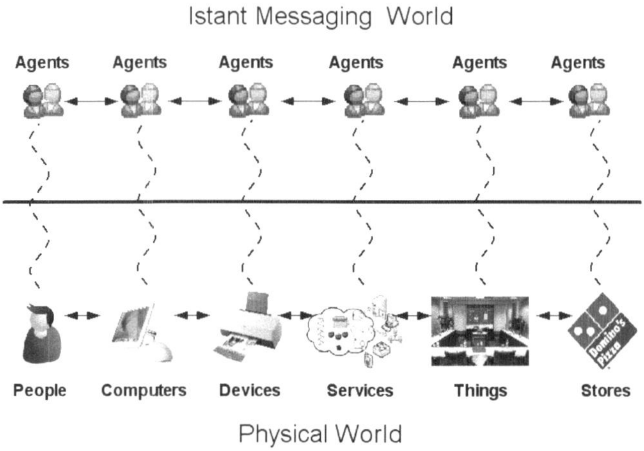

Istant Messaging World

Physical World

Fig. 3. Connection Physical World and Instant Messaging World

Besides the meeting the four requirements, instant messaging has one more significant advantage. By using instant messaging, we can utilize the existing instant messaging systems' facilities for communication and cooperation. For example, users can connect with things from mobile phone or PDA through CDMA or GSM networks without extra effort because these facilities are already established. Furthermore, users can also enjoy the voice interface supporting instant messaging [2].

3.2   Communication Scenario

Instant messaging removes the barrier between humans and things, and it allows people to communicate with computers, devices, services, things, etc by providing a simple and unified method. If we can communicate with devices, we can work more efficiently and easily. Here is a scenario that shows how to communicate by better utilizing our daily devices.

Scenario 1
A project manager, Choi, is far away from his office, and he notices that he needs to print his presentation handouts before the project meeting. He pulls out his mobile phone, and he selects "desktop" from the buddy lists of instant messaging on his phone. After the selection, a communication window opens to talk with the "desktop," and he then sends a message to print his presentation data before he comes to the office. A few seconds later, he gets a message from his "printer" through instant messaging, and the message is that it has run out of paper. He searches available coworkers in his office by checking the presence and availability marks in his instant messaging, and he asks one of them to fill his printer with paper. About five minutes later, his printer sends a message that it has finished the job.

Scenario 1 shows communication between human, computer, and printer. The real communication follows the sequence diagram illustrated in Fig. 4. In human-thing

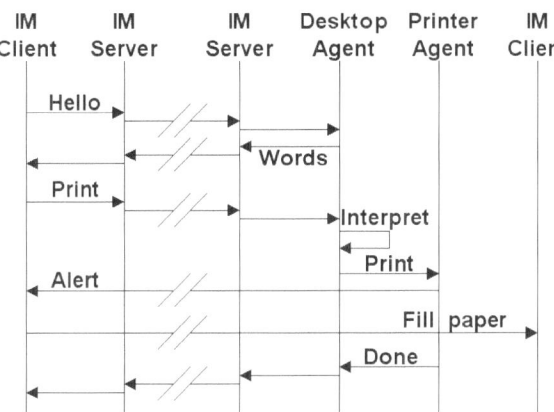

Fig. 4. Sequence Diagram for Scenario 1

communication, it starts with the "Hello" message. On receiving the "Hello" message, the thing agent sends its vocabularies and syntax for further communication.

Humans can communicate with things with instant messaging. For example, we can communicate with a conference room to reserve the room for the project meeting. Here is a scenario of communication with the room.

Scenario 2
Choi needs to reserve a conference room for the project meeting. He asks a confer-ence room called "Ocean," whether it is available at 3:00 p.m., and it replies that it is available until 5:00 p.m. He responds to it and reserves the room from 3:00 to 5:00. The project team members see that "Ocean" is marked in their instant messag-ing clients with a message about the project meeting.

Fig. 5 shows the sequence diagram for scenario 2. Room scheduler is an existing ap-plication for managing and scheduling rooms, and the room agent contacts it for schedule information.

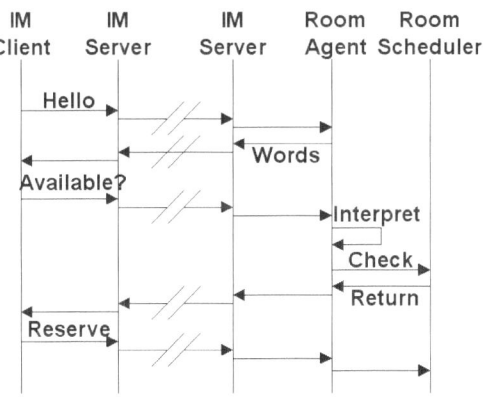

Fig. 5. Sequence Diagram for Scenario 2

Things can communicate and cooperate with other things using instant messaging. However, at least one of the communication and cooperation agents must have computing power. Here is a scenario for cooperation.

Scenario 3

After the reservation, Choi asks "Ocean" to prepare for the meeting. The room checks the availability of projector and computer, and talks to the computer to get the presentation file from Choi. Then the computer talks to him to get the file, and he sends the file to the computer. When everything for the meeting is prepared, the room talks him that the meeting room is ready.

Fig. 6 shows the sequence diagram for the scenario 3. The room agent must have the ability to check for the presence of a computer and projector in the room and the service availability of them.

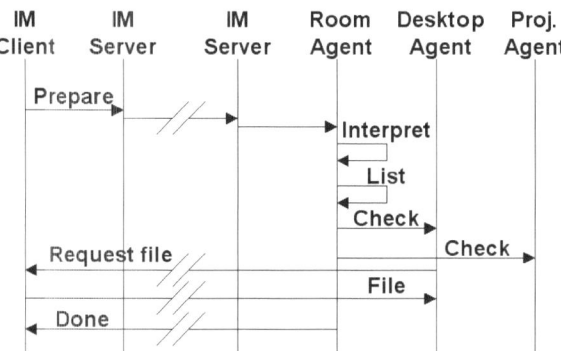

Fig. 6. Sequence Diagram for Scenario 3

3.3 System Architecture

An instant messaging system consists of servers and clients, and it follows peer-to-peer system architecture [19]. The main roles of the server are transferring messages and storing messages temporarily for clients. The message format follows Extensible Messaging and Presence Protocol (XMPP) [21]. An instant messaging client has direct connection with people or things, and we call it an instant messaging agent. Fig. 7 shows the architecture of instant messaging systems. In LAN, people use an instant messaging agent that runs on desktop computers, and they can connect with devices, services, and things through the instant messaging server. When people are outside, they can use a mobile instant messaging client that runs on mobile phone or PDA.

The instant messaging agent plays the central role in the communication façade because it enables people to connect with things. It consists of client module, interpreter module, and an external interface module. Fig. 8 shows an agent's architecture for a printer device.

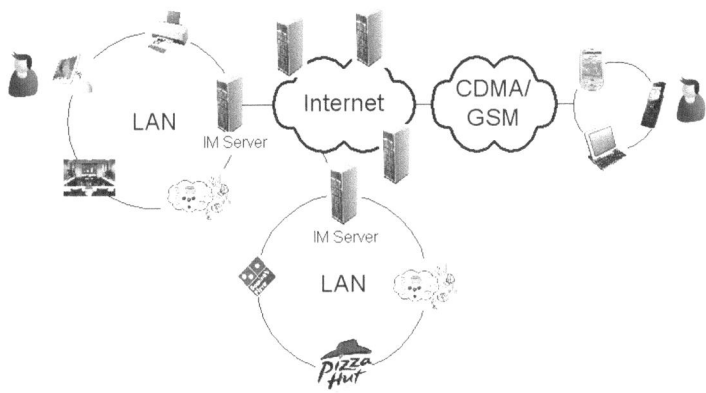

Fig. 7. Architecture of Instant Messaging System

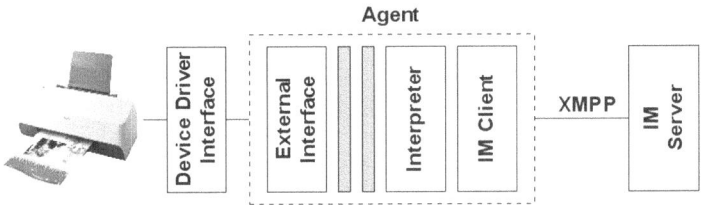

Fig. 8. Instant Messaging Agent's Architecture

In Figure 8, the client module is the most basic part for communication. It takes charge of the communication with an instant messaging server over the TCP/IP sockets, and it also manages the communication sessions. Another role is to parse the message in XML format (XMPP), extract the core message - we call it thing-specific message - from the whole message, and deliver it to the interpreter module.

Interpreter module is the brain of the instant messaging agent. It receives a thing-specific message from the client module. The message is the core of the data, and its format may be in application-specific XML format. For example, the message to a printer device is regarding sending requests to a printer or about getting information about the printer. Therefore, the interpreter module parses the thing-specific message transferred from the client module, understands the meaning of it, and translates it to the devices' commands or requests to devices. Most intelligent services are processed within this module.

External interface module is for connection with external things. The roles of this module are classified into two parts. First, it sends requests to things or devices, and the requests must be understood by the things. Second, it receives responses of requests and the presence information from the things. The response and status messages must be transformed into general terms in this module for corresponding parties.

The agent follows the pipe & filter pattern [20]. We can add more modules between the interpreter and the external interface module without recompiling the source. The filter module may support the interpreter module for more intelligent services.

3.4 Presence Model

Presence [18] is a very important feature in instant messaging, because it provides the presence information of other users. In communication façade, an instant messaging agent may have a variety of status information, because the agent may represent various things such as people, printer, room, etc.

In communication façade, the type of presence is classified into two: physical and logical. Physical presence means that a thing locates in a specific range so that it is available physically. Logical presence means the status of service availability. In communication façade, the importance of physical presence will increase. For example, in scenario 3, the room agent has to check the physical presence information of projector and computer. The physical presence information server can be implemented with location systems such as GPS or RFID [23]. Physical presence has one of two values: exist or non-exist.

The physical presence information of things is monitored and is provided by the presence monitor. Fig. 9 illustrates the architecture of presence server. The presence monitor checks the presence of things and their location information. If a thing is out of its location, the presence monitor reports that its physical presence is "non-exist."

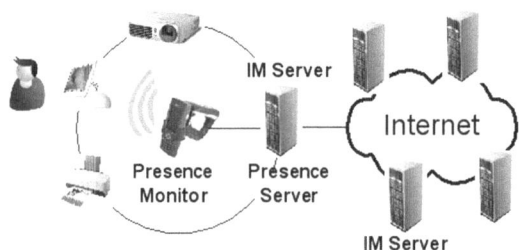

Fig. 9. Physical Presence Server Architecture

In contrast to physical presence, logical presence may have one of a variety of values. With regard to value, logical presence information has two features. First, some of the status values of presence are exclusive to each other. In this case, we say that the status values are in equal level. Second, the other status values are contained in another value. For example, an agent for a printer device can have off-line, on-line, printing, out-of-paper, and paper-jam. Then off-line and on-line are exclusive. However, printing, out-of-paper, and paper-jam are contained in on-line, and each of them is exclusive.

Logical presence is naturally modeled with UML statechart [22], because the status change is well modeled with statechart. All states in equal level have a transition of value among them, and the contained states are represented with internal states. Fig. 10 shows an example of a statechart for printer agent.

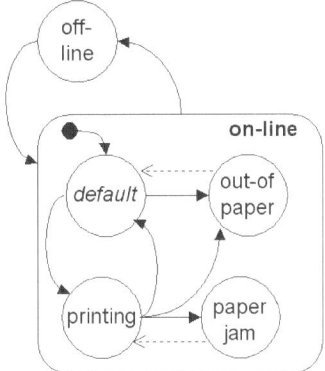

Fig. 10. Presence State Diagram

The statechart model can be transformed into a tree model. In the tree model, the root node is the name of a thing, and its child nodes are non-contained exclusive values. In addition, a node and its child nodes are determined by the containment of status values. The container is parent node and the containees are child nodes. Fig. 11 shows the tree model for a printer model.

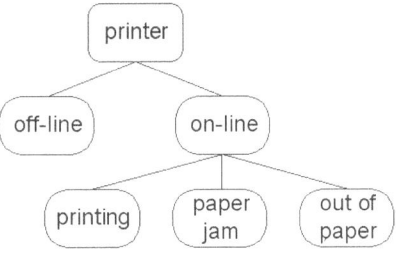

Fig. 11. Presence State Tree

4   Prototype System

We built first version of prototype system for communication façade paradigm. The main goal of this prototype is to implement scenario 1 and 2. We utilized Jabber [24] and XMPP protocol for implementing the system. Jabber is an open instant messaging and presence awareness system. This server is used to notify the state of people and agents, and to handle the interaction between people, agents and devices through XML messages. We used the existing instant messaging server for communication server, and we developed instant messaging agents with the Java language, especially with JDK 1.6 on Win 32 platform. Development of our prototype system was fairly straightforward because it utilized existing open source systems such as Jabber, and java shell [25].

The architecture of our prototype system is illustrated in Fig. 12. Our prototype system consists of five agents: Printer agent, Desktop agent, Room agent, Store agent, and BBS agent. The room agent uses database modules to check for room reservation information and to reserve a room by querying the database. Store agent uses order module to order goods, but we deployed it in another computer in our lab instead of real store for testing. Order System is so simple that it prints out order information to console, because it is developed just for testing. Desktop agent provides the ability to talk with a desktop computer, and it is based on an open source java shell [25]. It is able to execute commands by creating new processes. Printer agent has ability to print documents and to notify its status to users. To connect with a device driver, it uses JNI module.

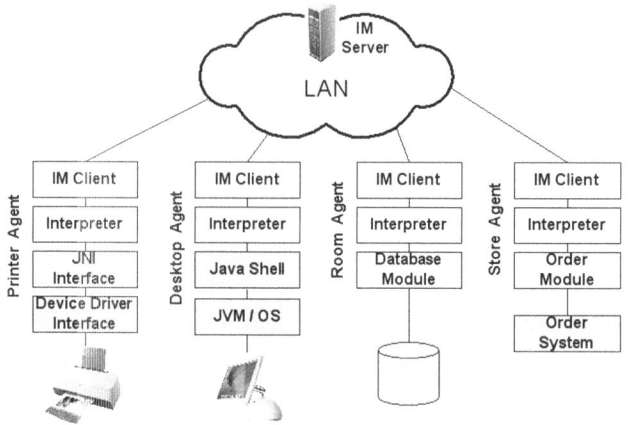

Fig. 12. Architecture of Prototype System

Out prototype system supports communication with co-workers, devices, applications, and things. Fig. 13 shows a snapshot of communication with a printer. When the printer runs out of paper, it notifies users about its status.

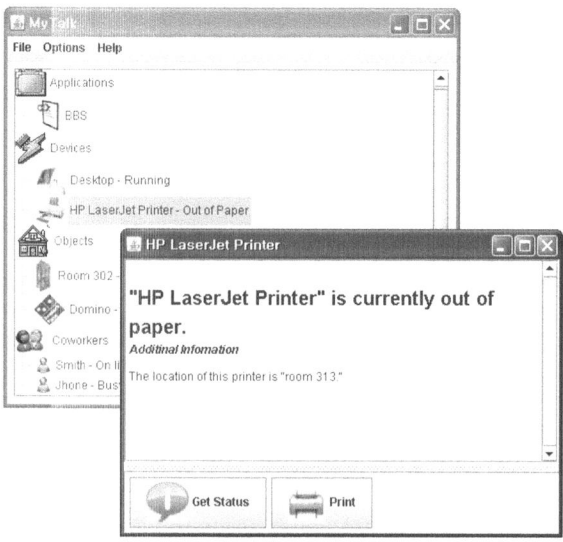

Fig. 13. Snapshot of Communication

Currently, we are developing the second version by extending the first prototype system to support scenario 3 and fix some problems. We re-implement Desktop agent and Printer agent with Microsoft Foundation Classes (MFC) library for more convenient control of the desktop computer and the printer. For supporting scenario 3, we build a presence server. For physical presence service, we use a 900M Hz RFID reader. The reader is able to check the physical presence of a beam projector and a notebook computer within the conference room by reading tag identifiers attached to them.

5 Conclusions

As technologies advance, we need more convenient and efficient methods for communication and cooperation with other people and even with things. There has been a great deal of work done for convenient human-thing and thing-thing communications, but no great success until now.

In this paper, we suggested a new paradigm for human-thing and thing-thing communication and cooperation by extending existing instant messaging. Instant messaging supports text-based real-time communication and cooperation. In addition, many additional features such as running on mobile phone or supporting voce communication are developed. The most important feature of instant messaging is its extensibility. We can add new features for human-thing and thing-thing communication façade with little efforts by adding messaging agents to the existing instant messaging systems.

We suggest instant messaging agent that enables things to talk and cooperate with people. An agent represents a thing or a device, and it keeps the information and status of the thing. Sometimes, it interprets the requests from users and executes the device on behalf of the users. Its architecture follows the pipe & filter pattern, and developers can add more features without rebuilding the system. It provide the presence model information for things, and for this, we suggested a presence model based on the statechart. We also introduced our prototype system that implements five types of agents.

In the near future, instant messaging will be used as the communication façade for everything, such as coworkers, devices, things, services, stores, and so on.

References

1. Eulynn Shiu and Amanda Lenhart: How Americans use instant messaging, Pew Internet & American Life Project (September 2004), http://www.pewinternet.org/
2. Wikipedia, http://en.wikipedia.org/wiki/Instant_messaging
3. Isaacs, E., Walendowski, A., Ranganathan, D.: Mobile instant messaging through Hubbub. CACM 45(9), 68–72 (2002)
4. Isaacs, E., et al.: The character, functions, and styles of instant messaging in the workplace. In: ACM Conf. on Computer Supported Cooperative Work, pp. 11–20 (2002)
5. Nardi, B.A., Whittaker, S., Bradner, E.: Interaction and outeraction: instant messaging in action. In: ACM Conf. on Computer Supported Cooperative Work, pp. 79–88 (2000)

6. Rebecca, E.: Grinter and Leysia Palen: Instant messaging in teen life. In: ACM Conf. on Computer Supported Cooperative Work, pp. 21–30 (2002)
7. Huang, E.M., Russell, D.M., Sue, A.E.: IM here: Public instant messaging on large, shared displays for workgroup interactions. In: Proc. of the SIGCHI Conf. on Human Factors in Computing Systems, pp. 279–286 (2004)
8. Peddemors, A.J.H., Lankhorst, M.M., de Heer, J.: Presence, Location, and Instant Messaging in a Context-Aware Application Framework. In: Chen, M.-S., Chrysanthis, P.K., Sloman, M., Zaslavsky, A. (eds.) MDM 2003. LNCS, vol. 2574, pp. 325–330. Springer, Heidelberg (2003)
9. Rovers, A.F., van Essen, H.A.: HIM: a framework for haptic instant messaging. In: CHI 2004 Extended Abstracts on Human Factors in Computing Systems, pp. 1313–1316 (2004)
10. Roychowdhury, A., Moyer, S.: Instant Messaging and Presence for SIP Enabled Networked Appliances. In: Proc. of Internet Telephony Workshop, pp. 73–79 (2001)
11. Aurell, S.: Remote Controlling Devices Using Instant Messaging. In: Proc. of the 2005 ACM SIGPLAN workshop on Erlang, pp. 46–51 (2005)
12. Foley, C., de Leaster, E., van der Meer, S., Downes, B.: Instant Messaging as a Platform for the Realisation of a true Ubiquitous Computing Environment. In: Innovation and the Knowledge Economy Issues, Applications, Case Studies, Part 2, pp. 1051–1060. IOS Press (2005)
13. Adams: You Have Mail! (2001), http://www.openp2p.com/
14. Rodríguez, M., Favela, J.: Autonomous Agents to Support Interoperability and Physical Integration in Pervasive Environments. In: Menasalvas, E., Segovia, J., Szczepaniak, P.S. (eds.) AWIC 2003. LNCS (LNAI), vol. 2663, pp. 278–287. Springer, Heidelberg (2003)
15. Rodríguez, M., Favela, J.: A Framework for Supporting Autonomous Agents in Ubiquitous Computing Environments. In: System Support for Ubiquitous Computing Workshop at UbiComp 2003 (2003)
16. Rodríguez, M., Favela, J., Gonzalez, V.M., Muñoz, M.A.: Agent Based Mobile Collaboration and Information Access in a Healthcare Environment. e-Health: Application of Computing Science in Medicine and Health Care 5, 133–148 (2003)
17. Murph, D.: Samsung prepping RFID-enabled refrigerator (2007), http://www.engadget.com/2007/01/04/samsung-prepping-rfid-enabled-refrigerator/
18. Day, M., Aggarwal, S., Mohr, G., Vincent, J.: Instant Messaging / Presence Protocol Requirements, RFC 2779 (2000), http://www.ietf.org/rfc/rfc2779.txt
19. Milojicic, D.S., et al.: Peer-to-Peer Computing, HP Laboratories, Palo Alto, Technical Report HPL-2002-57 (2002)
20. Buschmann, F., et al.: Pattern-Oriented Software Architecture, vol. 1. John Wiley & Sons, Chichester (1996)
21. Extensible Messaging and Presence Protocol (XMPP) (2004), http://www xmpp. org/
22. Booch, G., Rumbaugh, J., Jacobson, I.: The Unified Modeling Language User Guide. Addison-Wesley, Reading (1998)
23. Hazas, M., Scott, J., Krumm, J.: Location-Aware Computing Comes of Age. Computer 37(2), 95–97 (2004)
24. Jabber, http://www.jabber.org/
25. Jsh: OpenSource Java Shell, http://gerard.collin3.free.fr/

Developing a Wearable Assistant for Hospital Ward Rounds: An Experience Report

Kurt Adamer[4], David Bannach[1], Tobias Klug[2], Paul Lukowicz[1],
Marco Luca Sbodio[3], Mimi Tresman[5], Andreas Zinnen[2], and Thomas Ziegert[2]

[1] University of Passau, Germany
[2] SAP Research, Germany
[3] Hewlett Packard, Italy Innovation Center
[4] Steyr hospital, Austria
[5] Edna Pasher PhD. and Associates, Israel

Abstract. We describe the results of a three year effort to develop, deploy, and evaluate a wearable staff support system for hospital ward rounds. We begin by describing elaborate workplace studies and staff interviews and the resulting requirements. We then present a wearable system developed on the basis of those requirements. It consists of a belt worn PC (QBIC) for the doctor, wrist worn accelerometer for gesture recognition, a wrist worn RFID reader, a bedside display, and a PDA for the nurse. Results of evaluation of the system, including simulated (with dummy patient) ward rounds with 9 different doctors and accompanying nurses are given. The results of the evaluation have lead to a new system version aimed at deployment in real life 'production environment' (doctors and nurses performing ward rounds with real patients). The paper concludes by describing this next generation system and initial experiences from a first two week test deployment in a real life hospital setting.

1 Introduction

Hospitals are complex, yet highly structured environment where electronic information collection, processing, and delivery plays a central role in virtually all activities. While the introduction of electronic records has improved quality and enabled more efficient information exchange between institutions, it has in many ways actually complicated the work of the hospital personnel. The core of the problem lies in the fact that complex interaction with a computer system is not possible during most medical procedures. Clearly it is hard to imagine a doctor dealing with a notebook or a PDA during surgery. However even during simpler procedures access to computers is problematic. This is due to several factors. First of all, examinations are often performed on thigh schedules with doctors having just a couple of minutes per patient. Quoting one of the interviewed doctors they want: 'time for the patient, not time for the notebook'. Second, the patients expect doctors to devote their full attention to their problems. Retaining the 'human touch' and showing concern were major issues voiced by doctors

C. Floerkemeier et al. (Eds.): IOT 2008, LNCS 4952, pp. 289–307, 2008.

interviewed in our work. Thus doctors are reluctant to interrupt the examination and focus on a computer instead of the patient, even if is only for a very short time. Finally examinations require hands to be sterile. This means that hand based I/O (mouse, keyboard, touchscreen etc.) is not feasible in between examination steps. As a consequence of the above issues electronic information access is severely restricted during dealing with patients. Patient data is often simply printed out before a procedure and taken along in paper form. Information is also rarely entered directly into the system. Instead it is entered into dictating machines, written on paper by accompanying nurses, or even stored as 'mental notes'.

This paper describes the results of a three year long effort to develop, deploy and evaluate wearable information access system for hospital staff. The work has concentrated on so called *ward rounds*, which are crucial to all hospital activity, and impose particularly stringent requirements on mobility, unobtrusiveness and cognitive load issues. The work described in this paper has been developed within the EU funded wearIT@work project [1].

1.1   Related Work

The only wearable solution for a ward round scenario has been investigated by the E-Nightingale project. It has looked at wearable monitoring of nursing activities to prevent accidents [2]. A number of usability studies were performed with respect to the use of different mobile devices. A PDA vs. notebook for accessing patient's records study was undertaken by [3]. It shows that physicians are generally more satisfied using the notebook computer although pointing-and-clicking applications are faster on PDAs. In [4], PDAs were compared to notebooks for nursing documentation application. The results also indicate that with the exception of writing notes, the overall user satisfaction is very similar for the both systems. Another evaluation of hand-helds for nursing was done by [5]. Many mobile and tablet based solutions were proposed and implemented. Examples include a prototype mobile clinical information system [6] and the WARD-IN-HAND project [7]. With respect to novel interfaces a glove-based gesture interface is evaluated in [8]. Finally a commercial product in form of a tablet PC optimized for hospital environments has been introduced by Intel (Intel mobile clinical assistant [9]).

1.2   Paper Contributions and Organization

The focus of the paper is not as much on the technological solutions, but on the experience made with design and evaluation of pervasive computing system for and in a complex real world environment. Section 2 describes an elaborate (over 1 year long) workplace study that included observation of ward rounds and interviews with all involved persons. The derived system requirements exemplify the benefits that hospital staff expects from pervasive technology and the boundary conditions for its deployment. We then described a system that we have implemented on the basis of this requirements. Detailed evaluation of the system with doctors and nurses is then given. We conclude with a description of a next generation system based on this evaluation. We sketch initial experience with the deployment of this system in real life 'production' environment (ward rounds with real patients during normal hospital workday).

2   Application Design

The design and development of the wearable system for ward rounds followed an ISO 13407 compliant UCD (User Centered Design) process model. The different methodologies discussed in this section covered the first iteration of the process.

2.1   Field Observations

Early in the project a series of observations were conducted, where four doctors from two different departments were shadowed by two researchers during their daily ward round, two from the surgery department and two from the internal medicine department. The observations aimed at identifying typical activities during a ward round, and how these were distributed between doctor and nurse. During the observations, a number of challenges were encountered. First of all, the environment was completely unknown to the observer. When observing office users, the basic setting is already known in advance and the observer has first hand experiences with this kind of settings. As a consequence, it was difficult to assess the importance of events and environmental factors as well as note the right things. The speed at which unknown activities take place is another challenge the observer has to cope with. The ward round is a highly optimized and parallelized process. For example, a doctor might read the patient's file, order an examination, talk to the patient and document his visit, all in a matter of seconds. In a typical approach, one would interrupt the doctor and ask for clarification when things went too fast and were not understood. However this approach is prohibited by the environment that is studied. The doctor is not performing the ward round in his office, but in the hospital based on a strict schedule. In our scenario the hospital administration did not approve any technique that would have interrupted the ward round or would have made the patient aware of the fact that a study was in progress. For this reason and also for privacy reasons, audio and video recordings were not possible. Therefore the only option was to conduct retrospective interviews after the observations if the doctor's did not have other appointments which is typically the case after the ward round.

Jotting field notes on paper worked for the initial observations when trying to uncover the breadth of activities and environmental influences. However, during later observations when the goal was to capture the sequence and typical duration of events, manual note taking was not appropriate any more. During these observations, a Tablet PC running the custom TaskObserver software [10] was used to assist taking these kinds of notes. This solution allowed capturing of good data even though video recording was impossible.

In the following step the findings were used in order to develop first design ideas. These Mock-ups and the respective results of their evaluation are described in the following section.

2.2 Mock-Up Design and Evaluation

Based on observations and interviews, first design ideas were generated. Those ideas were evaluated using a mock-up system as described in [11]. The system allowed the doctor to view patient records on a Tablet PC and gave the nurse the opportunity to complete an x-ray request on a PDA. The studies, which were done using the mock-up system, pursued three goals:

1. Evaluate the feasibility of using the selected devices during the ward round.
2. Analyze whether the distribution of tasks between doctor and nurse was practical.
3. Compare speech and pen input as possible input modalities for the Tablet PC.

The evaluation was focused on following four ward round activities: (1) accessing patient data, (2) accessing individual findings, (3) organizing a laboratory analysis, and (4) organizing an examination.

The above mentioned activities form a representative subset of all the different tasks which have to be done during the ward round, assuming that the input and output modalities and the workflow of the other activities would be similar to those that have been tested.

The mock-ups were constructed using Macromedia Flash. The ward round activities obviously require collaborative work between doctors and nurses. In order to support collaboration, two mock ups were created. These were installed on a tablet PC and on a PDA for the doctor and nurse respectively. The tablet PC was remotely controlled (VNC) to simulate missing system functions such as: identification of the patient, automatic display of patient information as a response to identification and communication between the tablet PC and the PDA.

Medical data and findings of a fictitious patient were entered to assure a realistic feeling while interacting with the system. The realistic medical content of the system was created with the support of the medical staff of the Steyr hospital. The participants in the study were five doctors: two surgeons, one orthopedist and two internists. Two nurses were assisting, one for the surgeons and the orthopedist, and one for the doctors from the internal medicine department. The tests were performed in a patient room at the hospital, but without any patient being present. The participants were introduced to the use of the tablet PC and the PDA. Thereafter, the evaluation was carried out. Therefore, the doctor/nurse pairs were asked to role-play a typical ward round using the mock-ups available. Comments, opinions and suggestions from the participants were encouraged. Although discussions during the role-playing interrupted the workflow to a certain extent, it was important to allow immediate feedback to avoid incomplete response after the tests.

The mock-ups were complete enough to give a basis for discussion and answers to certain questions, but incomplete enough to allow room for comments and new ideas. The setup also included enough context to bring up relevant feedback about practical aspects of the workflow and devices. The Tablet-PC was regarded too heavy by most participants. Additionally, the device interfered with the physical patient-doctor interaction because there was no place to put it down in a patient room where it would not have been contaminated. These aspects would practically not have been discovered without doing tests within the considered context.

2.3   Functional Requirements

The field observations and the evaluation of the mock-ups resulted in a set of 57 requirements. In the course of the project, those requirements were iteratively adapted and specified in a more detailed way. As not all requirements could have been implemented within the scope of the wearIT@work project, the most important and manageable from an end user perspective ones were selected. These requirements are described in detail in the following:

Input Interface. Users must be able to switch to hands-free use for short periods of time. All time hands-free use is not necessary, but changing the interaction type has to be easy and fast. Doctors frequently perform examinations on patients which require them to use their hands. Once they have touched the patient, the hands are contaminated and have to be disinfected before using any device again. Finally, the interaction with the system must allow actions to be performed in a minimal amount of steps.

Content display. System must be able to display different types of content such as text (examination and lab reports), audio (recordings of examination questions and voice mails) or pictures (photos of wounds, ultrasonic examinations) - x-ray pictures (these require special treatment because of special requirements to resolution and size) - charts (vital data over time). It would also be desirable for the system to automatically adapt the information display to what the doctor is currently doing.

Information sharing. Doctor must be able to share and discuss information with colleagues and patients. The doctor often wants to show the patient for example x-ray pictures in order to help her understand her own condition better. The doctor often involves the junior doctors in diagnosing patients, and the information (the same display) should thus be visible to them all as a basis for discussion/ teaching.

Collaborative workflows. Users must be able to perform workflows cooperatively. Example: Several pieces of information are necessary to order an examination. The system should enable the doctor to start an order and fill in the desired information. The incomplete order is then automatically passed to an accompanying nurse to complete the request. Both doctor and nurse should be able to collaboratively edit and complete the order.

Authorization by proximity. The system needs to allow the following scenario: Person A and B are standing close together. A is authorized to do a specific task, but B is not. The proximity of B to A should authorize B to do the task as well, because it is assumed that A issued a vocal command to B thus indirectly authorizing B. In the healthcare scenario this maps to the relationship between doctor and nurse who can jointly fill in for example examination orders.

Patient identification. The system must be able to automatically identify patients that are close (1-3m) to the wearable user. Patient identification is needed to avoid mix-ups. Doctors cannot rely on the patient being able to state his name. A beacon on the patient should take the form of a wristband. The patient should not have to remove the beacon at any time (e.g. during showers).

Personal patient contact. The system should not hinder personal contact with the patient. The input/output modalities of the system should enable the doctor to give most of his attention to the patient, and not to the system. Therefore, voice commands to a computer system are inappropriate while a patient is present.

Sound recordings. The system should be able to record a dictation of free-text messages and enable the recipient to play them. Voice messages should also be available for listening in mobile situations. Shorter messages can be recorded without form restrictions.

Ability to make photos. Some device is necessary that can take photos of a reasonable quality. The system needs to attach context information to these photos and store them automatically. Example: When a doctor examines a wound she often wants to document the current state. A photo taken at this moment should automatically be associated with the patient and date and attached to the patient file.

Unobtrusiveness. It should be possible to wear the devices so that contamination is avoided. The system should enable interaction during a physical examination, i.e. when hands are contaminated, without risk of spreading contamination to other parts of the doctors body.

The physical properties of the system components should make use and transport easy and comfortable. System components should be light (lighter than a tablet PC), easy to store during movement (e.g. in a lab coat pocket), not become warm during use (when handheld), not contain too many small loose parts (these can get lost).

3   First Demonstrator System

This section will briefly describe the first demonstrator system which was built on the basis of the above requirements.

3.1   Scenario Description

Fig. 1 shows an overview of the first prototype. The doctor and nurse teams move around without paper, laptop or filing card. The doctor identifies the patient by means of his RFID reader on his wrist. The relevant patient files appear on the screen, attached to the bed, which at other times could be used as a television or an internet terminal. The doctor immediately gets an overview of all the important data about the patient, most recent results, current medication, x-rays or operation reports. With the inertial sensor attached to his wrist, he can navigate through the application without a mouse or keyboard. This is keeping his hands sterile and free to examine the patient at any time. Via his headset, the doctor can record new findings about x-ray requests related to his patients. The nurse, next to him, receives the information on her PDA, where she can view the diary of the x-ray department and directly arrange an appointment for the patient.

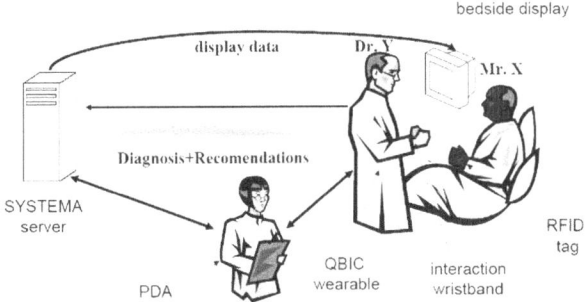

Fig. 1. The scenario behind our system

3.2   System Architecture

The first prototype consisted of three main subsystems; the doctor system, the bedside system and the nurse system (see Fig. 2).

Doctor Subsystem. The doctor subsystem consists of an unobtrusive wearable computer, an interaction wristband, and a headset for audio in- and output.

1. Wearable Computer. The wearable system is the doctors personal information gateway. It controls the I/O devices, all the sensors, it performs most context recognition tasks, and it provides network connectivity (with the nurse, the bedside system and all hospital databases). We chose the QBIC wearable computer, which features a 400 MHz Xscale processor, Bluetooth and USB interfaces, and a Debian Linux OS, all integrated in a wearable belt. A USB WiFi adapter is used for communication with the other subsystems.
2. Interaction Wristband. The interaction wristband is used for patient identification using RFID reader, and it provides a simple gesture interface as described in [12]. We used a small RFID-reader module attached to a USB-powered 'tmote sky' sensor node. For gesture recognition we used the MT9 motion tracker sensor from Xsens. It features accelerometers, gyroscopes, and magnetic field sensors, all on three axis and sampled with 100 Hz. The sensor was attached to a Bluetooth module including a battery pack.

3. Audio headset. For speech recording a wireless headset was used. Where a headset is not possible (because of the need for a stethoscope) a chest worn microphone will be used.

The doctor subsystem acts as the main gateway between sensors and the other subsystems. It gathers and processes the information from sensors, and sends suitable commands to the nurse- and bedside subsystem.

The gesture recognition and sensor processing software on the doctor subsystem was implemented using Context Recognition Network (CRN) Toolbox [13]. The CRN Toolbox basically consists of a set of parameterizable algorithms that can be applied on data streams. An application is defined by selecting the algorithms to be applied, defining the parameters of each algorithm, and connecting the input- and output streams to form the path of the data streams.

Nurse Subsystem. The nurse subsystem consists of an application server and a PDA as a thin-client frontend. It allows the nurse to enter and retrieve parts of the information relevant for her tasks, and to use those feature of the system, which are too attention-consuming and require too much manual interaction for the doctor to perform.

The nurse subsystem runs an application server which receives messages from the doctor subsystem and sends according forms to the PDA of the nurse which, in turn, can review the form, enter additional information and send it back to the application server. Also, if a dictation of the doctor is recorded, the application server can forward the audio file to the nurse's PDA. The nurse subsystem is based on MS Windows.

The Bedside Display. The bedside subsystem runs a Java application which is simulating the document browser of the hospital information system. The browser is able to display different types of content. Examples of the manageable content so far are text documents like examination and laboratory reports, audio recordings of examination questions on the PDA, pictures of wounds or x-ray pictures. An icon on the top right corner displays the current activation state of the gesture recognition. Our document browser can be controlled with four basic commands (scroll up, scroll down, open document, close document). Further commands set the enabled/disabled state of gesture control and the display of the browser itself.

Gesture Recognition. For being able to control the document browser by gestures our system on the doctor's wearable needs to recognize at least four different gestures. The gestures should be simple to perform on one side, and on the other side, they must be unique enough to be distinguishable from all other human movements. For easing the second requirement we introduced two special gestures for activating and deactivating the recognition of the other, simpler to conduct gestures.

Meeting these requirements we defined the following six gestures that can be conducted by swiveling the forearm up and down, left and right, or by rolling it left and right respectively:

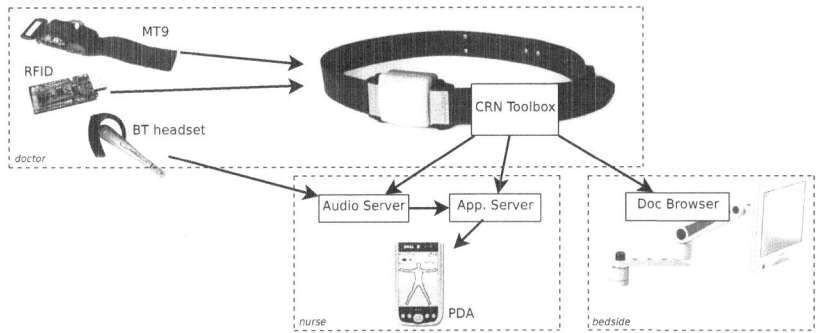

Fig. 2. System architecture of the first demonstrator

Forearm Gesture	Command
swivel up, then down	UP
swivel down, then up	DOWN
swivel left, then right	OPEN
swivel right, then left	CLOSE
roll right, then left, twice	ACTIVATE
roll left, then right, twice	DEACTIVATE

We use the three-axis gyroscope of the MT9 motion sensor to detect the swiveling and rolling of the forearm. The x axis of the gyroscope is aligned parallel to the main axis of the forearm, and the y axis is parallel to the plane of the hand. Rolling the forearm leads to either a positive or negative deviation of the angular velocity on the x axis, depending on the roll direction. Similarly, swivel left/right is measured on the y axis and swivel up/down on the z axis. The events of turning in either direction are extracted from the raw data stream by a threshold filter. The sequence of these events is analyzed and compared to pre-recorded sequences of valid gestures. Detected gestures with overlapping sequences are discarded.

4   Evaluation of the First Demonstrator System

The following section summarizes the results of a first system evaluation. The first subsection describes initial experiments with students. Afterwards, the system evaluation in a hospital in Steyr is described including the evaluation environment, training and test setup. Applied evaluation methodologies are introduced. The section ends with the main findings during the evaluation.

4.1   Student Experiments

Doctors are usually rarely available for in-depth user studies. From our experience this user group becomes very soon frustrated by a new system which is not functioning as expected. Therefore, we decided to first conduct user studies with students. The goal of these tests was to identify as many potential usability problems as possible, so that we were able to refine the prototype for the usability studies with our real end users.

Our test setup was designed to resemble the setting of the clinical ward round as closely as possible with respect to the interaction between user and system. The main activity that requires user interaction with the system during the ward round is a doctor browsing the patient health record. We have mapped the interaction necessary to navigate the patient health record into a domain where students that do not have a degree in medicine feel more comfortable with. Our test application lists headlines of news stories that can be opened to read the full text including related pictures. The interaction remains the same, but the news documents are meaningful to the test users, whereas medical records could have confused them. The computer display was positioned on a high desk, requiring the user to stand in front of the system.

The study required the user to switch frequently between system interaction and a real world task. The evaluation of the studies contained two parts. One was an analysis of video recordings towards obvious usability issues. The second part was an analysis of interviews we conducted with each person after performing the test. People were asked to explain what they did and did not like about the system.

In total 23 students from the local technical university took part in our tests. The analysis of these studies revealed a number of minor and some major usability issues. One of our findings was that the test participants felt that the *up* and *down* commands were easier to perform and more intuitive than the other gestures. This may relate to the fact that the up and down gestures are more naturally mapped to the visual feedback that the application gives (i.e. moving the hand upwards or downwards results in an upward or downward movement of the highlighting of the respective document on the screen). The gestures for opening and closing a document (i.e. moving the hand to the left and to the right), however, as well as activation and deactivation (i.e. turning the wrist twice to the right or twice to the left), are not naturally but rather semantically mapped to the application. In other words: these gestures were mapped to a specific function in the application and not to the same movement. These gestures were mainly chosen because of their simplicity, but partly also with the metaphor open and close book for open and close, and turn a simple electrical device on and off (using a knob) for activate and deactivate. These metaphors were not clear to all participants; therefore most participants found these gestures less intuitive than the more naturally mapped open and close. Another metaphor for open and close mentioned by one participant was the use of a door.

One of the most important insights from the results of the experiments was that it was very difficult for the test participants to understand exactly how they were to perform the gestures in order for the system to understand these correctly. One problem was that the time available for learning in the experiment was too short for the participant to have a chance to learn the gestures well enough to be able to perform the given tasks reasonably efficiently. The other problem was that our team, acting as test coordinators, had trouble communicating our knowledge on how to perform gestures to the test participants, even if we knew how to perform them effectively. The reason for this is that a large part of this knowledge is implicit, i.e. cannot really be described verbally since it is more or less tacit knowledge. The most effective way to gain tacit knowledge is to practice. For the following evaluation with doctors, we increased the training time. Furthermore, we applied the gained experiences in explaining and giving them metaphors to remember the gestures.

4.2   Setup of Experiments with Doctors

For evaluation, a patient room was prepared to test the prototype. The fixed installation included the bedside display attached to the patient's bed, and a patient doll from the medical training department, that was equipped with a RFID wristband for identification. The doll was used as a substitute for a real patient to simulate physical interaction with the patient. The technical infrastructure necessary to keep everything running was positioned on a table in the corner of the room. A video camera constantly captured the doctor during his interactions with the bedside display and the patient.

End users tested the system in doctor-nurse pairs. A total of 9 doctors from several departments took part in the study. The doctor was wearing the interaction wristband and a Bluetooth headset for audio in- and output. The nurse was given a PDA. Each evaluation session started with an introduction to the project. After a short explanation of the prototype, the doctors and nurses trained to operate with the system. The training session was split into two halves. During the first part, the doctor and nurse were introduced into the systems usage and asked to familiarize themselves with the interaction methods. After a certain degree of proficiency with the system was reached, a sequence of activities similar to that of the actual tests was explained and performed. During the training, instructions were given to the participants. Subsequently, an evaluation sequence was performed three times. After the training session, the doctors and nurses entered the room through the door. Next to the patient's bed, the doctor had to use the interaction wristband to scan the patient's identification wristband. The system brought up the patient's files on the bedside display. The doctors had to open a document and scroll to its end containing instructions about the next steps, e.g. opening another document. The last document contained instructions to perform an examination of a certain patient's body part. In a last step, the doctor issued an examination request that was completed by the nurse. Finally, after the practical part of the experiment, the participants filled out a questionnaire concerning their experience using the system and an interview was conducted.

4.3   Evaluation Method

The purpose of the test was to evaluate how well the nine participating doctors and nurses could interact with the system in the context of a typical ward round work flow.

In order to analyze the performance of the participants, identify problems, and capture spontaneous comments, each test sequence was entirely filmed. At a later point in time, practical interaction problems, conflicts between system interaction and work flow and comments were analyzed and documented.

Additionally, the participants filled out a questionnaire concerning their experience using the system. The questionnaires were constructed using Likert scales meaning that the test participants were shown a list of statements like *I found the gestures easy to perform* or *using this system could make my work more efficient*. The participants had to categorize on a scale between 1 and 5 to what extent this was true. The participants were also interviewed in order to find out more details about their opinions and ideas concerning the system and its use.

4.4 Results

In the following, the main results of the evaluation are summarized. In a first paragraph, overall system results are presented from doctors' and nurses' point of view. A second paragraph steps into the findings using gestures as an interaction method.

Overall System Results. When asked if they would use a system such as represented by the prototype if it was available, the doctors were hesitant, and gave on average neutral answers (3+/5 on Likert scale). The doctors were not fully convinced that such a system would make the ward round easier and more efficient. Most likely, the skepticism can be explained by the novelty of the gesture interaction, which is a central part of the prototype. Several doctors were reluctant to the idea of performing gestures in front of the patient, concerned that the patients would consider them crazy or strange, waving around with their hands in the air. According to the doctors, attention for interacting with the proposed system would decrease the patient contact. The doctors were only mildly thrilled (3+/5 on Likert scale) to perform extensive system training. However, several doctors could imagine using gesture interaction if the gestures were smaller and/or more elegant. Furthermore, some doctors commented that hands-free interaction was good because of hygiene reasons, and that the negative implications of the gesture interaction would most probably decrease with more practice.

The possibility to view the latest patient documents electronically, and to be able to enter fully authorized examination request immediately into the system, was seen as very positive and helpful for the efficiency of work, because it would reduce unnecessary paper work (post ward round) and duplication of information. The doctors on average (4/5 on Likert scale) found the audio feedback (start/end/error dictate audio recording) very helpful. Almost all doctors agreed that speech input was a suitable method to perform the examination request (in average 4+/5 on Likert scale). This probably has its cause in the fact that the doctors currently dictate the examination request to the nurses who write down the information. Basically, there is no great principal difference dictating to the system instead of the nurse.

Generally, the nurses were very positive. It should be said, however, that the nurses had quite small role in the experiment and their tasks were not very extensive, although still representative. Further, the time pressure that is often present during a real ward round was not part of the experiment setting. When asked if they would use the system if it was available, the nurses answered with a clear yes (4/5 on Likert scale). They felt that their ward round related work would go faster and be easier with the use of such system (4+/5 on Likert scale) connected to the hospital's backend. Today, in fact, the nurses process most of the information collected during the ward round only after it is finished; such situation could improve with the introduction of a fully functional version of the tested system. The nurses felt to have the system under control (4 /5 on Likert scale). The comments revealed that they found the interface easy to understand, and the touch screen interaction simple to perform. According to the nurses, they would have no problem controlling the interface perfectly (meaning a 5/5 answer to the control question) with a little more training.

The contact with the patient is to all nurses extremely important. Thus, several of them are worried that the new technology would take position in between themselves and the patient, and that they would be too busy with the PDA, especially at the beginning before the interaction becomes more automatic, but to a certain extent also after the initial stage. Especially the nurses from the pediatric department thought that the system would be a change to the worse and even unacceptable. Today, the only responsibility of the pediatric nurses is to talk to the children and comfort them during the ward round. Any further tasks that would take attention away from the children are not welcome. On the other hand, some of the nurses (non-pediatrics) felt that there was no greater difference using the PDA compared to taking notes like today, and that both tasks would require about as much attention. In average, the nurses answered slightly negatively to the question if the new system could help improve the patient contact (0-/5 on Likert scale).

Gesture Interaction Results. The gesture interaction was by far the most controversial aspect of the prototype, which is not surprising considering the novelty of the concept, and that none of the test participants had had experience with anything similar before. Below, the most frequently occurring issues and their potential causes are discussed. Most of the problems will most probably become less frequent when the test participants were given the possibility to practice the gestures for a longer period of time. Due to the short training-period included in the experiments, most of the participants did not reach a level of proficiency where they could work comfortably. However, a certain learning effect can already be observed from the tests results.

Shape: 6/9 doctors had problems performing a gesture without striking out in the other direction first in order to get power for the movement. This often led to a situation where the opposite gesture was recognized, i.e. the cursor moved up instead of down or a document that was meant to be opened remained closed (since the command close will have no effect on an already closed document). This issue was mainly caused by the fact, that doctors had no direct feedback, i.e. the effect of striking out is not perceived while doing it, only the end-result is seen. 4/9 doctors sometimes forgot to perform the second part of the gestures, i.e. the movement back to the original position. One reason for this mistake is that doctors found the back movement less intuitive. This is because the movement goes in the direction opposite of the desired effect. This is the case at least for the up-down gestures, i.e. in order to move the cursor up, one must move the hand up but then also back down.

Timing: 6/9 of the doctors had problems performing the gestures in the right tempo. Most frequently, they carried them out too slow, i.e. under the threshold speed defined in the gesture recognizer. However, a clear learning effect was observed from the beginning of the experiment to the end. In the initial practice phase, 5 doctors performed the gestures too slow, while during the third and last experiment task, only one doctor made this mistake. Some doctors also performed the gestures too fast to be recognized (3/9).

Conceptual: Most of the doctors at some point confused the gestures. 2/9 doctors at some point forgot completely how to perform a gesture, or at least had to think for a long time before remembering it. 7/9 confused the paired gestures with one another (for example open with close or activate with deactivate). After a bit of practice, however, the confusion decreased, and in the last phase only 1 doctor confused the gestures with each other. The up and down gestures were rarely confused at all.

A very frequent problem was that doctors completely forgot the gestures while activating/deactivating the system (i.e. doctors tried to interact with the system without activating it, or started with non-system-related activities, like examining the patient, without deactivating). A possible reason for this is the fact that doctors did not really see the benefit of being able to move their hands freely without their movements being randomly recognized. Probably, this was because the experiment context was not realistic enough in terms of activities involving the hands. For example the test did not contain a full patient examination (most doctors only symbolically touched the fictitious patient when asked to do so) or lively communication with either nurses, or colleagues, or the patient. Consequently, doctors did not really see a need for the de-/activation action and thus forgot about it. All 9 doctors at some point in time had this issue.

In summary, the use of arm gestures was the most controversial part of the tested system. As doctors had never used such an interaction technology before many beginners mistakes occurred, like performing gestures to fast or to slow, as well as confusing gestures or just forgetting how to perform a particular gesture.

Three issues regarding the gestures were found to be important for the further development of the prototype:

- Gestures have to be socially acceptable: There were several comments that doctors do not want to perform gestures in front of patients because of the fact that these might be distracting or doctors just felt stupid performing such gestures.
- Gestures should be easy and simple to perform in a way that those gestures require not to much of the doctors attention. This issue became apparent

when doctors were asked whether gesture interaction was comfortable and making their job easier.

- There should be some sort of feedback, whether a gesture was performed right or not: As there was no direct visual feedback showing what the systems interpretation of the performed gesture was (i.e. when performing a gesture in the wrong way) doctors felt not to be in control of the system.

5   Next Generation Prototype

Based on the above finding a next generation prototype has been designed and implemented. A major aim of the implementation has been to facilitate deployment in real life 'production' ward rounds with real patients, as opposed to simulated ward rounds in the evaluation described in the previous section. Such deployment is currently under way and an initial two week test has just been completed. This section outlines the main features of the new prototype and discusses the lessons learned from the initial two week test.

5.1   System Overview

Compared to the first generation prototype the system has been modified in the following way:

1. It has been integrated with the existing hospital information system and its back-end technology. This has involved a major software restructuring to make it compatible with the requirements (see software description below). It has also lead to a modified network structure and application distribution.
2. The hardware has been modified for increased robustness (see hardware description below)
3. The gesture recognition has been improved according to the findings of the previous prototype evaluation. The improvements include better senitivity to allow tiny inconspicuous gestures, improved acceptance of variations, and optimized recognition speed for more immediate feedback.
4. A proximity detection system has been integrated to provide the contextual information about which persons (doctors, nurses) are taking part at each ward round.
5. Pictures taken by the nurse (e.g. of wounds) with her PDA are automatically attached to the patient's files in the hospital information system (just like the sound notes in the old system).

Hardware Components. The main hardware modification can be summarized as follows.

- Redesign and re-implementation of the wrist worn sensor modules. This includes (1) a new RFID reader with improved range and higher reliability, which ensure an easier and more accurate retrieval of patients identifiers, (2) a re-design and re-implementation of the Bluetooth based communication module to include an improved battery, charging circuits, battery and function indicator and an external switch, and (3) re-packaging of the whole wrist band module in a more compact, easier to wear, and more nicely-looking component.

- The main processing unit worn by the doctor has changed from the QBIC system to a NOKIA 770, which, for our application, is only marginally more intrusive while having the advantage of being a 'mainstream', commercial device.
- An RF-based proximity detection system has been implemented and integrated, based on TecO Particles [14] which were attached to the badges of the hospital personnel.

Software Architecture. The overall software architecture of the prototype is shown in Figure 3. We can identify two main end-user applications the *MPA*

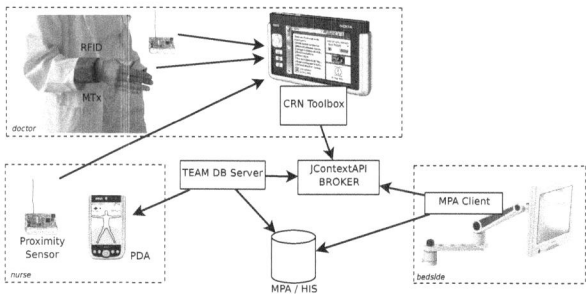

Fig. 3. The healthcare prototype system architecture

client application, which is used by the doctor to access the *Hospital Information System (HIS)* and the custom *PDA application*, which is used by the nurse to support her work during the ward round.

The MPA application (which is part of the hospital information system in Steyr) is provided by Systema (a partner of the wearIT@work consortium). It is a client-server application, which does not only allow the doctor to access all patients data, but also to order examinations, and to manage a number of other tasks related to daily activities. In our scenario, the front-end client of the MPA application has also been extended with two features: an interface to the gesture recognition system and an an embedded *Multimedia Server* (not shown in Fig. 3), which allows recording of the audio messages spoken by the doctor, and of the photos taken by the nurse.

To facilitate the integration with the existing HIS and MPA systems the distribution of context information is handled by *JContextAPI*, which is the Java implementation of the *Open Wearable Computing Framework (OWCF)* developed within the wearIT@work project. The core component of the *JContextAPI* acts as a broker to distribute context information to every registered consumer. In our prototype the JContextAPI broker sends proximity events, and patient identification events coming from the CRN Toolbox to both the MPA client and the PDA application. It also notifies the MPA client of gesture events, so that MPA can react to doctor's gestures in a context-sensitive way. Furthermore, it propagates events produced by the Multimedia Server to notify when and where a new audio recording is available.

Compared with the previous generation system, the PDA application used by the nurse has been modified to handle photos and to interact with the OWCF.

5.2   Test Deployment

The real life deployment experiments are currently in progress. An initial two week test was performed in August 2007. The aims of the test were to identify any remaining technical and/or organizational issue specific to the real world environment. The deployment has taken place in a two-bed room (see Fig. 4) that over the time period was occupied by a total of 4 patients. During this time the round has been performed 20 times by two different doctors and 4 different nurses.

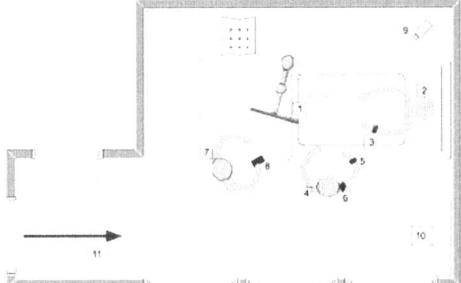 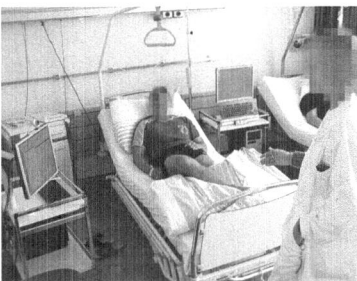

Fig. 4. Left: the evaluation setup for the first prototype. Right: The second generation prototype system in use in a two-bed patient room.

Deployment Issues. A number of issues emerged during planning for the initial deployment in the hospital. While they may seem trivial we choose to discuss them here as they are typical for moving from lab tests to real life deployment.

- Patient Considerations. To motivate the patients to participate in the experiments it has been suggested to allow internet access on the bedside systems. In addition our lab PCs had to be replaced by silent PC not to disturb the patients.
- Staff consideration. The staff had to be provided training on the use of the system including initialization, overall workflow and the use of gestures.
- Maintainability Considerations. During the deployment the system maintained by the hospital IT department. This meant that all system initialization procedures had to be properly automated and simplified. Protocols had to be established for daily maintenance including battery re-charging, rebooting systems and replacing components making problems. As will be described below the complexity and fragility of system setup components has been identified as a major source of problems.

Technical Lessons from Deployment Test. Despite intensive lab validation before the deployment, technical problems have invariably emerged during the test. Detailed analysis is currently underway. Key preliminary findings are:

1. The person independence of the gesture recognition was not satisfactory. While the gesture recognition worked fine for one of the doctors (as well as for test persons in our lab), the other had major problems handling it.
2. The procedures needed for system initialization and user registration were not robust enough. Obviously in many cases switching devices on/off, registering users, and coming into/out of range of Bluetooth nodes has been performed in sequence and at speeds which we did not consider in our lab trials. This often caused components to freeze.
3. Similar problems were caused by issues with wireless links. In particular when there were many people in the room (to view a demonstration of the system) Bluetooth links or the proximity detection signals would be blocked. In addition the way doctors were handling the wrist worn RFID reader combined with the new mounting often resulted in occlusions of the Bluetooth communication.

Initial User Feedback. Despite the focus on technical issues initial user evaluation was also undertaken. During this visit observations were conducted as well as interviews with two doctors, one of whom participated in the tests, and one who did not and spoke generally about working conditions and potential changes due to wearable computers. Additional interviews were conducted with two patients, who used the system during their stay, and with the hospitals technical specialist, who worked on servicing the system before and during the tests. The main results can be summarized as follows:

1. The staff reactions were strongly overshadowed by the technical issues described above. Clearly the problems with gesture training were, like in the first prototype, a key issue, and people suggested that speech control might be better. The fact that the system allows hands to remain sterile was nonetheless seen as positive.
2. The patients were positive about the system. They stated that seeing the results on the screen and having them explained in real time by the doctor increases patient's trust in staff. None of the patients interviewed felt such a system would have an effect on their relationship with staff, as the system will not prevent them from talking to the doctors and nurses. The patients interviewed stated that they would be happy to see such a system working in the hospitals in the future.

Acknowledgments

The authors would like to thank all those who have been involved in the work described in this paper. This includes in particular the wearIt@work partner companies TEAM and Systema, Steyr Hospital Team (Doctors, Nurses and IT department) and Peter Wanke from HealthInformatics R&D.

References

1. wearIT@work Project: http://www.wearitatwork.com/
2. Noma, H., Ohmura, A., Kuwahara, N., Kogure, K.: Wearable sensors for auto-event-recording on medical nursing - user study of ergonomic design. In: McIlraith, S.A., Plexousakis, D., van Harmelen, F. (eds.) ISWC 2004. LNCS, vol. 3298, pp. 8–15. Springer, Heidelberg (2004)
3. Rodriguez, N.J., Borges, J.A., Soler, Y., Murillo, V., Sands, D.Z.: A usability study of physicians' interaction with PDA and laptop applications to access an electronic patient record system. In: Computer-Based Medical Systems, 2004. CBMS 2004. Proceedings. 17th IEEE Symposium, pp. 153–160 (June 2004)
4. Rodriguez, N.J., Borges, J.A., Soler, Y., Murillo, V., Colon-Rivera, C.R., Sands, D.Z., Bourie, T.: PDA vs. laptop: A comparison of two versions of a nursing documentation application. In: Computer-Based Medical Systems, 2003 Proceedings. 16th IEEE Symposium, pp. 201–206 (June 2003)
5. Young, P., Leung, R., Ho, L., McGhee, S.: An evaluation of the use of hand-held computers for bedside nursing care. International Journal of Medical Informatics 62(2-3), 189–193 (2001)
6. Choi, J., Yoo, S., Park, H., Chun, J.: Mobilemed: A PDA-based mobile clinical information system. IEEE Transactions on Information Technology in Biomedicine 10(3), 627–635 (2006)
7. Ancona, M., Dodero, G., Gianuzzi, V., Minuto, F., Guida, M.: Mobile computing in a hospital: the WARD-IN-HAND project. In: ACM Symposium on Applied Computing (2000)
8. Tani, B.S., Maia, R.S., Wangenheim, A.v.: A gesture interface for radiological workstations. In: Computer-Based Medical Systems, 2007. CBMS 2007. Twentieth IEEE International Symposium, pp. 27–32 (June 2007)
9. Mobile Clinical Assistant, http://www.intel.com/healthcare/ps/mca/index.htm
10. Klug, T., Mühlhäuser, M.: Taskobserver: A tool for computer aided observations in complex mobile situations. In: Proceedings of the International Conference on Mobile Technology, Applications and Systems (Mobility), September 2007 (to appear)
11. Carlsson, V., Klug, T., Ziegert, T., Zinnen, A.: Wearable computers in clinical ward rounds. In: Proceedings of the third International Forum on Applied Wearable Computing, IFAWC 2006, pp. 45–53 (March 2006)
12. Carlsson, V., Klug, T., Zinnen, A., Ziegert, T., Levin-Sagi, M., Pasher, E.: A Comprehensive Human Factors Analysis of Wearable Computers Supporting a Hospital Ward Round. In: Proceedings of 4th International Forum on Applied Wearable Computing 2007, VDE (March 2007)
13. Bannach, D., Kunze, K., Lukowicz, P.: Distributed modular toolbox for multi-modal context recognition. In: Grass, W., Sick, B., Waldschmidt, K. (eds.) ARCS 2006. LNCS, vol. 3894, pp. 99–113. Springer, Heidelberg (2006)
14. TecO Particles, http://particle.teco.edu/

Social Devices: Autonomous Artifacts That Communicate on the Internet

Juan Ignacio Vazquez and Diego Lopez-de-Ipina

Faculty of Engineering
University of Deusto
Avda. de las Universidades, 24. 48007 Bilbao Spain
{ivazquez,dipina}@eside.deusto.es

Abstract. The Internet has boosted people collaboration, enabling new forms of exchanging knowledge and engaging in social activities. The Web 2.0 paradigm (also called the Social Web) has greatly contributed to this achievement. We believe that the next wave of smart devices and digital objects will leverage the pervasiveness of Internet connectivity in order to form ecosystems and societies of artifacts that implement Internet-based social behaviors and interact with existing Internet-based social networks. In this paper, we introduce the concept of *social device* and describe our experiences creating different kinds of augmented objects that use the Internet in order to promote socialization, look smarter and better serve users. We finally identify several challenges in the research of devices that behave in social ways.

1 Introduction

People are not isolated at their workplace or in public spaces: they meet, talk, exchange ideas and collaborate. In our modern world where no single person is likely to know everything about a particular domain, sharing information and cooperating is of foremost importance for the advance of the society.

In fact, complex tasks often require the combined contribution of different kinds of specialized knowledge and are also often more easily achieved as a collaborative activity. Nowadays, the concept of collaboration is itself undergoing a transformation driven by the new networked technologies. For example:

- Open Source communities have developed some of the most successful computer software applications in the last decade including GNU/Linux, Firefox, OpenOffice, Apache, Eclipse, and, more recently, the Java programming language that has also joined this trend.
- Collaborative content sites, created by the contributions of thousands of users, have been able to become the most important repositories of information in a surprisingly short time. The most popular examples are Wikipedia and YouTube.
- In P2P file exchange networks file sharing is promoted following the scheme "the more you share, the further ahead you are placed in download queues".

C. Floerkemeier et al. (Eds.): IOT 2008, LNCS 4952, pp. 308–324, 2008.

The Web has transformed from a place with many readers and few publishers, into a space where everyone can have a voice via tools such as blogs (weblogs), or a community of users can collaboratively create a body of knowledge about some concrete topic via wikis.

This evolution of the Web has been dubbed Web 2.0 or Social Web, and it promises to boost human collaboration capabilities on a worldwide scale, enabling individuals and organizations to face challenges that could not have been faced in isolation. In the very same way, there is clear evidence that digital and electronic devices should not be isolated in the places where they are deployed or spontaneously meet.

We are witnessing an unprecedented increasing number of digital objects that are populating and almost invading more and more environments in an imperceptible way. Most of times, they carry out simple but efficient tasks that help people in their everyday activities, simplifying their lives. However, when it comes to complex tasks involving a higher-level of intelligence and cooperation in unexpected situations or scenarios, they invariably fail, because *they have not been designed for collaboration.*

Just as Web 2.0 technologies have boosted human social capabilities on the Web, devices can also benefit from similar schemes, using the Internet in order to communicate, collaborate, use global knowledge to solve local problems and perform in exciting new ways.

In this paper, we present some concepts about devices powered with Internet-friendly social capabilities, and describe our experiences. In section 2, we refer to previous work on integrating Internet, especially Web-based technologies, into everyday objects. Section 3 is devoted to the analysis of desired features of social devices that communicate through the Internet. Section 4 includes a description of several prototypes of devices we have implemented that use Internet to communicate with other entities. Finally, in section 5 we identify different challenges associated with the design of everyday objects that behave in social ways, which must be addressed by the research community.

2   Related Work

Internet and, particularly, Web technologies have been used in the past to create exciting Ubiquitous Computing scenarios where objects are gifted with augmented capabilities. Cooltown [9] [10] [1] was a pioneer project applying Web technologies to support users of wireless, handheld devices interacting with their environment, anywhere they may be. WSAMI [5] enables the creation of Ambient Intelligence scenarios controlled by Web-based invocation mechanisms, Web Services.

Remarkably, the last few years have witnessed an increasing interest in the application of a particular type of Web technologies, Semantic Web technologies, to create connected societies of smarter artifacts. Semantic Web technologies (RDF [23] and OWL [22]) have been used in several projects in order to provide more intelligence in environments. In the Semantic Web, URIs are used for representing concepts, while HTTP [3] provides a natural means for retrieving RDF-based descriptions. Task Computing [13] [17] aimed at creating Semantic Web Services-based systems in which users could perform automatic composition of services, based on semantic service descriptions.

Other architectures such as CoBrA [2] provided both an architecture and an ontology (SOUPA) to create environments populated by smart devices that could communicate through the Internet using Semantic Web technologies. However, CoBrA requires a central server to be deployed where all the intelligence resides, acting the devices as simple slave entities out of the context-awareness process.

SoaM (Smart Objects Awareness and Adaptation Model) [18] [21] eliminated the need for a central component, and introduced the concept of collaborating *semantic devices*. The results demonstrated that spontaneous emergent intelligence could be achieved by the collaboration of individual objects that may access the Internet for retrieving data and ontologies.

Currently, one of the most popular approaches to implement Internet of Things experiences is the use of "touch computing" along with object tagging [16] [12] in order to obtain further information about a concrete object from the Internet. Mobile devices are generally used in these cases as service "activators".

However, the vision of a coherent and heterogeneous society of Internet-connected objects (*"from anytime, any place connectivity for anyone, we will now have connectivity for anything"* [4]) is in its first steps of realization. We think that current research must explore the possibility of embedding *social capabilities in objects and devices* in order to realize the "Internet of Things" vision as a replica to the current "Internet of People" reality. A lot of people use the Internet to find others' solutions to a particular personal issue; that is, they use global information for a local purpose.

In our vision, "social devices" should work very much in the same way. When facing a local problem, devices can "talk" with other artifacts that can provide their experience about that situation, or additional information that may help to come up with a solution. In the next section, we will describe the features that this kind of social devices must embody.

3   Features of Social Devices

Lassila and Adler [11] introduced the concept of *semantic gadget* to describe devices capable of performing *"discovery and utilization of services without human guidance or intervention, thus enabling formation of device coalitions"*. Vazquez et al. [18] [20] introduced the concept of semantic device as a system *"able to spontaneously discover, exchange and share context information with other fellow semantic devices as well as augment this context information via reasoning in order to better understand the situation and perform the appropriate reactive response"*.

The concept of *social device* emphasizes even more the benefits of the communication with other devices in order to determine the appropriate behavior. Internet is the enabling technology in this case, since the knowledge provided by local devices can be complemented with information obtained from the Internet (provided in turn by other devices, services, or humans). Further below, we will illustrate this situation with the example of a prototyped umbrella which is able to obtain current weather conditions through communication with surrounding rain sensors, but it is also able to download the weather forecast from the Internet (`weather.com`) in order to provide better advice to the user.

We have identified several characteristics that may help to define the concept of social device:

− Social devices are natively designed for collaboration
− Social devices integrate local and global information (obtained from the Internet) in order to apply community knowledge to particular problems
− Social devices are able to interpret all the exchanged information at a semantic level, whatever the vocabulary used

3.1   Designed for Collaboration

Social devices are inherently talkative: they are natively collaborative in the sense that they share all the information they can.

As already mentioned, the Web has transitioned from a basically "one publisher – many readers" model to a more collaborative "many publishers – many readers" model, in an approach that was called Web 2.0 [15]. The major representatives of this culture are weblogs, social bookmarking, wikis and RSS feeds.

We consider that this model can be also applied to devices, featuring a collaborative nature, sharing information, and creating a community of intelligent objects in the environment in order to better serve their users.

Higher and more useful knowledge can be obtained from the generous contributions of individual entities, rather than from selfishly not sharing information (of course, taking into account existing privacy concerns).

On the other hand, social devices can also interact with existing social networks (e.g., Facebook, YouTube, Flickr, Digg) in order to monitor some information on behalf of their user/owner. For example, a Flickr-enabled digital photo frame can periodically check whether new photos from the user's buddies have been published, download them and start a slideshow.

In this case, objects use existing social network infrastructure as a medium for retrieving user-related content. These social objects may also send data back to the social network (e.g., the user may vote for a photo in order to increase its popularity), thus becoming active participants in the community.

These scenarios illustrate how for some simple, well-defined activities users may interact with the social network via real objects instead of using a computer and a browser. These objects promote users' socialization in a transparent, unobtrusive way.

3.2   Integration of Local and Global Information

Users do not have the ability to connect their minds to the Internet, but devices can exhibit this feature. By managing more information they look more intelligent, as in the case of the weather-aware umbrella.

There are ubiquitous examples of this feature: a car may recommend the best route based on data provided by other surrounding cars and traffic information downloaded from the Internet; an eBay-aware ornament at the home entrance may project a green light beam if our bid was successful or a red one if not, as the user arrives home.

We have developed several prototypes of objects (described below in section 4) that simultaneously obtain context information provided by other local objects in the environment as well as remote services on the Internet, integrating all this knowledge to realize a higher level of context-awareness. In our prototypes local information was typically provided by wireless sensor network nodes, while global information was provided by public XML-based Web Services on the Internet. In both cases, we designed adapters that semantically annotated the data using RDF, and shared this information with all the entities in the environment, creating a distributed knowledge space, so existing objects could analyze the semantic context information and react in the desired ways [18].

3.3   Interpretation of Information at a Semantic Level

The vision of a myriad of social devices "talking" both at local and global level exposes the problem of the number of formats and vocabularies for exchanged information and messages. In our experiments [18] we used Semantic Web technologies, RDF and OWL, in order to support multiple vocabularies and dynamically learn relationships among concepts using ontologies.

We humans interpret the information on the Web at a semantic level, being able to understand, process and use it. Social devices should be able to perform in the same way: this is the main reason to apply semantic technologies.

Of course, this level of intelligence requires additional computation capabilities on devices (e.g., able to run a semantic reasoner), but we believe that semantic interpretation is a feature inherent to social communication, without imposing limitations in expressiveness or vocabularies.

The use of microformats [7] as an alternative provides the additional advantage that it is easier for humans to create microformats-enriched content than RDF-annotated content. This approach enables devices to interpret, at a basic level, user-generated content in blogs, wikis, and so forth, without demanding much effort from the contributors, thus bridging the communication gap between users and devices on the Internet.

4   Experiences Prototyping Social Devices

During the last years we have been researching the implications of Internet connected objects and their interaction with users. We have also developed several prototypes in order to evaluate their potential. The following subsections describe some of these prototypes.

4.1 Flexeo's SensorMap: A Wireless Sensor Network –Based Mash-Up

The goal of the project Flexeo was designing a system for remotely monitoring wireless sensor networks through the Internet, and integrating all the data into business processes in order to execute data-mining operations and determine correlations among data. Scenarios related to "health at home" and remote monitoring of industrial facilities were designed in order to test the resulting system.

As an additional outcome, a mash-up of sensor collected data and Google Maps was developed, called SensorMap. This subsystem illustrated the potential applications of publishing up-to-date sensor data on the Internet in order to dynamically integrate different sources of information for providing higher level services.

Figure 1 shows a sensorized chair with two pressure sensors (at the seat and at the back) connected to a wireless sensor network node in order to provide information about its use to local and remote objects. Figure 2 contains a screenshot of the SensorMap application showing the state of the chair (if someone is sat on and/or leaned backwards).

Fig. 1. Preparing the chair with pressure sensors and a WSN node

Flexeo's architecture was divided in to three different layers: the sensor layer, the concentrator layer and the business logic level.

The sensor layer was based on a wireless sensor network formed by Crossbow Mica2 nodes embedded into everyday objects such as a chair, a container, a

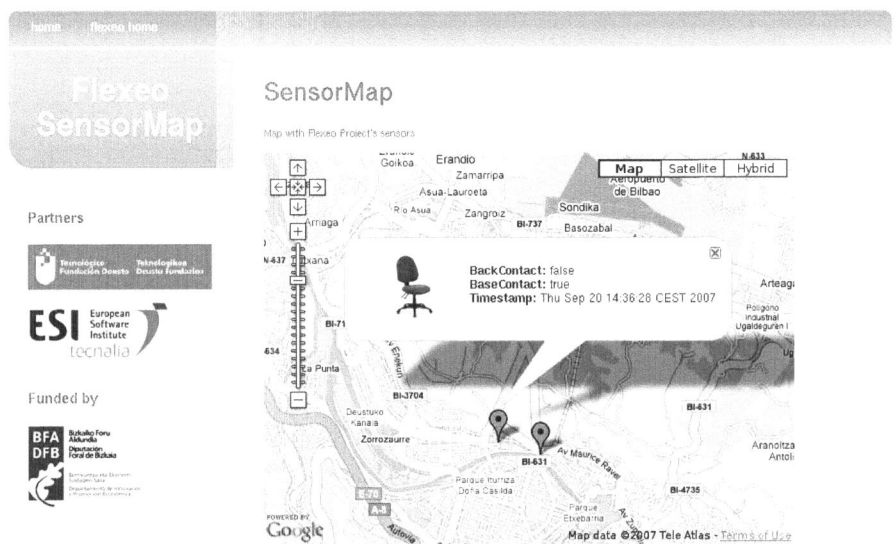

Fig. 2. Flexeo's SensorMap showing the state of a chair

wearable band and a wristwatch. The main goal was transforming different existing objects into wireless nodes, able to provide data about users' interactions.

The concentrator layer was implemented using an embedded computer, a Gumstix connex 400xm, connected both to the wireless sensor network and to the Internet via GPRS and/or Wi-Fi. The concentrator featured a middleware layer implemented using Java in the form of OSGi components, that was in charge of polling the sensor network, interpreting the data, evaluating rules in order to detect alarms, and grouping the data for sending them to the business logic servers that were on the Internet. In Flexeo, the concentrator was the architectural element in charge of connecting the sensor nodes, and thus, the real objects they were embedded in, to the Internet, acting as a communication gateway.

4.2   RealWidgets: From the Desktop to the Real World

Desktop-based operating systems are increasingly using some mini-applications called *widgets* or *gadgets* in order to provide small services or show concrete up-to-date information to users. Yahoo! Widgets, Apple Mac OS X Dashboard, Google Gadgets and Microsoft Windows Vista Gadgets are the most popular flavors of this form of interaction. Often, these widgets connect to online services in the Internet in order to provide weather or traffic information, most-popular YouTube videos, latests news, and so forth.

With RealWidgets we wanted to embody the functional capabilities of these digital entities into real world tiny wireless displays, in order to have small "windows" deployed everywhere opened to the information from the Internet.

The RealWidget is formed by an OLED display, with high resolution and contrast while small energy consumption, integrated with a Crossbow Mote2 wireless sensor network node. A computer acted as a gateway between the Internet and the wireless sensor network, running a RealWidget Management Application that connected to the required sites on the Web, downloaded the information, analyzed it and finally sent the content to the appropriate widget as configured by the user. Figure 3 illustrates a RealWidget showing information about the liquid level in a remote chemical container that was monitored by a wireless sensor network over the Internet.

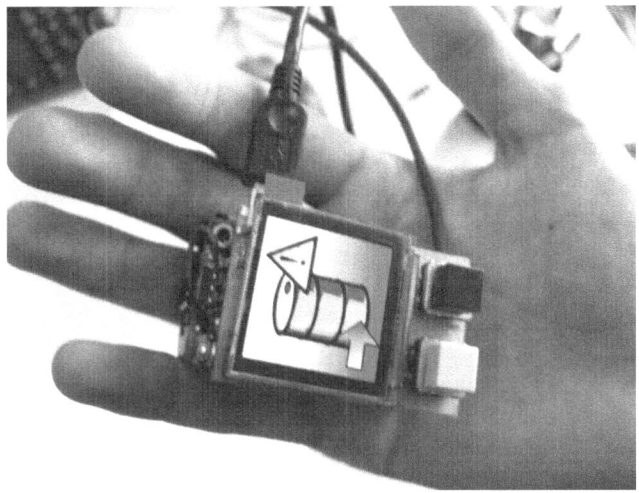

Fig. 3. RealWidget alerting about the liquid level in a tank

Two buttons were provided for interacting with the RealWidget, basically for managing the energy by postponing the information for a later time, or immediately discarding it.

One of the most popular widgets in computer desktops are those related to weather information, so one of the earliest examples of the RealWidget prototype involved obtaining the data from a weather information site on the Internet. We chose weather.com due to the availability of an open API that provided the ability to download an XML document with the weather forecast for any location in the world.

The RealWidget Management Application could host different adapters that acted as gateways between the Internet service and the physical RealWidgets. In the previous example, we created an adapter that was able to connect to the weather.com website, retrieve the XML document with the weather forecast of the configured location, and transformed it into a notification message that was sent to the RealWidget. If any change in the weather information forecast occurred, subsequent notification messages would be sent to the RealWidget. The Crossbow Mica2 wireless node on the RealWidget received the message, processed it, and displayed an icon on the OLED display representing the weather conditions (sunny, cloudy, raining, and so forth).

4.3   SmartPlants: Autonomous Objects That Interact with Their Environment

SmartPlant was one of the evaluation prototypes we designed for the SoaM (Smart Objects Awareness and Adaptation Model) [21] platform: a completely decentralized architecture for designing semantic devices that collaborate by sharing information about the environment, and spontaneously react to changes in the context. The two pillars of SoaM are a semantic discovery protocol called mRDP (Multicast Resource Discovery Protocol) [19], and proactive semantic information exchange among entities. The result is an architecture in which devices create dynamic communication flows, and perform semantic reasoning over context information contributed by all the entities in the environment in order to determine the most appropriate behavior.

One of the scenarios we envisioned at the beginning of the research was to create an artifact that could be attached to real objects, augmenting their perceptions and providing them with intelligent capabilities. An additional challenge was to attach this kind of artifact to living entities, such as plants, in a way that could result in intelligent behavior carried out by the entities from the user's point of view.

An important part of this smart behavior was realized by obtaining additional knowledge about the environment from the Internet. Creating this kind of "smart plants" raised several new important implications such as:

- They could become first-class citizens in the environment, rather than passive elements. They could influence temperature, humidity or lighting settings.
- They could be perceived as autonomic systems [6] in a twofold view: as normal living beings they try to survive and adapt to environmental conditions; but also, as augmented intelligent entities they can interact and communicate with surrounding objects to create a more suitable and healthy environment.

Figure 4 depicts the smart plant prototype. The plant communicated at a semantic level with a wireless sensor network that provided temperature and light measures about different nearby locations, asking the user to me moved to the most suitable place using a synthesized voice. The plant was also able to download ontologies from the Internet in order to interpret particular expressions and predicates provided by surrounding sensors.

For all the semantic devices in SoaM, including the SmartPlant, we designed a computing platform consisting on a Gumstix connex 400xm with Wi-Fi communication capabilities, and a semantic middleware layer. The location a concrete ontology had to be downloaded from was stored in a central server on the Internet that acted as ontology index for the deployed objects.

In order to process semantic information and ontologies, we implemented a minimal semantic engine in the Gumstix called MiniOWL, able to interpret

Fig. 4. The SmartPlant prototype

the most popular ontological relationships such as `rdfs:subClassOf`, `owl:sameAs`, `owl:TransitiveProperty`, `owl:SymmetricProperty` and `owl:inverseOf` [18]. Festival Lite was used as text-to-speech engine in the embedded computer.

The information captured by the wireless sensor network nodes was semantized by a software component running on a computer connected to the sensor network. We designed specific ontologies using OWL for this scenario in order to represent knowledge about lighting conditions, temperature, and location.

4.4 Aware-Umbrella: A Reactive Device Integrating Local and Global Communication

The Aware-Umbrella was also a semantic device designed for the evaluation of the SoaM architecture. As already mentioned, the ability to seamless integrate local and global information sources in order to augment local intelligence and knowledge is of foremost importance for social devices. The most probable, but not unique, source for this information is the Internet, and particularly, available dynamic web services.

Our goal in this scenario was to design some kind of social object that could be aware of both environment-provided and Internet-provided information in order to take decisions and look more intelligent from users' perspective.

Our choice was to create a smart umbrella that could obtain current weather information from surrounding sensors, as well as the weather forecast for the next hours from the Internet. The smart umbrella reacted when the user was leaving home without taking it by issuing a synthesized voice alert (see Figure 5).

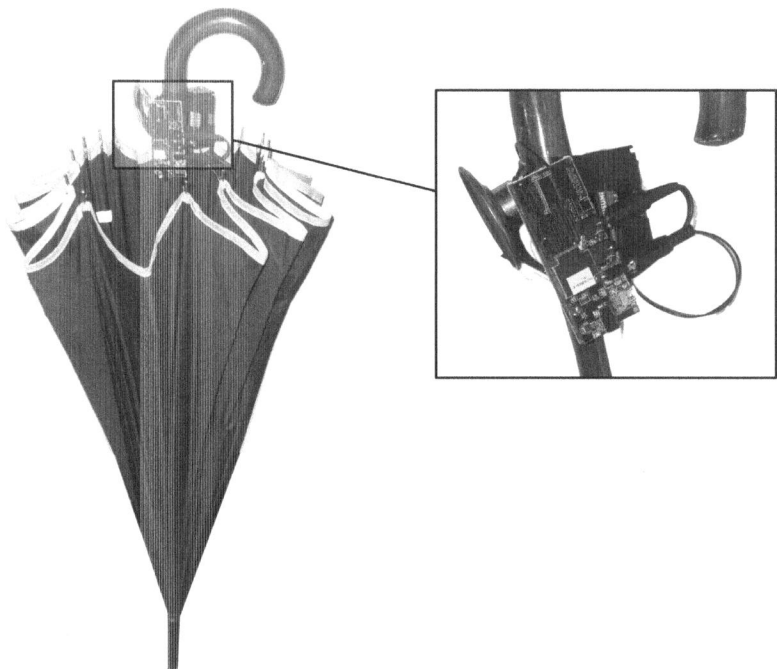

Fig. 5. The Aware-Umbrella prototype

The umbrella obtained context information from the Internet through a "virtual software sensor", a small piece of code that connected to the Internet to get weather information about the town (provided by `weather.com`) and semantized these data using a weather ontology we designed specifically for this case. The umbrella finally checked the state of the door in case the user was leaving when raining in order to decide whether to issue the voice alert.

In this scenario we used the same computing platform designed for the semantic devices in the SoaM project: a Gumstix connex 400xm with a Wi-Fi interface, a semantic middleware layer, and the Festival Lite TTS engine. The state of the door was provided by a proximity sensor (based on a magnetometer) on a Crossbow Mica2 at the door. As in the previous scenario, a computer acted as a gateway between the sensor network and the Wi-Fi network, semantizing the information provided by the node at the door.

5   Challenges for Web 2.0-Enabled Things

Web 2.0 is a synonym for collaboration, a natural habitat for the prosumer model (consumer + producer). Since web technologies have been the most popular infrastructure for designing objects that communicate on the Internet, are Web 2.0-enabled objects the next evolution of this paradigm? What are the interfaces and synergies between users participating in social networks and the Internet of Things?

Based on the experiences we have mentioned in the previous section, we have identified three main challenges on the integration of Web 2.0 collaborative models and Internet-enabled objects:

- Creation of social networks formed by collaborative devices
- Objects as consumers and producers of content in existing human-oriented social networks
- Objects as interfaces with Web 2.0-based services

Figure 6 depicts in a graphical way the relationships among these three challenges: devices collaborating in environment-specific social networks, information flows between devices and social websites, and objects as interfaces for people to interact with human-oriented social networks. In the next subsections, some discussion about these challenges is provided.

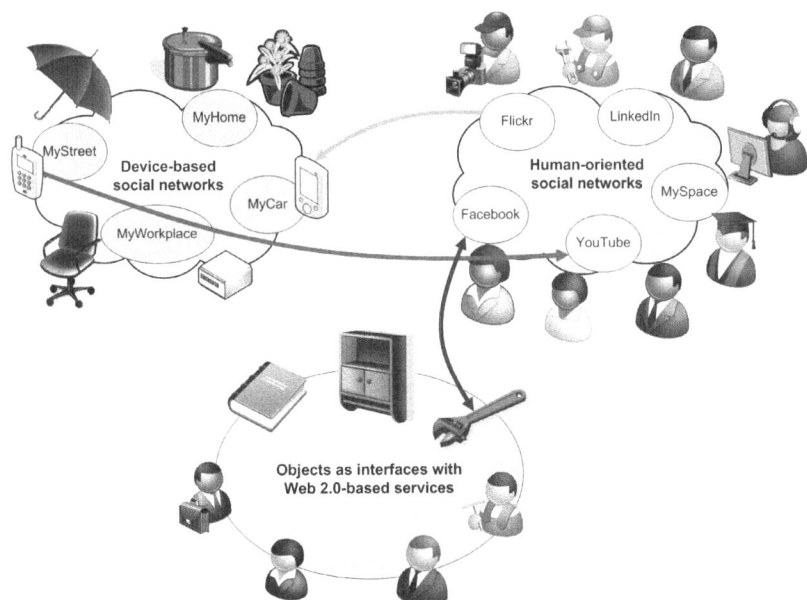

Fig. 6. Graphical representation of the three challenges for Web 2.0-enabled things

5.1   Creation of Social Networks Formed by Collaborative Devices

There are pervasive examples of Web 2.0 collaboration models that have boosted people participation: blogs, wikis, social tagging, social voting, content syndication, and so on. Are these models also suitable for boosting device collaboration on the Internet?

Devices can embrace Web 2.0-based communication models in order to implement a more human-like level of socialization. For people is relatively easy to understand the meaning of a blog entry, a photograph, or tags related to a website, since semantic interpretation is inherently human. However, when it comes to applying these social models to the Internet of Things, the problem of automatic content interpretation appears as a major technical challenge.

In our experiments we found Semantic Web technologies, mainly RDF and OWL, a suitable means for objects to express their knowledge of the environment, enabling fellow objects to interpret this information and behave accordingly. There is no limit in the number of vocabularies than can be applied, depending on the knowledge domains objects must be aware of: semantic technologies are the appropriate vehicle to interpret the information provided by different heterogeneous sources.

For instance, in the Aware-Umbrella prototype, the "virtual weather sensor" that connected to the `weather.com` website semantized the weather information using a vocabulary the umbrella was able to interpret. This semantized weather information was made available to any object in the environment upon request. A similar model was followed in the SmartPlants prototype where a software component semantized the information provided by the wireless sensor network using lighting, temperature and location ontologies, and made it available to interested parties upon request.

While the SoaM architecture used in these scenarios promotes social collaboration among devices in an *ad hoc* way, Web 2.0 models fit perfectly here in order to obtain similar results in a more structured form. Instead of having the objects in the environment creating multiple communication flows to exchange information, an "environmental wiki" could act as repository of information about the environment, with multiple contributors that continuously update this wiki based on their perceptions and interpretations. Annotations in the wiki must still be made using Semantic Web technologies, which makes them difficult for humans to read, but very easy for devices to interpret.

There are two major approaches for the implementation of this "environmental wiki": a centralized approach based on a server in the environment with the semantic wiki software storing all the information; and a decentralized approach based on the wiki being stored in a distributed way, different parts in different objects with some kind of distributed searching mechanism. The later approach is similar to Tuple Spaces [14] or even Semantic Tuple Spaces [8], but with a wiki orientation.

Similarly, existing objects and devices generally maintain a log file with all the incidences that occur during normal operation. Access to this log file is generally restricted to administrators or special users. Just as people have transitioned from secret diaries to public blogs, in which users want to share their experiences, knowledge and opinions with everyone, we think that entities in the Internet of Things should follow a similar approach: creating blog entries using Semantic Web technologies so that objects can share their "personal" interpretations and knowledge. Microblogging (Twitter- and Jaiku- style) may be even more appropriate due to its brevity.

For example, we are working on an extension of the SmartPlant in which the plant creates blog entries in both human-readable and semantically annotated form, in order to share *feelings* such as "I feel lonely", "I am thirsty", "I need more light", or "It is dark and nobody is here", based on its interpretation of the environmental information provided by surrounding entities.

5.2   Objects as Consumers and Producers of Content in Existing Human-Oriented Social Networks

The second challenge for Web 2.0-enabled things consists on connecting them to existing social networks in order to consume and/or produce content in behalf of their users. This approach would result in new ways of interacting with social websites, different from the traditional method of using the web browser.

For example, we are working on an additional extension to the SmartPlant in which the plant projects a blinking green light everytime the user has a Facebook friend request waiting in the queue. Thus, instead of having to check the social network homepage every now and then, existing objects in the environment can perform the task for their users and react in different ways.

Since the RealWidget prototype is basically a wireless OLED display, we are also working on a particularization of this device based on connecting to `Flickr.com` using the user's account, and start a photo slideshow everytime there are new photos in any of the user's groups.

While people are more and more involved in social networking on the Internet, these examples illustrate the powerful capabilities of using physical objects to relieve users from continuously checking up-to-date information from these websites. Designing things that actively contribute with content in behalf of the user seems a bit more complicated, but equally feasible, specially for simple activities.

For instance in the SmartPlant prototype, the user could automatically accept the pending Facebook friend requests by simply touching a part of the plant's pot where a small touch sensor resides. While we are not planning initially to augment the SmartPlant with a small display, it can be useful in this situation so that the user can visualize the name of the friend and accept or deny the request at his will. Precisely, user interactions with social websites-connected things is the subject of the third challenge we have identified for future research.

5.3   Objects as Interfaces with Web 2.0-Based Services

The major concern here is how to map user-object interactions to concrete actions on the social network. In the SmartPlant extension, is a green blinking projected light the most suitable means for alerting the user that he has some friend requests waiting? Is touching the pot the most appropriate means for indicating that the user accepts the requests?

Of course, a small display and keyboard in every object of the environment would eliminate any ambiguity about user interaction, but would also result in unnatural objects. Perhaps the most daunting challenge in having things connected to information sources on the Internet, and particularly to social websites, is how to map the information onto the object in a way that a user can intuitively understand, and even more difficult, how can user's interactions on the object be mapped onto concrete actions back to the website.

In the early experiments we carried out with the SmartPlant Facebook extension, we found that the user has to, obviously, have learned the interpretation of the blinking light previously as there is no natural mapping with its function. While this may be a problem if some concrete object has to express a plethora of different status or information updates using few interaction means, it is not such if there is only one simple function mapped onto the object as in the case of the mentioned Facebook extension. We consider this issue the most critical factor of success in order to design everyday objects that act as connectors with the user social life (on the Internet) in a natural way.

6   Conclusion

Devices should talk and share information without any geographical limitation, just as we humans do. Internet and the Web have demonstrated their ability for providing these capabilities to people, and also constitute the suitable means for enabling global-scale social devices.

This new wave of artifacts will exhibit social capabilities, e.g. promoting information sharing, recording individual opinions and interpretations about the context that can be easily accessed by others, tagging others' resources, or implementing voting and ranking strategies among themselves and the services they provide. As we described in this paper, Web 2.0 mechanisms are ideal candidates to be applied here.

We consider the concept of "social device" to be of foremost importance for creating smart environments. It comprises some of the fundamental aspects researchers have been looking for during the last years, especially intelligent and global context-awareness, and serendipitous collaboration even with remote objects and services.

In our research we investigated the foundations of this new wave of social objects that use the Internet for communication. We outlined their desired characteristics and designed experimental prototypes in several scenarios to demonstrate their functionality.

We also identified three main challenges in the research of social implications on the Internet of Things: the creation of social networks of devices in the environment based on semantic technologies, the connection of these objects with human oriented social websites, and the interaction challenges between virtual information and physical object characteristics.

This vision promotes the emergence of new scenarios in which surrounding objects become more interactive than ever before, being aware of the fact that the Internet is a natural extension of users' life.

Acknowledgments

This research has been partially supported by the Department of Innovation and Economic Promotion of the Biscay Regional Council (Spain), through the Ekinberri program funds for the project Flexeo. The authors would like to thank all the research staff that took part in the projects Flexeo (especially Joseba Etxegarai, Xabier Laiseca and Pablo Orduña) and RealWidget (Iker Doamo), as well as Joseba Abaitua for his contributions on the social vision of devices.

References

1. Barton, J., Kindberg, T.: The cooltown user experience. Hewlett-Packard, Technical Report HPL-2001-22 (2001)
2. Chen, H., Finin, T., Joshi, A.: An intelligent broker for context-aware systems. In: Adjunct Proceedings of Ubicomp 2003, UbiComp, pp. 183–184 (October 2003)
3. Fielding, R.T., Gettys, J., Mogul, J.C., Frystyk, H., Masinter, L., Leach, P.J., Berners-Lee, T.: Hypertext Transfer Protocol – HTTP/1.1. IETF RFC 2616 (1999)
4. International Telecommuncations Union. The Internet of Things. Executive Summary (November 2005)
5. Issarny, V., Sacchetti, D., Tartanoglu, F., Sailhan, F., Chibout, R., Levy, N., Talamona, A.: Developing ambient intelligence systems: A solution based on web services. Automated Software Engineering 12(1), 101–137 (2005)
6. Kephart, J.O., Chess, D.M.: The vision of autonomic computing. Computer 36(1), 41–50 (2003)
7. Khare, R., Çelik, T.: Microformats: a pragmatic path to the semantic web. In: WWW 2006: Proceedings of the 15th international conference on World Wide Web, pp. 865–866. ACM Press, New York (2006)
8. Khushraj, D., Lassila, O., Finin, T.: Stuples: Semantic tuple spaces. In: Proceedings of Mobiquitous 2004: The First Annual International Conference on Mobile and Ubiquitous Systems: Networking and Services, pp. 268–277 (2004)
9. Kindberg, T., Barton, J.: A web-based nomadic computing system. Computer Networks: The International Journal of Computer and Telecommunications Networking 35(4), 443–456 (2001)
10. Kindberg, T., Barton, J., Morgan, J., Becker, G., Caswell, D., Debaty, P., Gopal, G., Frid, M., Krishnan, V., Morris, H., Schettino, J., Serra, B., Spasojevic, M.: People, places, things: web presence for the real world. Mobile Networks and Applications 7(5), 365–376 (2002)
11. Lassila, O., Adler, M.: Semantic gadgets: Ubiquitous computing meets the semantic web. In: Spinning the Semantic Web: Bringing the World Wide Web to Its Full Potential, pp. 363–376 (2003)
12. de Ipiña, D.L., Vázquez, J.I., García, D., Fernández, J., García, I., Sainz, D., Almeida, A.: EMI^{2}lets: A Reflective Framework for Enabling AmI. Journal of Universal Computer Science 12(3), 297–314 (2006)
13. Masuoka, R., Labrou, Y., Parsia, B., Sirin, E.: Ontology-enabled pervasive computing applications. IEEE Intelligent Systems 18(5), 68–72 (2003)
14. Obreiter, P., Gräf, G.: Towards scalability in tuple spaces. In: SAC 2002, pp. 344–350. ACM Press, New York (2002)
15. O'Reilly, T.: What Is Web 2.0. Design Patterns and Business Models for the Next Generation of Software (September 2005, retrieved on January 2007), http://www.oreillynet.com-/pub/a/oreilly/tim/news-/2005/09/30/-what-is-web-20.html
16. Siorpaes, S., Broll, G., Paolucci, M., Rukzio, E., Hamard, J., Wagner, M., Schmidt, A.: Mobile interaction with the internet of things. In: 4th International Conference on Pervasive Computing (Pervasive 2006) (2006)

17. Song, Z., Masuoka, R., Agre, J., Labrou, Y.: Task computing for ubiquitous multimedia services. In: MUM 2004: Proceedings of the 3rd international conference on Mobile and ubiquitous multimedia, pp. 257–262. ACM Press, New York (2004)
18. Vazquez, J.I.: A Reactive Behavioural Model for Context-Aware Semantic Devices. PhD thesis, University of Deusto (2007)
19. Vazquez, J.I., de Ipiñato, D.L.: mRDP: An HTTP-based Lightweight Semantic Discovery Protocol. In: The International Journal of Computer and Telecommunications Networking ((to be published, 2007)
20. Vazquez, J.I., de Ipiña, D.L.: Principles and experiences on creating semantic devices. In: Proceedings of the 2nd International Symposium on Ubiquitous Computing and Ambient Intelligence - UCAMI 2007 (2007)
21. Vazquez, J.I., de Ipiña, D.L., Sedano, I.: SoaM: A Web-powered architecture for designing and deploying pervasive semantic devices. IJWIS - International Journal of Web Information Systems 2(3–4), 212–224 (2006)
22. World Wide Web Consortium. OWL Web Ontology Language Guide. World Wide Web Consortium, W3C Recommendation (February 2004)
23. World Wide Web Consortium. RDF Primer. World Wide Web Consortium, W3C Recommendation (February 2004)

Indoor Location Tracking Using Inertial Navigation Sensors and Radio Beacons

Pedro Coronel*, Simeon Furrer**, Wolfgang Schott, and Beat Weiss

IBM Research GmbH, Zurich Research Laboratory, Zurich, Switzerland
sct@zurich.ibm.com

Abstract. We consider the problem of real-time sensing and tracking the location of a moving cart in an indoor environment. To this end, we propose to combine position information obtained from an inertial navigation system (INS) and a short-range wireless reference system that can be embedded into a future "network of things". The data produced by the INS lead to accurate short-term position estimates, but due to the drifts inherent to the system, these estimates perform loosely after some time. To solve this problem, we also generate estimates with a wireless reference system. These radio-based estimates can be used as accurate long-term position estimates because their accuracy improves over time as the channel fading can be averaged out. We use a data fusion algorithm based on Kalman filtering with forward/backward smoothing to optimally combine the short- and long-term position estimates. We have implemented this localization system in a real-time testbed. The measurement results, which we obtained using the proposed method, show considerable improvements in accuracy of the location estimates.

1 Introduction

Real-time position localization of moving objects in an indoor environment is an important enabling technology for realizing the vision of the "internet of things" and will thus create numerous novel location-aware services and applications in various market segments. In the retail industry, for example, shopping carts equipped with shopping assistants enriched with additional location sensing means can guide customers through the store, provide them with location-based product information, and alert them to promotions and personalized discounts as they walk through the aisles. However, customer satisfaction, and thereby the success of this advanced shopping service, depends crucially on the achievable position accuracy. Moreover, the price of the shopping-cart tracking solution should be low enough to keep its attractiveness in the eyes of retailers.

Location tracking in an indoor environment can be done with various techniques based on mechanical, acoustical, ultra-sonic, optical, infrared, inertial, or

---

* Pedro Coronel is now with the Communication Technology Laboratory, Swiss Federal Institute of Technology (ETH), Zurich, Switzerland.
** Simeon Furrer is now with Broadcom Corporation, Sunnyvale, CA, USA.

C. Floerkemeier et al. (Eds.): IOT 2008, LNCS 4952, pp. 325–340, 2008.

radio-signal measurements. Localization algorithms in radio-based systems are frequently implemented measuring signal strength [1], angle of arrival, or time of arrival. As the radio beacons are deployed at known positions, triangulation or a signature method yield the location of the moving cart. As far as the radio technology is concerned, the short-range wireless personal-area and sensor network technology as specified by the IEEE 802.15.x standardization body and the ZigBee alliance are well-suited for obtaining low-cost implementations. However, the IEEE 802.11.x WLAN technology can also be attractive in buildings where an infrastructure is already deployed. The location estimates obtained with a radio-based system are stable on the long-term as they are updated rather infrequently, but they suffer from a large error variance due to channel fading.

On the other hand, location estimates can also be derived using inertial navigation sensors. Inertial measurements obtained with accelerometers, gyroscopes, and a compass are frequently applied in outdoor environments - often in conjunction with the global positioning system (GPS) - to track maneuverable vehicles like land crafts, aircrafts and ships [2]. Nevertheless, measurements obtained from an Inertial Navigation System (INS) can also be used for indoor tracking as they yield reliable short-term location estimates. Unfortunately, the localization algorithm has to be implemented with integrators and thus suffers from drifts that prevent keeping this accuracy over the long-term.

Various studies have shown that neither an inertial navigation system nor a radio-based positioning system on their own can achieve the target accuracy often required in an indoor environment. Therefore, we propose to merge both methods using a data fusion algorithm. This task can be solved in an optimal way by modeling the generation process of acceleration, bias, velocity, and position with a common state-space model and, posteriorly, estimating the position from measured data using an extended Kalman-filter algorithm that combines the short-term estimates produced by the INS and long-term estimates obtained from the radio positioning system. Fig. 1 shows a more detailed view of the three components of this new tracking system: the inertial navigation system, the wireless reference positioning system, and the data fusion system called location position estimator. In the sequel, we shall describe in detail each of these components and assess their localization accuracy performance by means of measurement results.

2 System Model

As shown in Fig. 1, the INS comprises an inertial measurement unit (IMU) with accelerometers and gyroscopes that measure acceleration and angular velocity in a right-handed Cartesian coordinate system (x, y, z), and a coordinate transformer that estimates the angle and rotates the coordinates system. The INS is mounted on a slowly moving cart in order to track its location. We assume that the z-axis is pointing in the opposite direction of the gravity vector; therefore, any movement of the cart on a plane floor can be tracked with one gyroscope and two accelerometers.

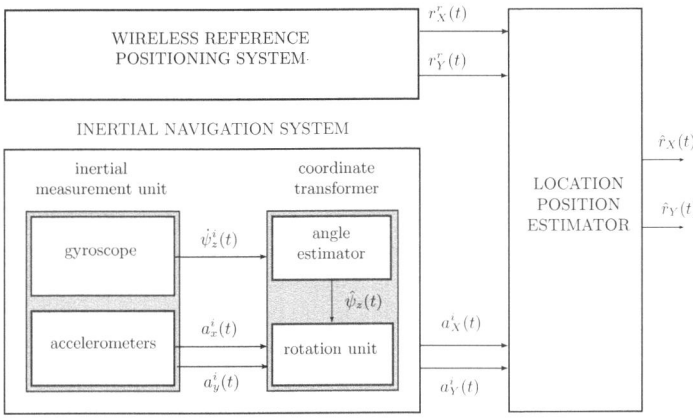

Fig. 1. Block-diagram of the new location tracking system. It comprises a wireless reference positioning system to produce long-term position estimates, an inertial navigation system that yields short-term estimates, and a location position estimator that optimally combines both using Kalman filtering.

2.1 Accelerometers

The accelerometers provide real-time measurements $a_x^i(t)$ and $a_y^i(t)$ on the acceleration of the cart in the x- and y-direction. Since these measurements are rather noisy due to cart vibrations, and often biased by a non-negligible, time-varying offset, we assume that each measurement $a^i(t)$ is related to the true acceleration $a(t)$ and the bias $b(t)$ according to

$$a^i(t) = a(t) + b(t) + v_a(t) \ , \tag{1}$$

where $v_a(t)$ is additive white Gaussian noise (AWGN) with a variance $\sigma_{v_a}^2$.

2.2 Gyroscope, Angle Estimator, and Rotation Unit

The gyroscope measures the angular velocity $\dot{\psi}_z^i(t)$ around the z-axis. Based on this noisy measurement, the angle estimator continuously estimates the rotation angle $\hat{\psi}_z^i(t)$. In a low-noise environment, this operation can be performed with an integrator. Its output value has to be initialized so that the x- and y-axis are aligned to the target coordinate system. After this calibration step, the rotation unit continuously transforms the measurements $a_x^i(t)$ and $a_y^i(t)$ to the target co-ordinate system by

$$\begin{bmatrix} a_X^i(t) \\ a_Y^i(t) \end{bmatrix} = \begin{bmatrix} \cos(\hat{\psi}_z^i) & -\sin(\hat{\psi}_z^i) \\ \sin(\hat{\psi}_z^i) & \cos(\hat{\psi}_z^i) \end{bmatrix} \begin{bmatrix} a_x^i(t) \\ a_y^i(t) \end{bmatrix} \ . \tag{2}$$

2.3   Process and Measurement Models

The acceleration $a(t)$ of the cart along one axis of the coordinate system is modeled by a first-order Markov process defined by

$$\dot{a}(t) = -\alpha a(t) + w_a(t) \ , \ \ \alpha \geq 0 \ , \tag{3}$$

where $w_a(t)$ represents white Gaussian noise with variance $\sigma^2_{w_a}$, α denotes the correlation between successive acceleration values, and, hence, it is inversely proportional to the time maneuver constant of the cart. Different α values can be used for modeling the acceleration in distinct directions. Similarly, a time-varying bias $b(t)$ introduced by sensor imperfections can be modeled by

$$\dot{b}(t) = -\beta b(t) + w_b(t) \ , \ \ \beta \geq 0 \ , \tag{4}$$

where $w_b(t)$ is AWGN with variance $\sigma^2_{w_b}$. The correlation coefficient β, however, takes on a larger value than α because the bias $b(t)$ changes with a much slower rate than the acceleration $a(t)$. The position $r(t)$ of the cart relates to its acceleration $a(t)$ according to

$$\ddot{r}(t) = \dot{v}(t) = a(t) \ , \tag{5}$$

where $v(t)$ denotes the cart velocity.

The linear differential equations (3)-(5) represent the continuous-time process model of the investigated system and can be written in the state-space form

$$\dot{\mathbf{x}}(t) = \mathbf{F}\mathbf{x}(t) + \mathbf{G}\mathbf{w}(t) \ , \tag{6}$$

where the state vector $\mathbf{x}(t)$, the process noise $\mathbf{w}(t)$, and the corresponding matrices $\mathbf{F}$ and $\mathbf{G}$ are given by

$$\mathbf{x}(t) = \begin{bmatrix} r(t) \\ v(t) \\ a(t) \\ b(t) \end{bmatrix} \ , \ \ \mathbf{F} = \begin{bmatrix} 0 & 1 & 0 & 0 \\ 0 & 0 & 1 & 0 \\ 0 & 0 & -\alpha & 0 \\ 0 & 0 & 0 & -\beta \end{bmatrix} \ ,$$

$$\mathbf{w}(t) = \begin{bmatrix} w_a(t) \\ w_b(t) \end{bmatrix} , \ \ \mathbf{G} = \begin{bmatrix} 0 & 0 \\ 0 & 0 \\ 1 & 0 \\ 0 & 1 \end{bmatrix} \ . \tag{7}$$

The measurement model is chosen in accordance with (1) and (2), and can be written as

$$\mathbf{z}(t) = \mathbf{H}\mathbf{x}(t) + \mathbf{v}(t) \ , \tag{8}$$

where the output vector $\mathbf{z}(t)$, the measurement noise $\mathbf{v}(t)$, and the matrix $\mathbf{H}$ are respectively given by

$$\mathbf{z}(t) = \begin{bmatrix} a^i(t) \\ r^r(t) \end{bmatrix} \quad , \quad \mathbf{H} = \begin{bmatrix} 0\ 0\ 1\ 1 \\ 1\ 0\ 0\ 0 \end{bmatrix} \quad ,$$

$$\mathbf{v}(t) = \begin{bmatrix} v_a(t) \\ v_r(t) \end{bmatrix} \quad . \tag{9}$$

The process and measurement model given by (7), (8) and (9) reflects the behaviour of the system if the cart is moving and location estimates are provided by the reference positioning system. If the cart is standing still, the dynamics are better described by incorporating the constraint $a(t) = v(t) = 0$ into the system matrix $\mathbf{F}$ of the process model (7). Similarly, if reference location estimates are not available, the matrix $\mathbf{H}$ in the measurement model (8) can be modified so that the signal $r^r(t)$ in the output vector $\mathbf{z}(t)$ is zero.

3   Wireless Reference Positioning System

The wireless reference positioning system provides estimates $r_X^r(t)$ and $r_Y^r(t)$ on the current position of the slowly moving cart in the target coordinate system (X, Y) using received signal strength indicators (RSSIs). For that purpose, we consider a wireless reference system, where K beacons (e.g. sensors equipped with a low-cost radio transmitter) are deployed at known positions over the location tracking area. Each beacon periodically broadcasts a radio packet containing the beacon identifier $k = 1, \ldots, K$. On the mobile cart, a radio receiver is mounted which continuously listens for incoming packets. When it receives a packet, the receiver determines the RSSI and decodes the packet to retrieve the originator's identifier k. The received signal strength indicator is used to approximate the distance d_k of the mobile to the corresponding beacon by $\hat{d}_k$. When the mobile has decoded radio packets from $K' \in \{3; K\}$ different beacons, the position of the mobile is given as the intersection of K' spheres centered on the packet originator position with radius $\hat{d}_k$. Various efficient methods can be employed to compute the position estimate on the basis of approximate distances [3]. As we assume that the cart is moving on the plane (X, Y), the computation takes into account the fact that the intersection lies in the plane $Z = 0$.

Next we shall discuss how to approximate the distance d_k on the basis of an estimate of the average RSSI derived from several decoded packets. The signal power received from the kth beacon, RSSI_k, is related to d_k by a path-loss model, i.e. a random function that describes the signal attenuation as a function of the propagation distance, whose model parameters are specific to the tracking environment. Before describing the RSSI estimation algorithm, we discuss the path-loss model and how to determine the model parameters in a specific tracking environment.

3.1   Path-Loss Model

The mean signal attenuation described by the path-loss (PL) model decays exponentially with distance. However, the signal strength varies considerably around its mean value due to the constructive and destructive superposition of multiple propagation paths between a beacon and the mobile. In an indoor environment, this effect is particularly important due to a typically obstructed line of sight, and reflections on walls and other objects (tables, shelves, window panes, ...). A general PL model that has been widely validated by experimental data [4] is given by

$$\mathsf{RSSI}(d) = \mathsf{RSSI}_0 + \alpha \; 10 \log_{10} \left(\frac{d}{d_0} \right) + X_\sigma \; , \tag{10}$$

where RSSI_0 is an RSSI value measured in dB at a reference distance d_0 in meters (close to the transmitter) and α is the path-loss exponent. The random variable $X_\sigma \sim \mathcal{N}(0, \sigma^2)$ accounts for log-normal shadowing. More specifically, X_σ models the fact that measured signal levels at the same transmitter-receiver separation have a normal distribution around a distance-specific mean.

On the basis of this model, the transmitter-receiver separation can be approximated as follows: Suppose that the mobile successfully decodes at least one packet while it stands at d meters from the originating beacon, and it computes an estimate of the signal strength denoted by RSSI. Then, d can be approximated as

$$\hat{d} = d_0 \; 10^{\frac{1}{10\alpha}(\mathsf{RSSI}(d) - \mathsf{RSSI}_0)} \; , \tag{11}$$

where we have used (10) and the model parameters tailored to the tracking environment.

3.2   Model Parameters

Depending on the environment, the parameters RSSI_0, α and σ in (10) take on different values (we assume that $d_0 = 1$ meter). For the purpose of identifying their values in a given environment, a large number of RSSI values are collected for multiple measurement locations *and* transmitter-receiver separations. The optimal parameters $\mathsf{RSSI}_0^\star$ and $\alpha^\star$ minimize the mean square error (MSE) between the measured data and the predicted values, that is

$$\arg\min_{\mathsf{RSSI}_0, \alpha} \sum_{n, m_n} \left(\mathsf{RSSI}_{m_n}(d_n) - \mathsf{RSSI}_0 - 10\alpha \log_{10} \left(\frac{d_n}{d_0} \right) \right)^2 \; , \tag{12}$$

where m_n is an index that runs over the RSSI values collected at different positions for the same transmitter-receiver separation d_n. The parameter σ^2 describing log-normal shadowing is found as the variance of the error between the RSSI measurements and the values predicted by the model with parameters $\mathsf{RSSI}_0^\star$ and $\alpha^\star$.

3.3   RSSI Estimation

In this section, we describe a simple method to estimate the RSSI. Suppose that the time-varying wireless channel has the following input-output relation

$$y_n = h_n d^{-\frac{\alpha}{2}} x_n + z_n \ , \tag{13}$$

where y_n, x_n and h_n are respectively the received signal, transmit signal and channel gain at time n. We assume an ergodic (across time) fading process with channel gains h_n distributed according to $\mathcal{CN}(0,1)$ for each n. The noise z_n is a white Gaussian random process with spectral density N_0. Using N signal samples taken every T_s seconds, a signal power estimate can be computed as

$$P_N = \frac{1}{NT_s} \left(\sum_{n=1}^{N} |y_n|^2 \right) \ .$$

Upon application of the strong law of large numbers (SLLN), the sample mean converges to the statistical average with probability 1 as $N \to \infty$. As the channel and input signal are mutually uncorrelated, and so are the signal and noise components in (13), we have

$$
\begin{aligned}
P &= \lim_{N \to \infty} P_N \\
&\xrightarrow{\text{w.p.1}} \frac{1}{T_s} \left(d^{-\alpha} \mathbb{E}\left\{|h_n|^2\right\} \mathbb{E}\left\{|x_n|^2\right\} + \mathbb{E}\left\{|z_n|^2\right\} \right) \\
&= d^{-\alpha} \mathcal{P}_x + N_0 \ ,
\end{aligned}
\tag{14}
$$

where $\mathcal{P}_x$ is the mean signal power. Equation (14) suggests that d can be obtained from P by either using an estimate of SNR $= \frac{\mathcal{P}_x}{N_0}$, or by assuming that the noise power is negligible compared to the power of the signal component, i.e. $d^{-\alpha}\mathcal{P}_x \gg N_0$. In practice, P is approximated by P_N for a suitable N.

Note that the SLLN theorem does not only hold for i.i.d. random variables, but also for correlated variables - as the channel gains are - provided that the correlation vanishes for increasing lags [5]. In indoor environments, the Doppler shift is typically small and the wireless channel changes very slowly, i.e. large channel coherence time, implying that the number of samples N has to be carefully chosen depending on the channel coherence time and T_s to get reasonable distance estimates.

3.4   Implementation of the Reference Location Positioning System

We have implemented a reference location positioning system consisting of multiple Crossbow MicaZ motes [6], which comprise a Chipcon CC2420 RF transceiver attached to an Atmega128L micro-controller. The radio transceiver is compliant with the ZigBee/IEEE 802.15.4 standard [7]. It is thus operated in the 2.4 GHz ISM frequency band using direct sequence spread spectrum (DSSS) and can reach transmission data rates of 250kb/s. MicaZ motes are programmed in NesC as TinyOS applications to transmit or receive radio packets according to the application needs. Since MicaZ motes have been designed for a small size, low power consumption, and lost cost, they are well-suited for the use as pervasive devices.

In our testbed, most of the motes are deployed as radio beacons that transmit packets containing the beacon identifier k at a rate of 10 packets per second. One mote, however, is attached to the mobile cart to listen to incoming beacon transmissions. For every incoming packet, the mobile node first retrieves the originator identifier and the link quality information, and then forwards them through a serial interface to an attached portable computer. On the computer, we use a component object model to read the data forwarded by the mobile, before the data is processed online by MATLAB scripts.

In a MicaZ receiver, the signal is demodulated as illustrated in Fig. 2. After 10-bit A/D conversion and filtering, the frequency offset is compensated and an RSSI estimate is obtained by averaging the signal amplitude over 8 successive symbol periods corresponding to the standard-defined packet preamble having a duration of $128\mu s$. In indoor environments, the channel does not change significantly during this time span, i.e., the channel coherence time is typically of the order of several milliseconds, and, therefore, we cannot assume (14) to hold for such a short averaging window. This suggests to average RSSIs across multiple packets to improve the distance estimate.

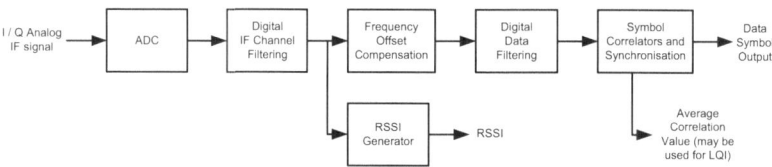

Fig. 2. Demodulator simplified block diagram [8]

Coming back to Fig. 2, we note that following digital filtering and before the data symbols can be recovered, synchronization is also performed based on the received packet preamble. The (peak) correlation value ρ obtained by matched-filtering the received signal with the preamble sequence is used to compute a link quality indicator (LQI). In general, the LQI is a function of both, the correlation value ρ and the RSSI. Using only the RSSI value as a LQI is not accurate enough as, for instance, a narrowband interferer inside the transmission bandwidth would increase the RSSI value although it actually reduces the true link quality. Hence, the correlation value provides a mean to determine the interference level in the received signal. Large correlation values correspond to low interference, while low correlation values evidence large interference. Corrupted RSSIs are thus be discarded as they will not lead to accurate distance estimates.

In summary, for each incoming packet, the mobile retrieves the originating beacon identifier k, the received signal strength RSSI_k value and correlation value ρ_k, and forwards them to the computer via the serial port. For $k = 1, 2, \ldots K'$, only the RSSI_k values corresponding to large correlation values ρ_k are used to approximate the distance d_k. When $K' \geq 3$, a position estimate is computed by triangulation.

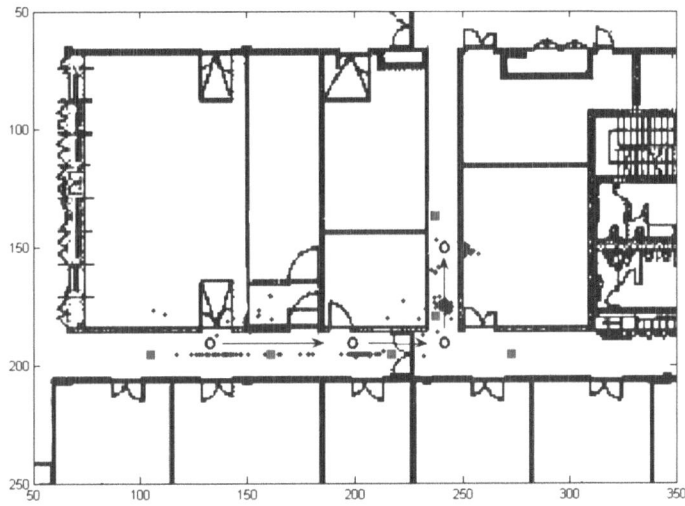

Fig. 3. Floor plan showing the beacon positions (red squares), the cart position at 4 stops (black circles) and the corresponding position estimates (blue dots). The dimensions are of approximately 30 x 20 meters.

3.5   Measurement Results

We have conducted several measurement campaigns in a commercial building to assess the performance of the proposed wireless reference positioning system. In a first measurement campaign, we derived the optimal PL model parameters for the given radio environment as described in Section 3.2. Then, we analyzed the accuracy of the position estimates obtained with the proposed system using the PL model with the optimized parameters.

Figure 3 shows the floor plan of the building where the measurements were performed. The arrows denote the path followed by the cart through the aisles of the building. The beacon positions are indicated by red squares. The mobile stops during 1 minute at each of the four different positions are shown by black circles in the figure. In blue, we plot the different position estimates derived from the RSSI values. Note that four dominating clouds can be distinguished close to the stops.

We are interested in evaluating two different alternatives to approximate d_k. The first method consists in generating a distance estimate $\hat{d}_k$ for each successfully decoded packet, i.e. the RSSI is computed over a time window of $128\mu s$. The second method averages the RSSI_k for multiple incoming packets from beacon k as long as the mobile is standing still. Note that if the averaging window is large compared to the channel coherence time (i.e. several orders of magnitude), this will improve the convergence of the sample mean in (14); in turn, better estimates of d_k and more accurate position estimates will be obtained. This intuition is confirmed in Fig. 4, where we plot the cumulative distribution of the

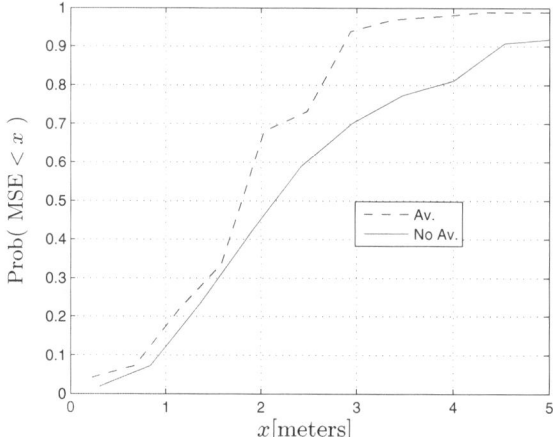

Fig. 4. Comparison of the cumulative distributions of the estimation error (MSE) with/whithout RSSI averaging

error obtained with and without RSSI averaging. Note that averaging results in a position error that is statistically smaller. For instance, the probability of a MSE smaller than 2 meters is about 0.45 without averaging and nearly 0.7 with averaging. When the RSSI values are averaged over multiple packets, the error is bounded to 3 meters in 95 % of the measurements, while this percentage drops to 70 % without averaging. In conclusion, computing an average RSSI using multiple packets improves significantly the accuracy of the position estimate as the sharp variations in the slow-fading channel can be averaged out.

4   Location Position Estimator

Short- and long-term position estimates obtained from the INS and the wireless reference system are combined using the location position estimator (see Fig. 1). The latter comprises a forward Kalman filter that yields optimal real-time location-position estimates $\hat{r}(t)$, a backward Kalman filter with a recording and smoothing unit for optimal smoothed estimates $\tilde{r}(t)$, a motion and reference signal detector, and a control unit for controlling the operation of the estimator. These components are illustrated in Fig. 5. Note that in this section the proposed location-positioning method and the corresponding algorithms are only given for one dimension of the target coordinate system (X, Y).

4.1   Forward Kalman Filter

The forward Kalman filter continuously monitors the noisy measured signals acceleration $a^i(t)$ and reference location $r^r(t)$, and computes in real-time a location-position estimate $\hat{r}(t)$ based on particular process and measurement

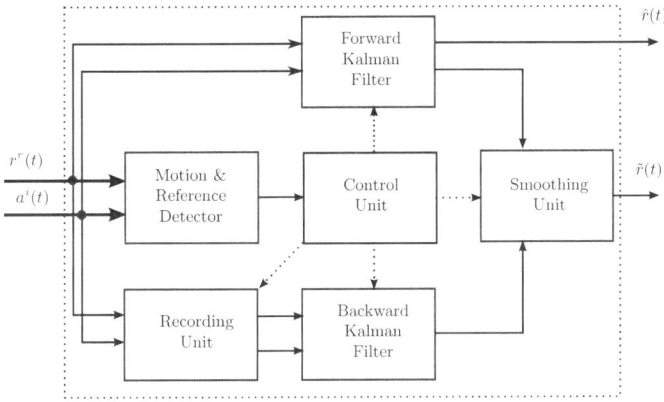

Fig. 5. Details on the various components of the location position estimator

models. This is achieved by, firstly, replicating the process and measurement model (without noise sources) in the filter to generate the state vector estimate $\hat{\mathbf{x}}(t)$ and the corresponding output vector $\hat{\mathbf{z}}(t) = \mathbf{H}\hat{\mathbf{x}}(t)$ and, secondly, continuously updating the state vector estimate according to

$$\dot{\hat{\mathbf{x}}}(t) = \mathbf{F}\hat{\mathbf{x}}(t) + \mathbf{K}(t)\left[\mathbf{z}(t) - \hat{\mathbf{z}}(t)\right] \ . \tag{15}$$

$\hat{\mathbf{x}}(t)$ is an optimal estimate of $\mathbf{x}(t)$ in the mean square error sense. The filter is driven by the error between the measured signal vector $\mathbf{z}(t)$ and the reconstructed output vector $\hat{\mathbf{z}}(t)$ weighted by the Kalman gain matrix $\mathbf{K}(t)$. To obtain the optimum Kalman filter gain settings, the covariance matrix of the state vector estimation error also has to be computed as described, for example, in [9].

A realization of the Kalman filter obtained for the process and measurement model given in (6)-(9) comprises eight time-varying Kalman gains as shown in Fig. 6. The estimates for the location $\hat{r}(t)$, velocity $\hat{v}(t)$, and accelaration $\hat{a}(t)$ of the cart, and accelerometer bias $\hat{b}(t)$ are provided at the output of four integrators. Since three of the four integrators are cascaded, the filter is especially sensitive to DC-offsets present at the input of the integrators leading to severe drift effects in the integrator output signals. As discussed below, additional bias and drift compensation techniques are applied to cope with this inherent deficiency.

4.2   Backward Kalman Filter with Recording and Smoothing Unit

While the forward Kalman filter provides real-time estimates $\hat{\mathbf{x}}(t)$ on the basis of measurement signals obtained up to time t, a better estimate $\tilde{\mathbf{x}}(t)$ can be obtained by forward-backward Kalman filtering. Intuitively, this technique is able to improve the estimate by exploiting past and future data. A drawback is that the smoothed estimate $\tilde{\mathbf{x}}(t)$ can no longer be computed in real-time, but only after some delay. However, this inherent disadvantage can be dealt with

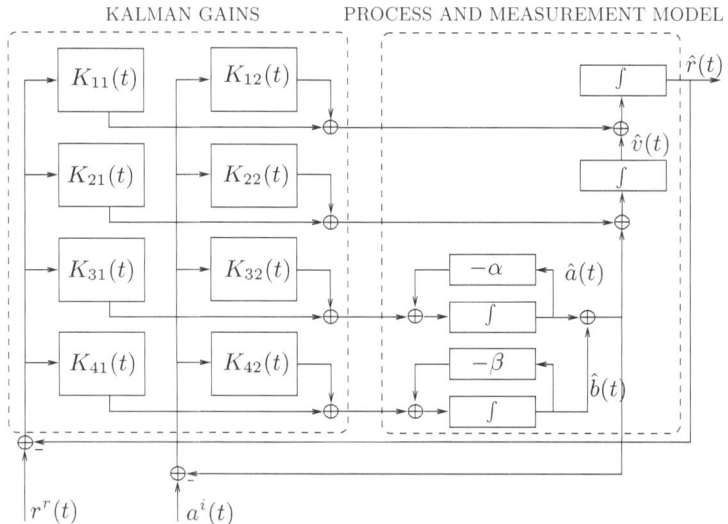

Fig. 6. Kalman Filter Designed for Moving Cart in Presence of Reference Signal

when the state vector does not change its value. Hence, we propose to apply forward-backward smoothing whenever the cart is not moving.

The forward-backward smoothing algorithm can be realized with a recording unit, a forward Kalman filter, a backward Kalman filter, and a smoothing unit as shown in Fig. 5. When a cart starts moving at any arbitrary time t_0, the recording unit starts storing the signals acceleration measurement $a^i(t)$ and reference location $r^r(t)$ until the cart stops at time t_1. At that moment, the location position estimator computes the smoothed estimate $\tilde{\mathbf{x}}(t)$ using forward-backward filtering over the data collected during the time interval $[t_0, t_1]$. The forward Kalman filter is implemented as discussed previously with the initial conditions $\hat{r}(t_0) = r_0$, $\hat{b}(t_0) = b_0$, $\hat{a}(t_0) = \hat{v}(t_0) = 0$. It runs forward in time from t_0 to t, and computes the forward state vector estimate denoted by $\hat{\mathbf{x}}^f(t)$. The backward Kalman filter has the same structure as the forward filter but it runs backward in time, i.e. from t_1 to t, and uses the following initial conditions: the acceleration $\hat{a}(t_1)$ and the velocity $\hat{v}(t_1)$ are set to zero because the cart is not moving, and the initial location $\hat{r}(t_1)$ and bias $\hat{b}(t_1)$ are modeled as Gaussian random variables. The estimate provided by the backward Kalman filter is denoted by $\hat{\mathbf{x}}^b(t)$.

Finally, the smoothing unit optimally combines the estimates $\hat{\mathbf{x}}^f(t)$ and $\hat{\mathbf{x}}^b(t)$ to generate $\tilde{\mathbf{x}}(t)$. Optimal combining is accomplished by weighting the estimates for each t using the covariance matrices of their corresponding state vector estimation error that are provided by the forward and backward filters.

4.3   Motion and Reference Signal Detector with Control Unit

The motion and reference signal detector continuously monitors the signals $a^i(t)$ and $r^r(t)$ for two purposes: firstly, to detect whether the cart is *in-motion* and, secondly, to recognize the *presence-of-reference* data or the *absence-of-reference* data. The latter may also be directly signaled to the location position estimator by the reference positioning system.

The motion detector continuously traces the magnitude of the acceleration measurement signal $a^i(t)$. As it strongly oscillates when the cart is moving, we sample it at rate $\frac{1}{T}$ and average N consecutive samples to obtain the $\bar{a}(kT) = \frac{1}{N}\sum_{l=0}^{N-1} |a^i(kT - lT)|$. Then, an *in-motion* event is declared if

$$\bar{a}(kT) > \mu \; , \tag{16}$$

where μ is a pre-determined threshold. Both N and μ have to be chosen so as to reliably detect short cart stops, but also to ensure that false decisions are rather unlikely.

Depending on the events *in-motion/no-motion* and *presence/absence-of- reference*, the control unit dynamically switches the location position estimator between the following basic modes of operation:

1. **Cart *In-Motion* and *Presence-of-Reference* Signal.** The control unit enables the operation of the forward Kalman filter. The filter provides estimates of the location, acceleration, and velocity of the cart, as well as an estimate of the accelerometer bias in real time. Moreover, the acceleration $a^i(t)$, the reference location $r^r(t)$, and the mode parameter are recorded in a storage device for posterior use (mode 3).

2. **Cart *In-Motion* and *Absence-of-Reference* Signal.** Since no reference signal is available, the location position estimator derives an estimate based on the accelerometer measurements $a^i(t)$ only. This change is modeled by setting all elements in the second row of the measurement matrix **H** to zero. The control unit reconfigures the Kalman filter so as to reflect this change of the underlying measurement model in the filter. Moreover, the control unit freezes the bias estimate $\hat{b}(t)$ in the Kalman filter to its current value when the mode of operation is switched from *presence-of-reference* to *absence-of-reference* signal. This freeze is recommendable to avoid observability problems due to simultaneous changes of acceleration and bias in the system model while no reference signal is present. Finally, the acceleration $a^i(t)$, the reference location $r^r(t)$, and the mode parameter are recorded in a storage device for posterior use (mode 3).

3. ***No-Motion* of the Cart.** Whenever the motion detector detects a *no-motion* event, the control unit overrides to zero the acceleration estimate $\hat{a}(t)$ and the velocity estimate $\hat{v}(t)$ in the Kalman filter. This calibrates the integrator output signals to zero, and thus prevents offset accumulation by the integrators. Moreover, an enhanced process model for the dynamic behavior of the cart is obtained by incorporating the constraint $a(t) = v(t) = 0$ into the process equations (6) and (7). The control unit reconfigures the Kalman filter to reflect this change of the underlying process model in the filter.

Since the measurements $a^i(t)$ and $r^r(t)$ were recorded while the cart was *in-motion* (see mode 1), the data stored up to the latest stop of the cart can be post-processed by forward-backward smoothing. The control unit initializes the execution of the smoothing when the event *no-motion* is detected. After this operation, the location position estimator outputs the optimal smoothed estimate of the cart location.

4.4   Measurement Results

We have implemented the inertial navigation system and the location position estimator by mounting a commercially available inertial measurement unit (IMU) to a mobile cart and attaching the IMU to a portable computer carried by the cart. As IMU, we selected the sensor unit MT9-B from Xsens Motion Technologies [10], which can easily be embedded in a cart due to its small size and light weight. The unit measures its own acceleration in the x- and y-direction, as well as its angular rate around the z-axis. The sensed data are digitized at a sampling rate of 100 Hz and forwarded in real-time through a serial interface to the portable computer. On the computer, MATLAB scripts are executed to derive from the sensed data and reference signals the current location position of the cart using the Kalman filter algorithm introduced above. On the mobile

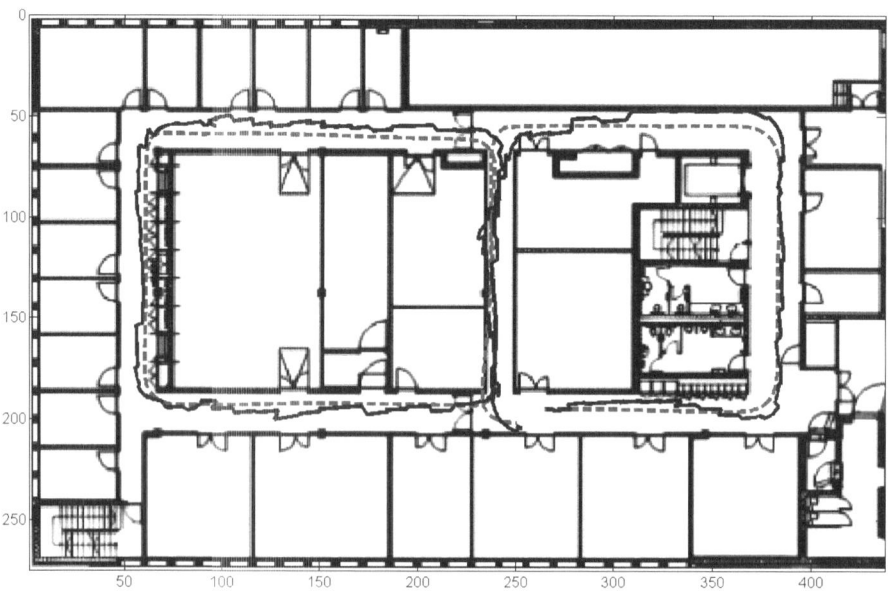

Fig. 7. Measurements conducted at an industrial building. The red dashed line represents the true path of the cart as indicated by the odometer; the blue solid line represents the position location estimates as indicated by the combination of the INS measurements and the reference signal using forward-backward Kalman filtering.

cart, we also mounted an odometer to track its true position and compare it to the estimates provided by the proposed location positioning system. With this prototype implementation of a location-aware mobile cart, a measurement campaign has been conducted in an industrial building to assess the performance of the location positioning system.

Fig. 7 shows the floor plan corresponding to the building where the measurements were performed. The mobile cart was moved in the aisles of the building along the red dashed path. This trajectory has been recorded with the odometer. The blue solid line indicates the location position estimates that have been obtained with the proposed location sensing and tracking solution. We observe that the Kalman filtering method presented by the blue trajectory compares favorably to the actual trajectory followed by the cart.

5   Conclusions

We have presented a novel indoor location tracking system that can be embedded into the envisioned "internet of things". The system uses an inertial navigation system and a wireless reference system for real-time sensing and tracking the location of a moving cart in an indoor environment. The wireless reference system uses multiple low-cost beacons deployed over the tracking area that periodically transmit radio packets. The signal strength of the received packets at the mobile is used to approximate the distance to its originator, and distances corresponding to multiple beacons are combined to determine the position of the mobile. We have used an experimental setup with MicaZ motes to demonstrate that a simple path-loss model and received signal power estimation algorithm can be used to derive a reference position with a reasonable accuracy. The PL model parameters have been derived in a preceding measurement campaign, before they have been used for online computation of the location estimates. By averaging the RSSI values over multiple packets, we obtained a more robust location estimate.

In order to improve the accuracy of the position estimate, we have proposed and investigated a merge of the measurements obtained by the INS and the reference positioning system. This is achieved by a data fusion algorithm that is based on Kalman filtering. The generation processes for the acceleration, bias, velocity and position are modeled using a common state-space model. The position estimates are obtained from measured data using an extended Kalman filter that exploits the reference position. Measurement results have shown the improvement that the proposed method is able to achieve.

References

1. Xiang, Z., Song, S., Chen, J., Wang, H., Huang, J., Gao, X.: A wireless LAN based indoor positioning technology. IBM Journal of Research and Development (2004)
2. Singer, R.A.: Estimating optimal tracking filter performance for maneuvering targets. IEEE Trans. on Aerospace and Electronic Systems (1970)

3. Murphy, W., Hereman, W.: Determination of a position in three dimensions using trilateration and approximate distances. Technical Report MCS-95-07, Mathematical and Computer Sciences Dept., Colorado School of Mines (1995)
4. Rappaport, T.S.: Wireless Communications. Prentice Hall, New Jersey (1996)
5. Serfling, R.J.: Approximation Theorems of Mathematical Statistics. In: Probability and Statistics, Wiley, Chichester (2002)
6. Crossbow Technology, Inc.: MPR-MIB User Manual
7. IEEE: Standard for information technology - Telecommunications and information exchange between systems - Local and metropolitan area networks - Specific requirements - Part 15.4: Wireless Medium Access Control (MAC) and Physical Layer (PHY) Specifications for Low-Rate Wireless Personal Area Networks (LR-WPANs), IEEE, Los Alamitos (2003)
8. Texas Instruments, Inc.: Chipcon Products - 2.4 GHz IEEE 802.15.4 / ZigBee-ready RF Transceiver
9. Maybeck, P.S.: Stochastic Models, Estimation and Control. Academic Press, New York (1982)
10. Xsens Technologies, B.V: MT9 Inertial 3D Motion Tracker

Tandem: A Context-Aware Method for Spontaneous Clustering of Dynamic Wireless Sensor Nodes

Raluca Marin-Perianu[1], Clemens Lombriser[2],
Paul Havinga[1], Hans Scholten[1], and Gerhard Tröster[2]

[1] Pervasive Systems Group, University of Twente, The Netherlands
[2] Wearable Computing Lab, ETH Zürich

Abstract. Wireless sensor nodes attached to everyday objects and worn by people are able to collaborate and actively assist users in their activities. We propose a method through which wireless sensor nodes organize spontaneously into clusters based on a common context. Provided that the confidence of sharing a common context varies in time, the algorithm takes into account a window-based history of believes. We approximate the behaviour of the algorithm using a Markov chain model and we analyse theoretically the cluster stability. We compare the theoretical approximation with simulations, by making use of experimental results reported from field tests. We show the tradeoff between the time history necessary to achieve a certain stability and the responsiveness of the clustering algorithm.

1   Introduction

Wireless sensor networks, smart everyday objects and cooperative artefacts represent all different facets of the ubiquitous computing vision, where sensor-enabled devices become integrated in the environment and provide context-aware services to the users. Various systems have already been demonstrated to be able to retrieve the context, such as the physical context (e.g. position, movement [5]), the situation (e.g. meeting [17]) and even the emotional context (e.g. mood detection [4]). One step further is to have sensor nodes that interact and use common contextual information for reasoning and taking decisions at the point of action. Such a "networked world" opens perspectives for novel applications in numerous fields, including transport and logistics [10], industrial manufacturing [13], healthcare [8], civil security and disaster management [9].

In this paper, we explore a non-traditional networking paradigm based on context sharing. Previous work showed that sensor nodes can recognize online a common context and build associations of the type "moving together" [7,10]. Starting from this result, we address the problem of organizing the nodes sharing a common context into stable clusters, given the dynamics of the network and the accuracy of the context-recognition algorithm.

The contributions of this paper are as follows. Firstly, we propose Tandem, an algorithm for spontaneous clustering of mobile wireless sensor nodes. The

C. Floerkemeier et al. (Eds.): IOT 2008, LNCS 4952, pp. 341–359, 2008.

algorithm allows reclustering in case of topological or contextual changes, and tries to achieve stable clusters if there are no changes in the network, by analysing the similarity of the context over a time history. Secondly, we approximate the behaviour of the algorithm using a Markov chain model, which allows us to estimate the percentage of time the clustering structure is correct given that the topology is stable. Thus, we are able to analyse the cluster stability both theoretically and through simulations, using experimental results reported from real field tests. Thirdly, we study the tradeoff between the time history necessary to achieve a certain stability and the responsiveness of the clustering algorihm. As a result, we estimate the delay induced by the time history, given a desired cluster stability.

2   Application Scenarios

We describe two applications where wireless sensor nodes are able to extract and communicate the general contextual information for creating a dynamic cluster. The clusters then provide services such as reporting the group membership, analysing the cluster activity, recognizing fine-grained events and actions.

2.1   Transport and Logistics

Transport and logistics represent large-scale processes that ensure the delivery of goods from producers to shops [10]. Using the wireless sensor networks technology in transport and logistics is particularly interesting for dynamically locating the goods, generating automatic packing lists, as well as for monitoring the storage condition of a product (e.g. temperature, light) or its surroundings.

The delivery process starts at a warehouse, where the transport company personnel assemble rolling containers (Returnable Transport Items - RTIs), pick the requested products from the warehouse shelves, and load them in the RTIs. Next, the RTIs are moved on the *expedition floor*, a large area used for temporary storage (see Figure 1). From the expedition floor, the RTIs are loaded into trailers according to the shop orders. In order to avoid the errors made during loading of items in the RTIs, a node equipped with a movement sensor can be attached to each product. At the moment the RTIs are pushed on the expedition floor, the sensors from each container correlate their movement and report as a group to the devices carried by the company personnel. In this way, a missing or wrong item can be detected before arriving on the expedition floor. The same solution can be applied for checking the correct loading of the RTIs into the trailers that deliver the goods to the shops.

In a similar manner works the list generator for automatic packing [1]. Various order items are packed in a box and an invoice is generated. In order to find out where a certain item is, nodes with movement sensors can be attached to each good. When the box is moved around, the items inside can correlate their movement and decide that they form a group. When the objects are grouped, checking on the goods and packing lists can be done automatically.

Fig. 1. Transport and logistics scenario

2.2 Body Area Networks (BAN)

Wearable computing aims at supporting workers or people in everyday life by delivering context-aware services. One important aspect is the recognition of human activities, which can be inferred from sensor integrated into garments and objects people are interacting with. The usage of context information enables a more natural interaction between humans and computers, and can assist workers with complex tasks with just now relevant information. Examples include training unskilled workers for assembly tasks [13], monitoring the health and activity of patients [8], assisting firefighters engaged in rescue operations in unknown environments with poor visibility.

Clustering of nodes based on the movement of persons simplifies the selection of relevant sensors in the environment which can contribute to the recognition of the currently performed activity. The advantage is that the recognition processing can be kept within the cluster, which is important for environments where multiple people are present. This provides: (1) identification of the body wearing the sensors, (2) a better recognition stability when selecting sensors moving with a person for the recognition task, and (3) potentially a trusted network where private data can be communicated only to nodes within the cluster.

3 Related Work

Clustering in ad-hoc and sensor networks is an effective technique for achieving prolonged network lifetime and scalability [14]. Parameters such as the node degree, transmission power, battery level, processor load or degree of dynamics usually serve as metrics for choosing the optimal clustering structure. Nevertheless, recent initiatives address the problem of grouping based on application-specific attributes. For instance, Bouhafs et al. [3] propose a semantic clustering algorithm for energy-efficient routing in wireless sensor networks. Nodes join the clusters depending on whether they satisfy a particular query inserted in the network. The output of the algorithm is called a *semantic tree*, which allows for layered data aggregation.

The Smart-Its project [6] first introduces the notion of context sharing: two smart objects are associated by shaking them together. Using this explicit interaction between the two devices, an application-level connection can be established. For example, the two devices can authenticate using secret keys that are generated based on the movement data [11]. Siegemund [12] proposes a communication platform for smart objects that adapts the networking structure depending on the context. A clusterhead node decides which nodes can join the cluster, based on a similar symbolic location. Strohbach and Gellersen [15] propose an algorithm for grouping smart objects based on physical relationships. They use associations of the type "objects on the table" for constructing the clusters. A master node (the table) has to be able to detect the relationships for adding/deleting the nodes to/from the cluster. The solution described is suitable for static networks, where the master node is stationary for a long period of time. In our work, the network is dynamic, the context is permanently changing and every pair of nodes is capable of understanding the physical relationships, and thus the common context.

We now give two examples of shared-context recognition algorithms, where the practical results help us evaluate the clustering algorithm. First, Lester et al. [7] use accelerometer data to determine if two devices are carried by the same person. The authors use a coherence function to derive whether the two signals are correlated at a particular frequency. Second, in our previous work [10], we proposed a correlation algorithm which determine whether dynamic sensor nodes attached to vehicles on wheels move together. We use the real-world experimental results reported by both algorithms to analyse the performance of Tandem.

4 Algorithm Description

The goal of Tandem is to organize the nodes that share the same context, so that they can subsequently collaborate to provide a service. Tandem assumes that each node runs a *shared-context recognition algorithm*, which provides a number on a scale, representing the *confidence value* that two nodes are together. This algorithm can be for example a *coherence* function, which measures the extent to which two signals are linearly related at each frequency [7] (on a scale from 0 to 1), or the *correlation coefficient*, which indicates the strength and direction of a linear relationship [10] (on the scale from -1 to 1). Such an algorithm permanently evaluates the context, so that at each time step every node has an updated image of the current situation, reflected in a new set of confidence values (one for every neighbour). Since the context recognition algorithm has a certain accuracy, the perceived context determined by the confidence values may vary in time.

4.1 Requirements

Following the scenarios from Section 2, the nodes sharing the same context are within each-others transmission range, so we consider only one-hop clusters. The environment is dynamic, where the topology and context can change at any time. The requirements for the clustering algorithm are the following:

1. *Incorporate dynamics.* The clusters can merge or split, depending on the context changes. Nodes can join and leave the cluster if the topology or context changes. For example, in the BAN scenario, people can pick up and use different tools, and then return or exchange them with other people. In this case, the nodes attached to the tools have to join and leave the BAN clusters. Contextual and topological changes cannot be predicted, so the algorithm cannot assume a stable situation during cluster formation.

2. *Stability.* If there are no contextual or topological changes, the clustering structure has to be stable. Following the remark that every node periodically re-evaluates the shared context with its neighbours, the fluctuations of the confidence values may lead to unwanted changes in the cluster structure. Therefore, the cluster has to cope with these fluctuations in order to keep the structure as stable as possible. A solution to increase the stability is to analyse the similarity of the context over a larger time history. In this sense, a tradeoff has to be found between the spontaneity in accommodating changes and the desired cluster stability.

3. *Energy-efficiency.* The communication overhead should be kept to a minimum, for prolonging the lifetime of the wireless network.

4. *Facilitate service provisioning.* The clusters have to easily interact with the higher-layer applications and provide context-aware services to the user.

4.2   Cluster Formation Algorithm

For the sake of simplicity, we assume that a node is aware of its neighbours (every node within the range) and the one-hop communication is reliable (this can be achieved for example by using a simple stop-and-wait ARQ protocol).

Each node v periodically computes the confidence of sharing the same context with its neighbours. If the confidence with a neighbour u exceeds a certain threshold, then v considers that it shares the same context with u for the given time step. The final decision for sharing the same context with u is founded on the confidence values from a number of previous time steps, called the *time history* (see Section 5.1).

The fact that the perception of the shared context may vary from one node to another leads to nodes having different views of the cluster membership. To provide a consistent management of the membership list and cluster organization, a *clusterhead* or *root* node is dynamically elected among the nodes that share the same context. This also assists the service provisioning, by having a single point of interaction with the higher-layer applications (see Requirement 4). In order to allow merging of clusters and to facilitate the election process, the candidate clusterheads dynamically generate unique *priority numbers*, either based on the hardware addresses, the capabilities of the nodes based on the resources available or as a context-dependant measure. A regular node subscribes to the clusterhead with which it shares a common context and has the highest priority number.

Algorithm 1. Tandem - node v (events/actions)

Initialization:
1. $r(v) \leftarrow \bot$, $r(u) \leftarrow \bot$, $\forall u \in \Gamma(v)$

GetContext:
1. $r_0(v) \leftarrow r(v)$ // Store the root of v
2. Update $h(v, m), \forall m \in \Gamma(v)$ // Update the history
3. $M \leftarrow \{m \in \Gamma(v) \mid h(v, m) > h_{min}\}$ // Select the nodes sharing the same context with v
4. **if** $M \neq \emptyset$ **then**
5. $M_0 \leftarrow \{m \in M \mid r(m) = m\}$ // Select from M the nodes that are clusterheads
6.   **if** $M_0 \neq \emptyset$ **then**
7.   Choose $u \in M_0$ such that $pn(u) = max\{pn(m) \mid m \in M_0\}$
8.   **if** $(r(v) = \bot) \vee (r(v) \notin M_0) \vee (r(v) = v \wedge pn(v) < pn(u))$ **then**
9. $r(v) \leftarrow u$ // Choose u as the root of v
10.   **end if**
11.   **else if** $r(v) \neq v$ **then**
12. $M_1 \leftarrow \{m \in M \mid r(m) \neq \bot\}$ // Select from M the nodes that have a valid root
13.   **if** $M_1 \neq \emptyset$ **then**
14. $r(v) \leftarrow \bot$
15.   **else**
16. $r(v) \leftarrow v$ // Become root
17.   Generate $pn(v) > 0$
18.   **end if**
19.   **end if**
20. **else**
21. $r(v) \leftarrow \bot$ // v is unassigned
22. **end if**
23. **if** $r_0(v) \neq r(v)$ **then**
24.   Send *SetRoot* $(v, r(v), pn(v))$ to neighbours // Announce root change
25. **end if**

We use the following notation:

- V is the the set of nodes in the network.
- n is the number of nodes in the network, $n = |V|$.
- $r(v)$ is the root or clusterhead of node v.
- $pn(v)$ is the priority number of node v.
- $\Gamma(v)$ is the neighbourhood of node v.
- $h(v, u)$ represents the number of times v and u are sharing a common context over a total time history of H steps.
- h_{min} is the minimum amount of times when two nodes are sharing a common context, such that they can safely be considered part of the same cluster.

The algorithm constructs a set of one-hop clusters, based on the context information shared by the nodes. A node v can be: (1) *unassigned*, where v is not part of any cluster, (2) *root*, where v is clusterhead, or (3) *assigned*, where v is assigned to a cluster where the root node is one of its neighbours.

Algorithm 1 gives the detailed description of the cluster formation and update of knowledge among neighbouring nodes. Every node has the following information about its neighbours: the root, the priority number and whether it shares a common context for a specified time history. Let v be an arbitrary node in the network. At each time step, node v changes or chooses its root node in the following cases: (1) v is unassigned, (2) v does not share a common context with its root, (3) the root of v is no longer a root or (4) v is root and there is another neighbour root, sharing the same context with v, that has a higher priority

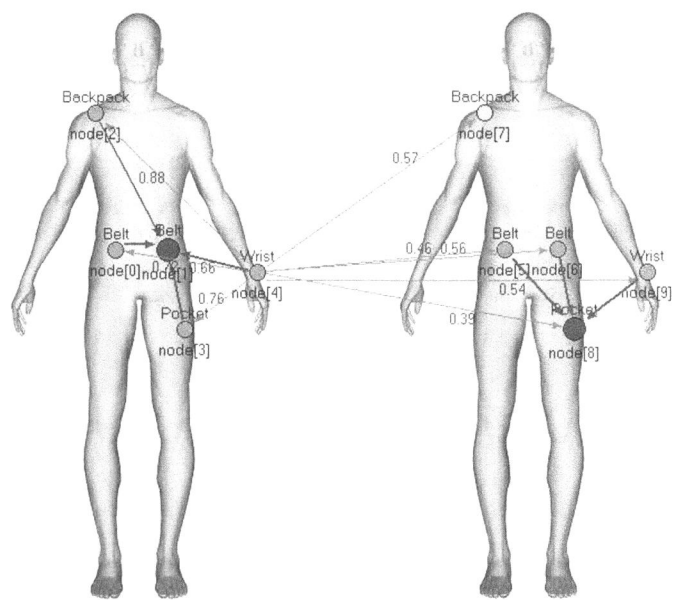

Fig. 2. Graphical simulation of the clustering algorithm on BANs

number. In any of these cases, v chooses as root node the neighbour root u with which it shares a common context and which has the highest priority number. If such a neighbour does not exist, v competes for clusterhead or becomes unassigned. The decision is based on the current status of the neighbours and tries to minimize the effect of the following erroneous situation: due to context fluctuations, an assigned node v may loose its root node and cannot join another cluster because none of its neighbours is root. Therefore, v may become root, form a new cluster and attract other nodes in that cluster. To avoid this undesirable outcome, a node declares itself root only if all its neighbours with which it shares a common context are unassigned. If there exists at least one neighbour u with which v shares a common context and u has a valid root node, then v becomes unassigned.

Node v announces the changes in choosing the root node by sending a local broadcast message *SetRoot* to its neighbours. In case of topological changes, this message is also used to announce the new neighbours of the current structure. The algorithm thus allows cluster merging or splitting, which meets the Requirement 1 from Section 4.1. We make the observation that there is no additional strain in terms of communication overhead on the clusterheads compared to the other nodes.

Let us consider the example from Figure 2. A BAN is associated with each person, consisting of five nodes placed in various locations: backpack, belt, pocket and wrist. The clustering structure is seen from the perspective of node 4, which is attached to the wrist of the left person. The clusterheads are represented with big circles (nodes 1 and 8). The dark-coloured arrows indicate the assignment of the other nodes to the current clusterheads. Node 7 is unassigned, as the shared-context recognition algorithm did not associate this node with any of the neighbouring clusterheads at the previous time step. The light-coloured arrows show the confidence values computed by node 4 at the current time step in this particular case. The confidence values for the nodes on the same body with node 4 range between 0.66 and 0.88, while for the other body they lie between 0.39 and 0.57. Because the confidence of sharing the same context with the root node 1 is 0.66 and above the threshold, node 4 keeps the current root. Otherwise, it would have become unassigned (node 4 has some neighbours with the same context, having a valid root node), or assigned to the other cluster, if the confidence value for the neighbouring root node 8 was higher than the threshold.

5 Cluster Stability Analysis

Several algorithms for context sharing have been proposed in the literature, using various sensors and providing different accuracies (see Section 3). However, none of them gives a measure of the overall accuracy of the system, when multiple nodes sharing different contexts come together. We would like to analyse the cluster stability from both the theoretical point of view, by giving average approximation, upper and lower bounds, and through simulations. In addition, we are interested in the tradeoff between the time history necessary to achieve a certain stability and the responsiveness of the clustering algorithm. First, we compute the probabilities of correctly assessing the context, given the distribution of the confidence values. Second, we model the algorithm using Markov chains and we derive an approximation for the proportion of time the clustering structure is in a correct state.

5.1 Determination of Common Context

In this section, we give an example of how the probabilities of correct detection of the shared context can be computed.

Let v be a node in the network and u a neighbour of v. If v does not share the same context with u (e.g. they represent sensor nodes attached to different persons), we model the confidence value computed by the shared-context recognition algorithm with the random variable $X_1(v, u)$. If v shares the same context with u (e.g. they are attached to the same person), we model the confidence value as a random variable $X_2(v, u)$. We take the distribution encountered during the experiments as the reference Probability Density Function (PDF): we associate the random variables $X_1(v, u)$ with the PDF φ_1 and the corresponding Cumulative Distribution Function (CDF) Φ_1. Similarly, we associate the random variables $X_2(v, u)$ with the PDF φ_2 and CDF Φ_2.

Node v selects the subset of its neighbours with which it shares a common context based on a threshold value T. We choose T as the intersection point of the two PDFs φ_1 and φ_2, which minimizes the sum of probabilities of an incorrect determination. We denote p as the probability of a correct detection of a *common*

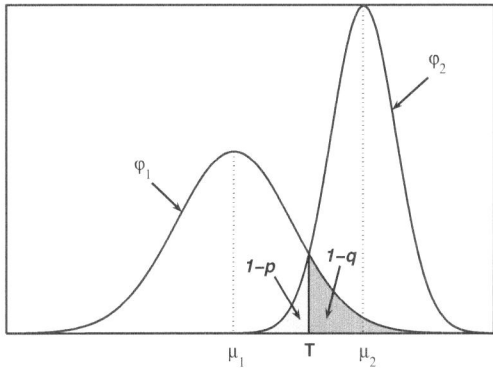

Fig. 3. The calculation of the threshold value T and the probabilities p and q

context and q as the probability of a correct detection of *different* contexts. The probabilities p and q are computed in the following way (see Figure 3):

$$p = 1 - \Phi_2(T), \quad q = \Phi_1(T) \tag{1}$$

We compute the threshold value for the case where the distributions are normal, which is valid for the applications described in Section 2 (see the experimental data reported by [7,10]). Let us consider two normal distributions, $\varphi_1(\mu_1, \sigma_1)$ and $\varphi_2(\mu_2, \sigma_2)$. The intersection point of φ_1 and φ_2 which lies between μ_1 and μ_2 is the following:

$$T = \frac{\mu_1 \sigma_2^2 - \mu_2 \sigma_1^2 + \sigma_1 \sigma_2 \sqrt{(\mu_1 - \mu_2)^2 + 2(\sigma_2^2 - \sigma_1^2)ln(\sigma_2/\sigma_1)}}{\sigma_2^2 - \sigma_1^2} \tag{2}$$

Using Eq. 1 and 2, it is straightforward to compute p and q, knowing the characteristics of φ_1 and φ_2. We are now interested in how these probabilities change if we involve the time history in the decision process. The probability p_h of the correct detection that two nodes share a common context for a minimum time history h_{min} out of a total of H time steps is given by the CDF of the binomial distribution:

$$p_h(h_{min}, H) = \sum_{k=h_{min}}^{H} \binom{H}{k} p^k (1-p)^{H-k} \tag{3}$$

Similarly, the probability q_h of the correct detection of different contexts for a minimum time history h_{min} out of a total of H time steps is:

$$q_h(h_{min}, H) = \sum_{k=h_{min}}^{H} \binom{H}{k} q^k (1-q)^{H-k} \tag{4}$$

We have therefore $p = p_h(1, 1)$ and $q = q_h(1, 1)$.

5.2   Modelling with Markov Chains

We approximate the behaviour of the algorithm with a Markov chain, which allows us to estimate the global probability of having a correct cluster. We stress on the difference between a *time step* and a *Markov chain step*. A time step is related to the periodic update of the context information by the shared-context recognition algorithm which runs on every node. For improving the probabilities of a correct detection of a shared context, the algorithm looks over a time history H, composed of a number of time steps (see Section 5.1). A Markov chain step is the "memoryless" transition from one state to another, which happens on intervals equal to the total time history H.

We define a *group* G as the collection of nodes that share the same context in reality. We define a *cluster* C as the collection of nodes which have the same root node (as a result of Agorithm 1). The goal of the clustering algorithm is that for any group of nodes G, there exists a cluster with the root $r_0 \in G$ such that $\forall v \in V, r(v) = r_0 \Leftrightarrow v \in G$. Taking the example from Figure 2, we have two groups: $G_1 = \{0,1,2,3,4\}$, $G_2 = \{5,6,7,8,9\}$ and two clusters: $C_1 = \{0,1,2,3,4\}$ with root node 1, $C_2 = \{5,6,8,9\}$ with root node 8.

We define the following states for a cluster C:

1. *Correct:* The cluster has exactly one root node from the group and all the members of the group are part of the cluster[1].
2. *Has root:* At least one group member is root.
3. *No root:* None of the group members is root.
4. *Election:* After reaching state 3, members of the group start an election process for choosing the root node.

For example, cluster C_1 from Figure 2 is *Correct*, while C_2 is in the state *Has root*, since node 7 is unassigned.

The transition matrix that determines the Markov chain is the following:

$$P = \begin{pmatrix} p_{11} & p_{12} & p_{13} & p_{14} \\ p_{21} & p_{22} & p_{23} & p_{24} \\ p_{31} & p_{32} & p_{33} & p_{34} \\ p_{41} & p_{42} & p_{43} & p_{44} \end{pmatrix}$$

Let $m \geq 0$ be the number of root nodes with higher priority than the current root and $k \geq 1$ the number of nodes in the cluster. If the cluster has a root, let r_0 be the root node. The probabilities p_{ij} are evaluated in a worst case scenario, by minimizing the chance to get in the *Correct* state.

The conditions which determine the probabilities p_{ij} are the following:

– p_{11}: (a) r_0 remains root in the next step, as r_0 does not share the same context with other root nodes with higher priority, and (b) all the nodes in the group share the same context with r_0.

---

[1] We intentionally take $G \subseteq C$. The nodes from $C \setminus G$ are part of other groups, which have the corresponding clusters in an incorrect state.

Table 1. Transition probabilities p_{ij}

Probability	Value	Probability	Value
p_{11}	$q^m p^{k-1}$	p_{21}	$q^m p^{k-1} q^{m(k-1)}$
p_{12}	$q^m(1 - p^{k-1})$	p_{22}	$q^m(1 - p^{k-1} q^{m(k-1)})$
p_{13}	$1 - q^m$	p_{23}	$1 - q^m$
p_{14}	0	p_{24}	0
p_{31}	0	p_{41}	0
p_{32}	0	p_{42}	$q^{mk}(1 - (1-p)^{k(k-1)})$
p_{33}	0	p_{43}	0
p_{34}	1	p_{44}	$1 - q^{mk}(1 - (1-p)^{k(k-1)})$

- p_{12}: (a) r_0 remains root in the next step, and (b) there exists at least one node in the group that does not share the same context with r_0.
- p_{13}: r_0 shares the same context with a root node with higher priority, so that it gives up its role and joins another cluster.
- p_{21}: (a) r_0 remains root in the next step, (b) all the nodes in the group do not share the same context with other root nodes with higher priority and (c) all the nodes in the group share the same context with r_0.
- p_{23}: r_0 shares the same context with a root node with higher priority, so that it gives up its role and joins another cluster.
- p_{34}: from state *No root* the system goes at the next step to state *Election*.
- p_{42}: (a) all the nodes in the group do not share the same context with any root node with higher priority, and (b) there are at least two nodes in the group that share the same context.

Table 1 gives the computed probabilities for each transition of the Markov chain. We notice that the *Correct* state can be reached only from the *Has root* state. If $m > 0$, the probability p_{21} is minimized, so that in the stationary distribution of the Markov chain, the probability to be in the *Correct* state is lower than the real probability. Calculating the fixed row vector of the Markov chain yields the following result:

$$p_1(m, k, q, p) = \frac{p_{21} p_{42}}{(1 + p_{21} - p_{11})(p_{42} + p_{42} p_{13} + p_{13})} \tag{5}$$

We define the *cluster stability* P_S as the probability of a cluster to be in the *Correct* state. Given that there are c clusters in the network, we have the following lower and upper bounds:

$$p_1(c - 1, k, q, p) \le P_S \le p_1(0, k, q, p) \tag{6}$$

An estimation of the average case is given by:

$$P_S \approx p_1\left(\frac{c-1}{2}, k, q, p\right) \tag{7}$$

6 Results

We analyse the performance of Tandem by running a series of simulations in the OMNeT++ [16] simulation environment. As the cluster formation and stability is affected only by the nodes in the one-hop neighbourhood, we simulate a network where the nodes are attached to different mobile objects or people, and where there is a wireless link between any two nodes. We focus on a mobile scenario with clustered nodes moving around and passing each other, and we analyse how the cluster stability changes when we vary the number of groups. First, we compare Tandem with a traditional clustering method and point out the factors that determine the improved behaviour of our algorithm. Second, based on the experimental results from previous work, we evaluate Tandem with respect to cluster stability and communication overhead.

6.1 Comparison to Traditional Clustering

Compared to traditional clustering methods, conceived to run only at the network layer, Tandem groups the nodes with similar semantic properties, namely the shared context. The uncertainty in determining the shared context and the resulting variation in time lead to a high instability of the clustering structure. Therefore, an algorithm that generates stable clusters distinguishes itself from generic algorithms designed for ad-hoc networks by paying attention especially on how to minimize the effect of the variation of confidence values.

To illustrate this difference, we present an alternative clustering algorithm following the idea of DMAC [2], denoted by DMAC*. In DMAC*, each node v takes into account every change in the confidence value. DMAC* has the following two main differences compared to Tandem:

1. If v detects in its neighbourhood a clusterhead r_1 with a higher priority value than the current clusterhead r_0, and v shares a common context with r_1, then v directly chooses r_1 as clusterhead.
2. If none of its neighbours sharing a common context is clusterhead, then v declares itself a clusterhead.

The first condition may lead to unwanted effects, where a single error in the detection of different contexts between v and r_1 changes the cluster membership of v. Likewise, an error in the detection of the common context between v and r_0 results through the second condition into the erroneous situation of having two clusterheads for the same group.

We show in Figure 4 a comparison between Tandem and DMAC* from the stability and energy efficiency points of view. Each group is composed of 10 nodes and we vary the number of groups between 2 and 10. We take $p = q = 0.99$. The energy spent by Tandem on communication is proportional to the number of *SetRoot* messages sent by every node in one time step. Compared with the results obtained for DMAC*, Tandem improves the stability up to 18% on an absolute scale and the energy efficiency up to 32% on a relative scale, thus satisfying Requirements 2 and 3 from Section 4.1.

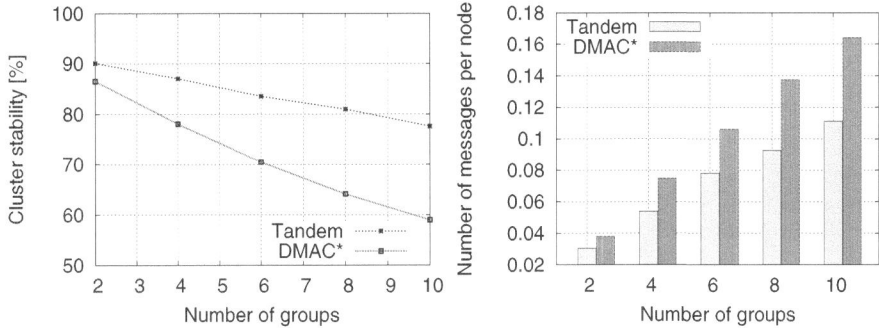

Fig. 4. Comparison between Tandem and a clustering alternative

6.2 Evaluation of Tandem

A typical example of context sharing is the similarity of movement, which we analyse in this section using experimental results corresponding to the scenarios described in Section 2. In general, the movement information is extracted from accelerometers. Simpler sensors such as tilt switches can be also used, but with less accurate results. We have the following two concrete examples of wireless objects moving together:

1. **RTI** - wireless sensor nodes used in a transport scenario, which correlate their movement pattern; the shared-context recognition algorithm computes a correlation coefficient between the movement data of two neighbouring nodes, which is extracted using both tilt switches and accelerometers [10].
2. **BAN** - smart devices that decide whether they are carried by the same person; the shared-context recognition algorithm uses a coherence function of the movement data provided by accelerometers [7].

Table 2 shows the characteristics of the normal distributions derived from the concrete experiments conducted in both application examples, together with the computed threshold from Eq. 2 and the probabilities p and q. Contrary to the RTI scenario, where the nodes moving together experience exactly the same movement pattern, in the BAN scenario different parts of the body are engaged in different types of movements during walking. For a realistic evaluation, we choose the worse case experimental results from the BAN scenario, where the nodes are attached to the pocket and wrist of the subjects (Pocket/Wrist trial [7]).

Table 2. Statistics from the experiments and computed probabilities

Application	μ_1	σ_1	μ_2	σ_2	T	p	q
RTI - tilt switch	0.641	0.087	-0.017	0.249	0.438	0.9902	0.9662
RTI - accelerometer	0.817	0.106	0.009	0.124	0.442	0.9998	0.9998
BAN - Pocket/Wrist	0.757	0.065	0.519	0.069	0.640	0.9636	0.9607

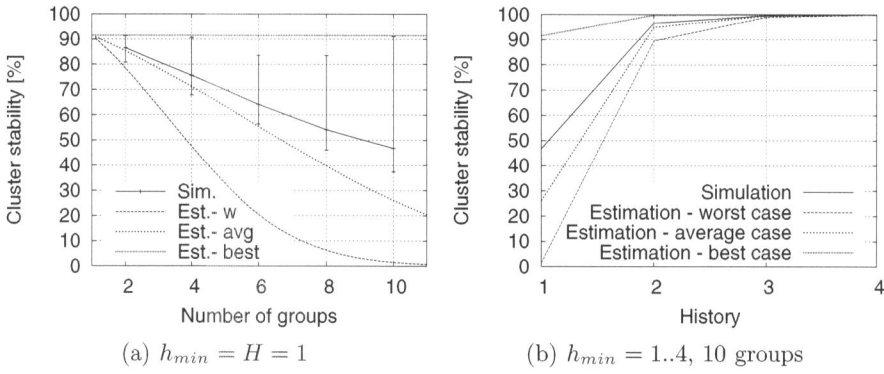

(a) $h_{min} = H = 1$ (b) $h_{min} = 1..4$, 10 groups

Fig. 5. Cluster stability in the RTI with tilt switches scenario

The scalability analysis of the movement correlation method proposed for the RTI scenario indicate a maximum network density of 100 nodes [10], imposed by the shared wireless medium. Because the same periodic transmission of the movement data is needed for the BAN scenario [7], we use in our simulations the same maximum network density for both applications. We have 10 nodes in each group moving together and interacting with other groups, and we vary the number of groups between 2 and 10 and also the time history. We recall from Section 5.2 that the cluster stability is the probability that the cluster is in the *Correct* state. We represent the cluster stability in percentage, for the following cases: (1) average simulation results, (2) estimation of the worst case, derived from Eq. 6, (3) estimation of the average case, derived from Eq. 7, and (4) estimation of the best case, derived from Eq. 6.

For each point on the simulation plots we run up to 10 simulations of $10^4 - 10^5$ time steps. In order to study the influence of the history size, we take $H = 2h_{min} - 1$ and we vary h_{min} from 1 to 4.

RTI with tilt switches scenario. Figure 5(a) shows the cluster stability depending on the number of groups present in the network, given that the time history is $h_{min} = 1$. The error bars represent the absolute minimum and maximum stability recorded during the simulations. We notice that the results respect the upper and lower bounds calculated theoretically. The estimation of the average case is close to the simulations for a small number of groups. However, increasing the number of groups decreases the precision of the approximation, due to the minimization of the transition probabilities to get in the *Correct* state (see Section 5.2).

Figure 5(b) shows the cluster stability depending on the time history, for a network composed of 10 groups. We notice that increasing the time history considerably improves the cluster stability and the theoretical approximation. For a time history $h_{min} = 4$ ($H = 7$), a stability of 99.92 is achieved, while the lower bound is 99.89. Therefore, for achieving a stability close to 100%, the necessary delay is $H \times 16 = 112$ seconds (the size of the data sequence is 16 seconds [10]).

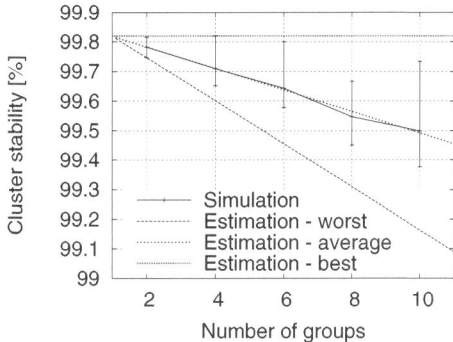

Fig. 6. Cluster stability in the RTI with accelerometers scenario ($h_{min} = H = 1$)

RTI with accelerometers scenario. The solution using accelerometers is more reliable, resulting in higher probabilities for a correct detection of the context (see Table 2) and consequently, higher cluster stability. Figure 6 shows the cluster stability depending on the number of groups present in the network, given that the time history is $h_{min} = 1$. We also represent the error bars for the absolute minimum and maximum values. We notice the high stability obtained even for large number of groups (99.5 for 10 groups). Due to the fact that the clusters stay in the *Correct* state for most of the time, the approximations are close to the simulation results. For this scenario, a high cluster stability can be achieved even considering the time history 1, reaching a delay of only 16 seconds.

BAN scenario. Figure 7(a) shows the cluster stability in the BAN scenario, depending on the number of groups, given that the time history is $h_{min} = 1$. Similarly with the two scenarios presented above, we notice that the results respect the upper and lower bounds calculated theoretically. The average stability is lower than in the previous cases, with a maximum of 67% and less than 50% for a network composed of more than 6 groups.

Figure 7(b) shows the cluster stability depending on the time history, for a network composed of 10 groups. The time history significantly improves the cluster stability: for the time history $h_{min} = 4$ ($H = 7$), the stability is 99.84 and the lower bound is 99.74. For achieving this, the delay is $H \times 8 = 56$ seconds (the window size is 8 seconds [7]).

Communication overhead. The communication overhead is induced by the *SetRoot* message, sent by every node when the current root node changes. In case the lower networking layers use a periodic message in order to maintain for example the synchronization (e.g. in case of a TDMA MAC protocol), Tandem can use this heartbeat to piggyback the small piece of information present in the *SetRoot* message (root address and priority number). Assuming that there is no such periodic message, we count the average number of *SetRoot* messages sent by each node in one step of simulation on average.

Figure 8(a) shows the number of messages on a logarithmic scale, depending on the number of groups. The results correspond to each of the three scenarios,

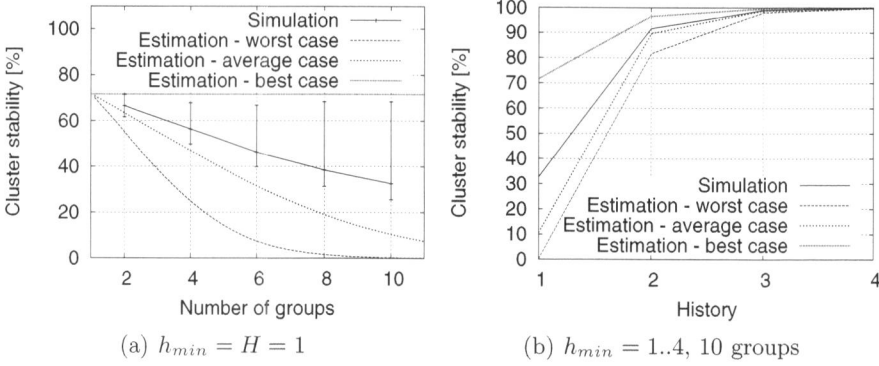

(a) $h_{min} = H = 1$ (b) $h_{min} = 1..4$, 10 groups

Fig. 7. Cluster stability in the BAN scenario

where the time history is 1. We notice that the more stable the structure is, the less communication overhead is needed. For the RTI with accelerometers scenario, less than 1 message is sent per node in 100 time steps, even for a large number of groups. The overhead is increasing as the number of groups increases, due to the diminishing cluster stability.

Figure 8(b) shows the number of messages depending on the time history, for the RTI with tilt switches and BAN scenarios. Increasing the time history improves the stability and thus reduces the communication overhead. For the time history 4, the overhead is less than 10^{-3} messages per node.

7   Discussion and Conclusions

We presented Tandem, a context-aware method for spontaneous clustering of wireless sensor nodes. The algorithm allows reclustering in case of topological or contextual changes. By analysing the similarity of the context over a time history, Tandem tries to achieve stable clusters. We approximate the behaviour of the algorithm using a Markov chain model and we analyse the cluster stability theoretically and through simulations, using experimental results reported from real field tests. The analysis gives the possibility to theoretically estimate the stability of the structure and the responsiveness of the algorithm. Computing the worse case stability via the Markov chain approximation, we can deduce the time history necessary to achieve stable clusters. In what follows, we discuss the main advantages and limitations of the proposed clustering method.

Advantages

- *Responsiveness.* The clustering structure reacts quickly to topological and contextual changes: nodes decide based only on the current situation of their neighbourhood, without the need of any negotiation with other parties.
- *Small-scale experiment required.* For computing the probabilities p and q that can be used to estimate the cluster stability, only a small-scale reproducible

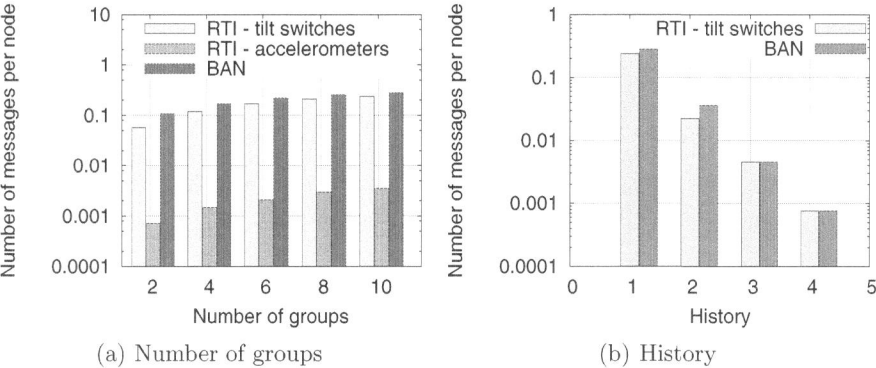

Fig. 8. Average number of *SetRoot* messages sent per node in 100 steps

experiment is required. For example, two nodes moving together and another two moving separately are enough to generate the statistical distributions of the confidence values.

- *Delay estimation.* By deducing the time history required to achieve a certain stability, the delay in accommodating the topological or contextual changes can be easily estimated.

Limitations

- *Rough approximation for many groups.* As we notice from Figures 5(a) and 7(a), the difference between the approximation that we derive using Markov chains and the real situation is increasing with the number of groups. However, the model offers a good approximation in case of highly accurate context detection methods (see Figure 6). Therefore, the approximation can be successfully used for deducing the minimum time history for a cluster stability close to 100%.
- *Multihop clusters.* The method that we propose is valid only for one-hop clusters, which is justified taking into account the scenarios from Section 2. Nevertheless, other applications may require multihop clusters, even several layers of clustering. For example, groups of people skiing together, forming multihop clusters, where each person is wearing a BAN that is a one-hop cluster. The algorithm can be easily extended to accommodate multihop clusters: instead of choosing directly the clusterhead node, every node selects a parent and thus joins the cluster associated with the parent node.

For future work, we intend to extend the algorithm for multihop clusters and to investigate the cluster stability in this case. We also plan a series of experiments involving clustering of nodes on body area networks. Subsequent to clustering, a task allocation mechanism distributes various tasks to the nodes which are part of the cluster, depending on their capabilities. The final goal is to have a distributed activity recognition algorithm running on a dynamic, context-aware BAN.

References

1. Antifakos, S., Schiele, B., Holmquist, L.E.: Grouping mechanisms for smart objects based on implicit interaction and context proximity. In: Dey, A.K., Schmidt, A., McCarthy, J.F. (eds.) UbiComp 2003. LNCS, vol. 2864, pp. 207–208. Springer, Heidelberg (2003)
2. Basagni, S.: Distributed clustering for ad hoc networks. In: ISPAN 1999, pp. 310–315. IEEE Computer Society, Los Alamitos (1999)
3. Bouhafs, F., Merabti, M., Mokhtar, H.: A semantic clustering routing protocol for wireless sensor networks. In: Consumer Communications and Networking Conference, pp. 351–355. IEEE Computer Society, Los Alamitos (2006)
4. Gluhak, A., Presser, M., Zhu, L., Esfandiyari, S., Kupschick, S.: Towards mood based mobile services and applications. In: Kortuem, G., Finney, J., Lea, R., Sundramoorthy, V. (eds.) EuroSSC 2007. LNCS, vol. 4793, pp. 159–174. Springer, Heidelberg (2007)
5. Gu, L., Jia, D., Vicaire, P., Yan, T., Luo, L., Tirumala, A., Cao, Q., He, T., Stankovic, J.A., Abdelzaher, T., Krogh, B.H.: Lightweight detection and classification for wireless sensor networks in realistic environments. In: SenSys, pp. 205–217. ACM Press, New York (2005)
6. Holmquist, L.E., Mattern, F., Schiele, B., Alahuhta, P., Beigl, M., Gellersen, H.-W.: Smart-its friends: A technique for users to easily establish connections between smart artefacts. In: Abowd, G.D., Brumitt, B., Shafer, S. (eds.) UbiComp 2001. LNCS, vol. 2201, pp. 116–122. Springer, Heidelberg (2001)
7. Lester, J., Hannaford, B., Borriello, G.: Are You with Me? - using accelerometers to determine if two devices are carried by the same person. In: Ferscha, A., Mattern, F. (eds.) PERVASIVE 2004. LNCS, vol. 3001, pp. 33–50. Springer, Heidelberg (2004)
8. Lorincz, K., Malan, D.J., Fulford-Jones, T.R.F., Nawoj, A., Clavel, A., Shnayder, V., Mainland, G., Welsh, M., Moulton, S.: Sensor networks for emergency response: Challenges and opportunities. IEEE Pervasive Computing 03(4), 16–23 (2004)
9. Marin-Perianu, M., Havinga, P.J.M.: D-FLER: A distributed fuzzy logic engine for rule-based wireless sensor networks. In: Ichikawa, H., Cho, W.-D., Satoh, I., Youn, H.Y. (eds.) UCS 2007. LNCS, vol. 4836, pp. 86–101. Springer, Heidelberg (2007)
10. Marin-Perianu, R.S., Marin-Perianu, M., Havinga, P.J.M., Scholten, J.: Movement-based group awareness with wireless sensor networks. In: LaMarca, A., Langheinrich, M., Truong, K.N. (eds.) Pervasive 2007. LNCS, vol. 4480, pp. 298–315. Springer, Heidelberg (2007)
11. Mayrhofer, R., Gellersen, H.: Shake well before use: Authentication based on accelerometer data. In: LaMarca, A., Langheinrich, M., Truong, K.N. (eds.) Pervasive 2007. LNCS, vol. 4480, pp. 144–161. Springer, Heidelberg (2007)
12. Siegemund, F.: A context-aware communication platform for smart objects. In: Ferscha, A., Mattern, F. (eds.) PERVASIVE 2004. LNCS, vol. 3001, pp. 69–86. Springer, Heidelberg (2004)
13. Stiefmeier, T., Lombriser, C., Roggen, D., Junker, H., Ogris, G., Tröster, G.: Event-based activity tracking in work environments. In: Proceedings of the 3rd International Forum on Applied Wearable Computing (IFAWC) (March 2006)

14. Stojmenovic, I., Seddigh, M., Zunic, J.: Dominating sets and neighbor elimination-based broadcasting algorithms in wireless networks. IEEE Transactions on Parallel and Distributed Systems 13(1), 14–25 (2002)
15. Strohbach, M., Gellersen, H.: Smart clustering - networking smart objects based on their physical relationships. In: Proceedings of the 5th IEEE International Workshop on Networked Appliances, pp. 151–155. IEEE Computer Society, Los Alamitos (2002)
16. Varga, A.: The omnet++ discrete event simulation system. In: ESM 2001, Prague, Czech Republic (June 2001)
17. Wang, J., Chen, G., Kotz, D.: A sensor fusion approach for meeting detection. In: MobiSys 2004 Workshop on Context Awareness (June 2004)

Stream Feeds - An Abstraction for the World Wide Sensor Web

Robert Dickerson, Jiakang Lu, Jian Lu, and Kamin Whitehouse

Department of Computer Science
University of Virginia
rfdickerson,jklu,jl3aq,whitehouse@cs.virginia.edu

Abstract. RFIDs, cell phones, and sensor nodes produce streams of sensor data that help computers monitor, react to, and affect the changing status of the physical world. Our goal in this paper is to allow these data streams to be first-class citizens on the World Wide Web. We present a new Web primitive called *stream feeds* that extend traditional XML feeds such as blogs and Podcasts to accommodate the large size, high frequency, and real-time nature of sensor streams. We demonstrate that our extensions improve the scalability and efficiency over the traditional model for Web feeds such as blogs and Podcasts, particularly when feeds are being used for in-network data fusion.

1 Introduction

The *Internet of Things* (IOT) refers to the vision in which computers on the Internet are able to monitor, react to, and affect the changing status of objects in the physical world. The two major forces driving the IOT are (i) the widespread adoption of RFIDs that allow physical objects to be uniquely identified, and (ii) rapid proliferation of portable devices like cellphones and wireless sensor nodes that push computation, communication, and sensing deeper into the physical world. Both of these new technologies produce streams of data that must be stored and processed.

In this paper, we address the issue of connecting these data streams with the World Wide Web. Although the terms *Web* and *Internet* are often used interchangeably, they are quite distinct: the Internet is a set of computers connected by the IP protocol, while the Web is a set of data resources that are connected by hyperlinks. Our goal in this paper is to allow data streams to be first-class citizens on the Web: every data stream has a URL, is manipulated through the HTTP protocol, and is hyperlinked to other objects on the Web. This would allow data streams to easily interoperate with other web applications and enable crawling and indexing by search engines. In addition, streams will inherit other Web features such as authentication and encryption.

There are currently several abstractions for stream-like objects on the Web. For example, a Web *feed* is an XML document that contains a dynamically changing sequence of content items, such as blog entries or news headlines. Feeds are transported as a single file using the *pull* mechanism provided by HTTP

C. Floerkemeier et al. (Eds.): IOT 2008, LNCS 4952, pp. 360–375, 2008.

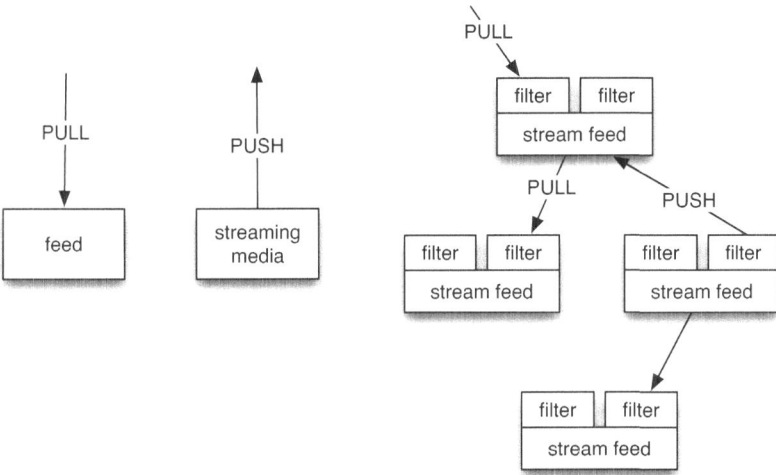

Fig. 1. Web feeds (left) only support a PULL abstraction while streaming media (middle) only supports a PUSH abstraction. The stream feed abstraction (right) must support both types of data transport, must provide server-side filtering, and must support data fusion trees.

requests. On the other hand, multimedia streams such as Internet radio stations are transported in real-time as the content is generated, using a *push* mechanism provided by protocols like the Real Time Streaming Protocol (RTSP). This protocol allows operations like "play" and "pause" on streaming media objects.

The problem we address in this paper is that sensor streams do not fit into any of the stream abstractions currently available on the Web. Clients will sometimes want to *pull* the historical data of a sensor stream, similar to a Web feed, but other clients will want real-time updates, similar to a media stream. Furthermore, many clients will only want a small part of the total stream. Finally, unlike Web feeds or media feeds, sensor streams will commonly be *fused* to create new sensor streams.

We hypothesize that we could provide an abstraction that meets the special needs of sensor streams by adding a few simple extensions to Web feeds. To test this hypothesis, we implement a new Web abstraction called the *stream feed*. Just like normal feeds, stream feeds can be accessed through a URL and are accessed through a Web server over HTTP. As such, they can be both the source and the target of hyperlinks. They are dynamic objects that are retrieved using non-streaming transport, so updates to the content can only be observed by retrieving the entire object again. However, stream feeds also support two extensions not supported by normal feeds: server-side filtering and streaming transport. These extensions allow stream feeds to be very large objects that change with very high frequencies and are served to clients with real-time updates.

Stream feeds can be *fused*, processed and filtered to create new stream feeds, and this process can be repeated to create a *fusion tree*. For example, data from multiple sensors on a highway can be combined to create new streams that indicate vehicle detections on the road, and these streams can in turn be fused to create a new stream with traffic conditions over time. This is different from the *aggregation* of other feeds in which, for example, news headlines from multiple sources are combined into a single feed. We demonstrate that the traditional model of serving feeds does not scale well with the number of clients and the size of the feeds, particularly with deep fusion trees, and that our stream feed extensions improve scalability, latency, and the total amount of bandwidth consumed. These results support the conclusion that stream feeds can serve as the basis for a World Wide Sensor Web in which sensor data from around the world can be shared and processed within a single global framework.

2   A Motivating Application

MetroNet is a sensor system that we are currently designing to measure foot traffic near store fronts using commodity sensors such as motion sensors and thermal cameras in downtown Charlottesville where shops are densely packed and there is high pedestrian traffic at points in the day. This simple system will be used by shops to identify the effectiveness of window displays, signs, or other techniques designed to attract foot traffic into the shop. It will measure (i) how many people walk by the shop (ii) how many people enter the shop. The ratio of these values is extremely valuable to retail stores. In addition, many stores consider the *conversion rate*, the store's number of retail transactions during a certain time period compared to its traffic, as good indicator of its performance. Measuring foot traffic can be helpful when evaluating proper labor usage, advertising and promotional programs, prime selling hours, visual merchandising strategies, future opportunities or challenges, new store concepts, store location analysis, and merchandise assortment and stock levels.

The MetroNet sensors provide valuable data to the shopkeepers, but there are also a number of compelling reasons to *share* the data. For example, shopkeepers could use data from neighboring shops to normalize their own data, allowing them to differentiate the effect of a bad storefront display from the effect of bad weather, vehicular traffic, and downtown events such as fairs or concerts. Once in place, the sensors and their data would also be extremely valuable to the residents and planners in the city. For example, potential residents that would like to buy a new house could query for aggregate activity levels in each part of the city to find a location that is close to the city's night life. This type of query could be augmented in a subsequent phase by collecting supplemental data such as average levels of noise, sunlight, humidity, or air pollutants. Potential business owners would have empirical data upon which to base the value of commercial real estate, and city planners could gear zoning and traffic decisions to increase business and quality of life for residents. In fact, one motivation for the MetroNet testbed is to provide empirical data for an ongoing debate in Charlottesville about the effect on business of vehicular traffic crossing the downtown pedestrian zone.

Because MetroNet data can be useful to many parties, it could be used to form *fusion trees*. For example, the shopkeepers might publish streams of raw sensor data, but this sensor data may be *fused* by third parties to produce the counts of people walking in front of and into the shops. These third parties may then publish this information on their own servers as streams of people detection events. These new streams may then be fused again to produce streams with commercial real estate values or real-time estimates of overall pedestrian activity levels. A client query at the top of a tree would cause a sequence of cascading queries that consume network bandwidth and increase the latency of the query response. Stream feeds must be able to support queries over fusions trees quickly and efficiently.

There are many different requirements under which MetroNet data and the data streams that are created from it could be shared. Some data consumers, such as city planners who are moving a road to minimize effects on pedestrian traffic, may want the sensor streams as a bundle of historical data. Sometimes, the sensor stream objects could be very large (on the order of Terabytes), and so the city planners would like to query for only a small part of the entire stream, such as the data collected in the past year. Other data consumers, such as those looking at a web page with current estimates of pedestrian activity, may require real-time updates to the sensor streams. Furthermore, some data streams could be completely private, shared with other shopkeepers, or shared only with storefronts that are owned by the same company. This type of sharing may require authentication and encryption. Data could also be shared publicly, in which case the data consumer must be able to find the data through some sort of index, such as a search engine.

This case study demonstrates that stream feeds must accommodate several needs:

– *Historical queries*: many applications demand analysis over past data, and should be accessible if a server has stored it.
– *Real-time updates*: some events have low frequency but real-time importance upon firing.
– *Server-side filtering*: only a subset of data should flow from source to destinations to lessen network load.
– *Authentication*: some aggregates should be made from authorized users
– *Encryption*: some branches of the fusion tree should remain private to a certain group.
– *Indexability*: keywords from the stream meta data should be indexable along with the URL of the stream itself.
– *Scalability*: The system performance should not deteriorate dramatically as the number of clients increases, or as the depth of a fusion tree increases.
– *Autonomy*: a server should not be required to store its data on a third-party's server, and vice versa.

3   Background and Related Work

We can define a *sensor stream* to be a set of data that is generated over time by a single sensor. For example, a stream could be the set of temperature data generated by a temperature sensor over time. A stream is distinct from a *connection*, which is a software abstraction through which one device may send data to another at any time. For example, a TCP socket is a connection, but is not a stream. *Multimedia streaming* is a common technique in which a media object is transported such that it can be played before the entire file is downloaded. This is a distinct concept from a sensor stream: one property of steam feeds is that sensor streams can be transported using either streaming or non-streaming connections.

Another important distinction for this paper is the difference between the World Wide Web and the Internet. The *Web* is a set of objects that are uniquely addressed by a uniform resource locator (URL), are hosted by a *web server*, and are accessed via the HTTP protocol. Objects on the web refer to each other by their URL's, creating *links* that form a directional graph. Links in HTML documents are called *hyperlinks* while links in XML documents are called *XLinks*. The directional graph formed by these links help users to locate objects on the Web in a process called *surfing*, allow automated bots to *crawl* the Web, and form the basis for indexing and search algorithms such as Google's PageRank algorithm [1]. Although the Web is often also called the Internet, it is a distinct concept: the *Internet* is a set of computers connected by the IP protocol, while the Web is a set of objects connected by links. Most Web objects can be accessed through the Internet, but not all Internet applications are on the Web. As we describe below, most architectures that have been proposed for the World Wide Sensor Web exploit the globally-connected nature of the Internet, but do not actually put sensor streams on the Web: data streams do not have a URL, are not served by Web servers over HTTP, and cannot be the subject or object of a link. The main contribution of the stream feeds abstraction is that high data rate, aperiodic, real-time data streams become first-class citizens on the Web.

3.1   XML Schemas for Sensor Streams

The problem of sharing sensor data streams is currently becoming an active area of research. The Open Geographic Consortium (OGC) is an organization of 352 companies, government agencies, and universities collaborating to design publicly available interface specifications. The SensorNet [2] project primarily uses the OGC's schemas to achieve web interoperability. One example schema is the SensorML specification, which is an XML schema used to define processes and components associated with measurement of observations. These schemas are being created to define the a standard model for Web-based sensor networks, which will enable the discovery, exchange, processing of sensor observations, and tasking of sensor systems. However, they are not associated with a proposed architecture to store, cache, filter, and fuse data streams, and to translate between the real-time streaming nature of data streams and the static, connectionless nature of the Web.

Other data formats include GeoRSS, which is used for encoding geographic objects in RSS fees and KML (Keyhold Markup Language), a similiar format especially designed for GoogleEarth and Google Maps [3]. However, these groups are only proposing data formats, they are not proposing an architecture or abstraction that will integrate these XML formats with the Web. Stream feeds are agnostic with respect to XML Schemas for sensor streams and any XML format may be used. Some architectures have been designed to integrate SensorML files with the Web, for example, in the Davis Weather System, AIRDAS airborne sensor, and the SPOT satellite. However, these implementations do not propose solutions to the criteria we laid out in Section 2, since they do not define complex interrelationships between data sources that are owned by several entities.

3.2   Web Feeds

A Web *feed* is an XML document that contains a dynamically changing sequence of content items, such as blog entries, Podcasts, or news headlines. Content providers *syndicate* their web feeds and content consumers *subscribe* to that feed. A subscription is usually done through a feed *aggregator*, which continually queries for the feed at regular intervals and processes it or notifies the user when the feed content has been updated. Using this model, there is no coupling between the publisher and the consumer: subscriptions can be broken by removing the feed from the user's aggregator. This model has worked well for feeds until now because traditional feeds are relatively small data objects and have little or not realtime requirements. This small size also makes it possible to filter the feed content within the client aggregator, instead of on the server side. Stream feeds differ from traditional feeds because they may often be much larger data objects, they may change with orders of magnitude higher frequency than traditional feeds, and clients that request stream feeds may require immediate update. This combination makes periodic requests for the entire data object inefficient. Another difference is that the set of operations that is typically performed on the data content of traditional feeds is limited to concatenation, sorting, and truncation of content entries. The content of sensor streams, however, can be processed, filtered, and *fused* with content from other sensor streams. There are systems that use XML files available on the web for disseminating sensor data both realtime and historical, resulting in a virtual sensor abstraction that provides a simple and uniform accesss to heterogeneous streams [?]. However, these systems do not provide a common interface for pushing and pulling for the data.

Research is interested in finding ways to search for data coming from these feeds. One example is Cobra which crawls, filters, and aggregates RSS feeds [5], and another is the popular search engine Technorati, designed to search for keywords in web feeds.

3.3   Streaming Media

The HTTP protocol was designed to retrieve an entire document at once, using a *pull* model for file transport. This model causes problems for large multimedia files that can take a very long time to download. To address this problem, several protocols and techniques have been designed to *stream* media files in order to allow the file to be played by the client before the file has been completely downloaded. For example, some systems stream data through multiple sequential HTTP responses while others use new protocols like RTP (realtime protocol) and RTSP (realtime streaming protocol) [6]. Protocols like Jabber (XMPP) [7] support streams of instant messages and can be used to stream realtime data past firewalls. The WHATWG is drafting a Web Applications 1.0 specification [8] which server-side pushing of content to the client.

Streaming file transport is used for finite-sized media files, but it also enabled the creation of *streaming media*, a type of data object that has no end, like an Internet radio station. This type of data object cannot be used with the traditional *pull* model for file transport. Sensor streams are similar to streaming media in this respect. However, the streaming transport protocols that have been designed until do not meet all of the requirements for stream feeds. They support the abstraction of a streaming media object, and provide functions such as 'play' and 'pause'. Sensor streams do not fit under this abstraction, and require other commands such as the downloading of historical data and server-side data filtering.

3.4 RFIDs

RFID tags can uniquely identify a physical object as it moves through a supply chain that may involve dozens of different vendors and shippers. For this reason, several architectures have been designed to make RFID data available from multiple administrative domains. For example, EPCglobal has created a global system for managing and sharing RFID and sensor data [9]. This system uses an asynchronous messaging protocol based on Web service standards to make Electronic Product Codes (EPC) data available in real time to hundreds of thousands of simultaneous users. However, this work differs from ours in that we are creating an abstraction based on URLs, so sensor data can be linked to by other Web objects, and can be crawled and index by search engines.

RFID applications have a *high fan-in* architecture where there are many devices at the edge of the network sending their data to a small number of more powerful processing nodes. High fan-in systems can suffer from congestion, and a system called HiFi [10] addresses this problem by performing data cleaning, event monitoring, stream correlation, and outlier detection among the mid-tier nodes.

3.5 Other World Wide Sensor Web Architectures

There have been several designs for an infrastructure for global sensor data sharing. Most architectures propose either putting all sensor data into one large centralized database or distributing data across a distributed database running the same software stack. Centralized databases have advantages that come with simplicity, but centralization can lead to high server loads and require others to trust the server with private information. Centralization runs tangent to the web's laissez-faire philosophy. The problem with database-oriented solutions is that they force the data producer to conform to a pre-specified implementation, and do not allow for diversity of software stacks. They also do not provide rank-ordered query results so they cannot deal with issues of data fidelity and popularity.

IrisNet is a distributed database that has been proposed as an architecture for the World Wide Sensor Web [11]. Sensor data is stored in an XML database which is hierarchically partitioned and stored in different parts of the network to minimize response time and network traffic during queries. IrisNet restricts data positioning to a sensor ontology based on a hierarchy of locations. Many other systems have also been designed for distributed stream processing using distributed hash tables[12] or by moving the data to a centralized location and running continuous queries [13,14,15].

The SensorMap project from Microsoft Research uses a database-oriented approach where users post queries to a central database by interacting with a graphical map interface [16]. SensorMap is similar to our project in that the thrust is to make publishing data especially easy. SensorMap has explored using *screen scraping* by enabling a web crawler to automatically index adjacent text descriptions next to embedded webcam content on a webpage.

Hourglass is a system that provides circuit-based connections for real-time persistent streams [17]. It emphasizes intermittent connection scenarios for mobile or poorly connected nodes. Circuits push data through a series of links and can be processed along the way. However, because all data is pushed, Hourglass only supports real-time data streaming and does not allow searching through the history of a data stream. In addition, Hourglass does not support indexing streams and all data queries must be made through their registry/query service. Hourglass does not require a particular data schema, like database systems, but it does require users to use a particular software stack.

4   The Stream Feed Abstraction

Stream feeds provide the abstraction of a dynamic content object containing a stream of sensor data. This data object can be located with a URL, can be linked to from other data objects, and can be accessed over the HTTP protocol. For example, to access a stream of temperature data from the front door of the Whitehouse, one would visit:

```
http://www.whitehouse.gov/streams/frontdoor/temperature
```

When the client requests this URL, the client opens a TCP socket to the server and sends a HTTP GET request, as usual. The server responds immediately with all data that is already contained in the sensor stream, followed by updates about the sensor stream in real-time as new data objects are added to the stream.

A parameter can be appended to this URI to indicate that the query should take place over a different time region. For example, the client can request:

```
/streams/frontdoor/temperature/historical
```

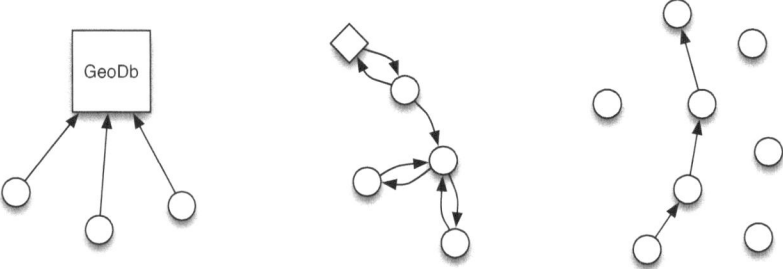

Fig. 2. SensorMap (left) collects all data into a central data base. IrisNet (center) provides a distributed XML database abstraction that can be queried by location. Hourglass (right) allows the user to setup *circuits* along which sensor data is pushed.

to restrict the query only to data that has already been stored, or can request

```
/streams/frontdoor/temperature/incoming
```

to restrict the query only to data that has not yet been received. Thus, the client can pose the query over three temporal scopes: (i) all time (default), (ii) all time before now, and (iii) all time after now. These three scopes are illustrated in Figure 3.

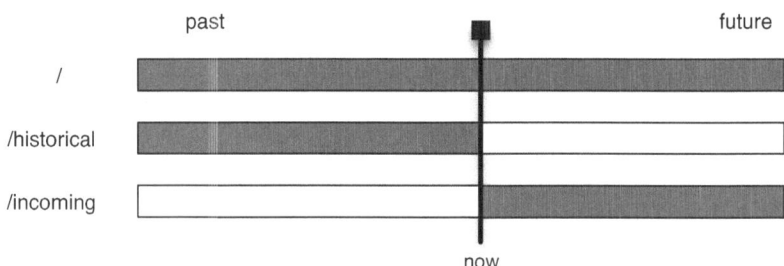

Fig. 3. Every query over a stream feed must take one of three temporal scopes. The default (top) is a query over all time, but queries over historical data (center) and incoming data (bottom) are also possible. Historical data is returned immediately while incoming data is returned as it arrives.

The user can also *filter* the sensor stream by passing filter parameters as follows:

```
/temperature/filter/?value_lower_bound=100&day_equal=Tuesday
```

This query restricts the query response to only contain the values in the data stream that have a "value" attribute greater than 100 and a "day" attribute equal to 'Tuesday'. The values that can be filtered depend on the XML schema that the sensor stream is using, so the client must be familiar with this schema. The values in the sensor stream are then filtered using syntactic filtering of the data points based on the XML tag names and values. This can be performed automatically for any XML schema. filters can be applied to a query with any of the three temporal scopes shown in Figure 3. Any tag can be used with four types of filters:

```
TAG_equal
TAG_not_equal
TAG_lower_bound
TAG_upper_bound
```

The URL-based stream feed interface conforms to REST (Representational State Transfer) principles describing how resources are defined and addressed. Unlike RPC-based web services, where a vast set of verbs to be applied to a set of nouns, the RESTful philosophy promotes the use of a simple set of verbs that operate on a vast collection of nouns. The advantage to using a RESTful interface is that there is an inherent standardization of the operations that can be applied to the resources, without needing to explicitly define descriptions of the methods that can be applied to them using the web services description language (WSDL). The URL contains all the information that is needed to return to a particular state of a web service. By including the name of the sensor, the temporal scope, and the filter parameters all in the URL, stream feeds make it easy to link to stream feeds, or even to link to partial stream feeds.

5   Implementation of the Stream Feed Abstraction

Our motivating application, MetroNet, was implemented using the Ruby on Rails framework. MetroNet is served using the lighttpd web server with the FastCGI module, and the MySQL database. The ActionController under Rails handles the complex routing requests required by StreamFeeds, so that XML feeds can be generated on demand based on the URL structure.

In MetroNet, all data is stored in a central database and a XML file is generated on the fly for each query. Thus, queries over historical data are performed as normal HTTP queries, and the generated file is returned to the client. Filters over historical queries are executed when the XML file is first generated.

Queries over incoming data are currently handled through a publish/subscribe mechanism. Every stream holds a list of subscriptions, and a client can add a receiver address by posting a URL to the location

http://.../sensorName/subscription

using the HTTP PUT command. Incoming data is then sent using a POST to the target URL. This interface is slightly different than that described in Section 4, although it achieves the same goals. The interface implementation is currently being changed to match that of Section 4.

The implementation described above is not fundamental to the steam feeds abstraction, and other implementations are possible as long as they provide the

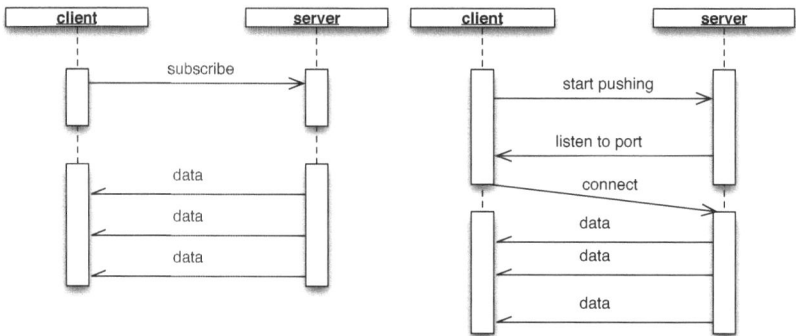

Fig. 4. A publish/subscribe mechanism can be used to efficiently push aperiodic to the client (left) while a new connection can be opened and remain open to push data to a client behind a firewall (right)

same interface and abstraction. For example, a stream feed could be stored in a flat file and served by a normal Web server. Filters could be implemented using syntactic filtering of entries with XML tags that match the filter. Incoming data could be appended to the file by a program that is independent of the Web server software, and the server could server queries over incoming data by observing changes to the file.

One limitation of the publish/subscribe implementation that we describe above is that clients cannot be behind firewalls or network address translation (NAT) devices. An alternative implementation is to negotiate a new long-lived TCP socket between the client and server, and to serve all HTTP responses over that socket. This scheme is compared with the protocol that we implemented for MetroNet in Figure 4. It would be able to stream data to clients behind NATs and firewalls, but has the disadvantage that a socket must be open during the entire query response, which could limit the number of queries that can be supported by the server at a time.

6   Providing Other Interfaces for Stream Feeds

The URL interface is not the only type of interface that can be provided to steam feeds. The URL interface is necessary to make sensor streams first-class citizens on the Web because they can be linked to and they can be crawled and indexed by search engines. However, we can also provide other interfaces by wrapping the URL-based interface. for example, we implemented the interface shown in Figure 1 using SOAP objects, providing a programmatic interface to stream feeds that can be used in any language. In our MetroNet server, the SOAP object for each stream can be located at a URL relative to the location of the stream itself. For example:

```
/streams/frontdoor/temperature/service.wsdl
```

The SOAP interface we provide contains three functions corresponding to the three temporal scopes over which queries can be posed. It also contains a member variable for each TAG in the XML schema used for that stream, so that the programmer can define a filter based on that XML schema. The filter defined through these variables is applied to all queries that are posed with the SOAP object, with any temporal scope. This interface provides the same abstraction as our URL-based interface and therefore provides the same abstraction over a sensor stream, but has the disadvantage that it is not as easily crawlable. Other similar APIs could be defined for stream feeds as necessary by wrapping the URL-based interface.

Table 1. A SOAP interface provides the same functionality as our URL-based interface in a programmatic way. Other interfaces can also be created for stream feeds.

API	Meaning
get()	query over historical data
getHistorical()	query over incoming data
getIncoming()	query over incoming data
TAG	a member variable through which to apply filters
TAG.eq	return only content where TAG equals a value
TAG.ne	return only content where TAG does not equal a value
TAG.lb	return only content where TAG is greater than a value
TAG.ub	return only content where TAG is less than a value

7 Evaluation

We evaluate the efficiency of the stream feed abstraction by measuring the scalability of the stream feed abstraction when applied to a query over a deep fusion tree, and compare our results from using standard Web feeds. We perform this evaluation in two parts. First, we empirically evaluate our server implementation to establish the trade-off between posing repeated queries for historical data and posing a single query for incoming data. Then, we use these empirical results to drive a simulation in which we pose queries over large fusion trees and measure the latency.

7.1 Repeated Historical Queries vs. Incoming Queries

In our first experiment, we empirically decide between repeatedly querying for the sensor stream and pushing the sensor stream. We setup a Web server that serves a sensor stream that is updated with a frequency of 1Hz. We also setup a set of clients that must receive the data with a frequency from 0.1Hz to 10Hz. The clients filter the query response by indicating the sequence number of the last value received, so only new data is sent in response by the server. We measure the total number of streams that can be served when all clients use push and when all clients use pull. The total values averaged over 100 runs can be seen in Figure 5.

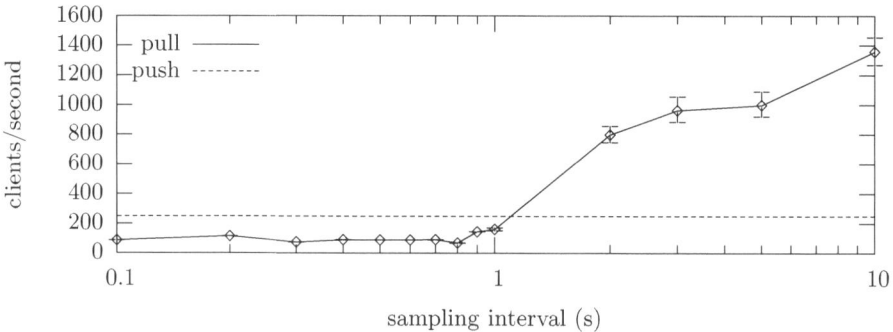

Fig. 5. When clients pull data from the server, the number of clients that can be served increases as the the frequency of requests increases. At a query rate of approximately 1 second, the number of clients served begins to exceed that supported when the server pushes data to the clients in real time.

In our experiment, we used the SOAP implementation to measure the performance of the query over incoming data, because the SOAP implementation keeps a socket open for the duration of the push, just as our proposed abstraction would due to problems with firewalls and NATs.

The results show that in the server we implemented, the PUSH queries can support a total of 250 clients at a time, independent of the frequency with which the client needs the data. This is because our web server can only support a maximum of 250 sockets at a time, and the PUSH query keeps the TCP socket open the entire time. This number may differ from servers to servers, due to the capability of the servers.

Theoretically, if each query over historical data only takes 0.2 seconds on the server side, then the server can support a total of more than $5 * 250$ queries in one second because the socket connection of a PULL request lasts less than 0.2 seconds. However, during the experiments we found that the TCP connection number is not the only factor that limits the number of clients we can support per second. As the query frequency increases, with 200 clients, the latency for each query increases due to the speed of database access. For queries over incoming data, a database access only occurs once when any new reading of the stream comes in. Then the server duplicates the new value and pushes it out to all clients. For repeated historical queries, on the other hand, each query requires access to the database because the server doesn't know whether the value of that sensor stream has changed since the previous client's query. This increases the latency of each query and brings down the number of queries we can support per second.

Our results show that more streams can be served using repeated historical queries if the query frequency is less than 1.2Hz. Therefore, since the data was being generated in this experiment at a frequency of 1Hz, we empirically validate that in our system the client should use running queries over incoming data when the sampling frequency is 1.2 times the data generation frequency or more, and should use repeated historical queries otherwise.

7.2 Scalability and Efficiency

In our second experiment, we use the results from our first experiment to test the efficiency of stream feeds when used with deep fusion trees in simulation.

With each run, we randomly generate a network of servers with a branching factor of 3 as the depth increases. Although the network resembles a tree, some nodes can have multiple parents assigned to random nodes in the layer above. The nodes are connected in multiple layers from 2 to 15 as shown in Figure 6. The nodes on the bottom are sensors and generate data which is requested by several clients through a series of servers. 5 clients are attached to each server node, and feed the network with queries. The clients are initialized with a random query frequency, and each node has a query rate equal to the sum of the parent's query rates. The latency results reported are the averaged for every client in the network over 5 runs.

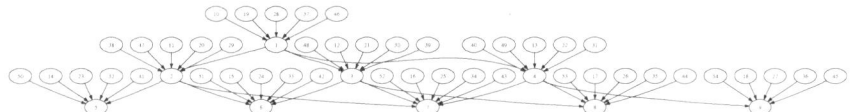

Fig. 6. A randomly generated *fusion tree* of 54 nodes randomly distributed and connected in three layers, with five clients per server

In our first case, we demonstrate the standard web feed abstraction by restricting servers only to pull for data without server-side filtering from a fusion tree of varying depth. The results from Figure 7 show that the average latency for each query grows exponentially with the depth of the tree. This is a problem, and demonstrates that normal Web feeds cannot efficiently support deep fusion trees.

In our second case, we enable servers to use filtering to prevent responses from returning results that are already known by the client. In other words, each client stores the sequence number of the last retrieved data point and passes that as a filter to the server in each request. This experiment shows that the latency grows proportionally with the depth of the network because the filtering significantly reduces the network traffic among all layers during the fusion process, and the latency is dominated by the roundtrip time rather than the transmission time.

In our third case, we evaluate the model with the two streamfeeds extensions, filtering and the ability to push. Each server dynamically decides between requesting historical queries or running queries over incoming data based on its observed query rate and data generation rate. We expect this optimization, by combining the advantages of pushing and pulling, to reduce the average latency of all queries even further than the reduction observed from filters. Because of this dynamic switching there is a balance point at a certain depth in the network, where all the nodes beneath that level begin to push their data to the

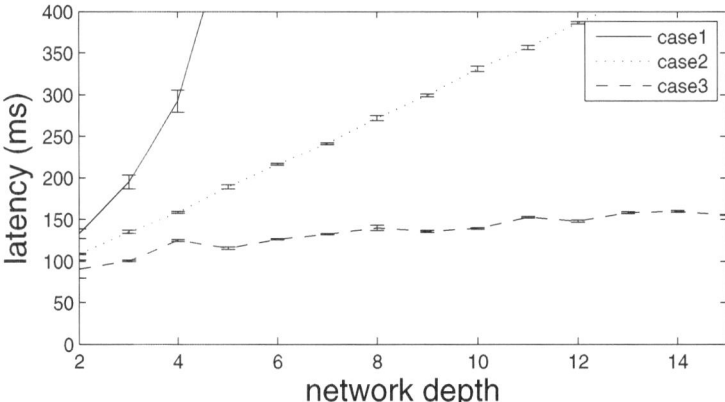

Fig. 7. As the network gets deeper, filtering with the ability to push outperforms the alternatives

parents. Because of this region of low latency, the queries suffer most latency when pulling for data in the upper levels of the topology, however more clients can be served in these upper levels. The result is an improvement in the the average latency for all the client's requests. We expect that with larger files, lower sampling rates, and/or deeper fusion trees, this benefit would be further increased.

8   Conclusions

In this paper, we showed that we can provide an efficient abstraction for sensor streams that is a first class citizen on the Web by combining the advantages of Web feed and multimedia streaming paradigms. Each sensor stream has a URL, and parameters encapsulated in this URL allow the user to query for historical data using a PULL model, to query for incoming data using a PUSH model, or both. Other parameters to this URL allow the user to filter the data on the server side, to reduce the size of the data response in repeated queries to the large streaming data objects.

Because this abstraction is a first-class citizen on the web it inherits security, authentication, and privacy mechanisms. Stream feeds can also be crawled and indexed by search engines. The servers are all autonomous, so the server administrators do not need to worry about storing data on the servers of others, or having other data stored on this domain's servers, as with the IrisNet and Hourglass systems. Thus, in this paper, we show that the stream feed abstraction meets the requirements of and can provides the basis for the World Wide Sensor Web, where sensor data can be shared, processed, and published within a single global framework.

References

1. Page, L., Brin, S.: The anatomy of a large-scale hypertextual web search engine. WWW7 Computer Networks 30, 107–117 (1998)
2. Gormon, B., Shankar, M., Smith, C.: Advancing sensor web interoperability. Sensors (April 2005)
3. Geller, T.: Imaging the world: The state of online mapping. IEEE Computer Graphics and Applications 27(2), 8–13 (2007)
4. Abrrer, K., Hauswirth, M., Salehi, A.: A middleware for fast and flexible sensor network deployment. In: VLDB 2006: Proceedings of the 32nd international conference on Very large data bases, VLDB Endowment, pp. 1199–1202 (2006)
5. Rose, I., Murty, R., Pietzuch, P., Ledlie, J., Roussopolous, M., Welsh, M.: Cobra: Content based filtering and aggregation of blogs and rss feeds. In: 4th USENIX Symposium on Networked Systems Design and Implementation (2007)
6. Schulzrinne, H., Rao, A., Lanphier, R.: Real time streaming protocol (rtsp). Technical report, Network Working Group (1998)
7. Saint-Andre, P.,, E.: Extensible messaging and presence protocol. Technical report, Jabber Software Foundation (October 2004)
8. Hickson, I.: Html 5. Technical report, Google (September 2007)
9. Armenio, F., Barthel, H., Burnstein, L., Dietrich, P., Duker, J., Garrett, J., Hogan, B., Ryaboy, O., Sarma, S., Schmidt, J., Suen, K., Traub, K., Williams, J.: The epcglobal architecture framework. Technical report, EPCglobal (2007)
10. Cooper, O., Edakkunni, A., Franklin, M.J., Hong, W., Jeefrey, S., Krishnamurthy, S., Reiss, F., Rizvi, S., Wu, E.: Hifi: A unified architecture for high fan-in systems. In: Proceedings of the 30th VLDB Conference (2004)
11. Gibbons, P.B., Karp, B., Ke, Y., Nath, S., Seshan, S.: Irisnet: An architecture for a world-wide sensor web. IEEE Pervasive Computing 2(4), 22–33 (2003)
12. Huebsch, R., Hellerstein, J.M., Boon, N.L., Loo, T., Shenker, S., Stoica, I.: Querying the internet with pier (2003)
13. Cherniack, M., Balakrishnan, H., Balazinska, M., Carney, D., Cetintemel, U., Xing, Y., Zdonik, S.: Scalable Distributed Stream Processing. In: CIDR 2003 - First Biennial Conference on Innovative Data Systems Research, Asilomar, CA (January 2003)
14. Chandrasekaran, S., Cooper, O., Deshpande, A., Franklin, M.J., Hellerstein, J.M., Hong, W., Krishnamurthy, S., Madden, S., Raman, V., Reiss, F., Shah, M.: Telegraphcq: Continuous dataflow processing for an uncertain world (2003)
15. Chen, J., DeWitt, D.J., Tian, F., Wang, Y.: NiagaraCQ: A scalable continuous query system for Internet databases, ACM SIGMOD, pp. 379–390 (2000)
16. Liu, S.N.J., Zhao, F.: Challenges in building a portal for sensors world-wide. In: First Workshop on World-Sensor-Web: Mobile Device Centric Sensory Networks and Applications (WSW 2006) (2006)
17. Shneiderman, J., Pietzuch, P., Ledlie, J., Roussopolous, M., Seltzer, M., Welsh, M.: Hourglass: An infrastructure for connecting sensor networks and applications. Technical report, Harvard University (2004)

.

Author Index

Lecture Notes in Computer Science

Sublibrary 4: Security and Cryptology

For information about Vols. 1– 3786
please contact your bookseller or Springer